End-to-End Mobile Communications: Evolution to 5G

About the Authors

Syed S. Husain is an Industry Standards expert specializing in mobile system architecture and 5G core network standards. He is president of SSH Consulting, Inc. His company provided consultancy services to NTT DOCOMO for over ten years representing them in 3GPP and oneM2M standards organizations. Prior to consulting, he worked at Motorola and AT&T Bell Laboratories as a systems architect in advanced technology. He has over 35 years of telecommunications experience in the areas of Digital Switching, Intelligent Networking, Next Generation IP Networks, and Wireless Communications. He has written numerous conference publications, and he is a Life Member of IEEE. He received his BSc and MSc degrees in Electrical Engineering from Syracuse University, in upstate New York.

Athul Prasad received his MBA from Massachusetts Institute of Technology (MIT), and his MSc (Tech.) and DSc (Tech.) degrees in Communications Engineering from Aalto University in Finland. He is also a graduate of the year-long executive education (LEAD) program at Stanford University Graduate School of Business. Most recently, he was part of Nokia's end-to-end 5G leadership program, based in Helsinki, Finland. Previously, he has worked with Huawei and NEC in various roles related to standardization research and product development. He is the co-author and co-inventor of over 90 international publications, patent applications (including granted patents), technical reports, and standards contributions. He has won the best paper award at IEEE VTC-Spring 2013, best 5G paper award at IEEE BMSB 2018, and best paper finalist at IEEE 5G World Forum 2018.

Andreas Kunz received his diploma degree in Electrical Engineering in 2001 and his Ph.D. degree in 2006 from the University of Siegen in Germany. His main research topic involved MAC-layer packet scheduling under QoS aspects in several radio access technologies. He joined NEC Network Laboratories in 2005 and was part of the Next Generation Networks research team focusing on 3GPP standardization, mainly in the system architecture working group SA2. From the beginning he was a permanent member of the NEC 3GPP SA2 delegation and was involved in various topics from the IP multimedia subsystem over machine-type communication to the current 5G core network, as well as selected system security aspects. Besides 3GPP, he also participated in other standards organizations such as GSMA and ETSI. In 2017 he joined Lenovo Germany as member of the research and technology team, representing Lenovo/Motorola Mobility in the 3GPP security group SA3.

JaeSeung Song received his BSc and MSc degrees in Computer Science from Sogang University, South Korea, and his Ph.D. degree from the Department of Computing at Imperial College London, United Kingdom. He is an Associate Professor at Sejong University, leading the Software Engineering and Security Laboratory in the Computer and Information Security Department as the oneM2M Technical Plenary Vice Chair. Between 2012 and 2013, he worked for NEC Europe Ltd., as a leading standards senior researcher where he actively participated in IoT-related R&D projects. From 2002 to 2008, he worked for LG Electronics as a senior researcher leading a 3GPP system architecture standards team. He is the co-author and co-inventor of over 100 international publications, conference papers, and patent applications. He also holds leadership roles in several journals and conferences, such as IoT series editor of IEEE Communications Standards Magazine and TPC co-chair of IEEE Conference on Standards for Communications and Networking.

Adrian Buckley has worked on wireless systems since the early 1990s, starting his career by helping launch the Orange network in the UK. He also helped launch the Pacific Bell network in the late 1990s before being responsible for the worldwide R&D GPRS operations for SBC Telecommunications. He spent a brief time at a startup before spending the next 16 years representing BlackBerry in various 3GPP (SA1, SA2, CT1, and CT6 working groups), OMA, GSMA, and other industry forums. He authored over 175 granted USA patents, many in industry standards. Now he runs his own consulting firm advising on standards, IPR, and disruptive technologies. He received his BSc (Engr.) and MSc (Telecommunications Technology) degrees from Aston University UK, and JD Law degree from Purdue University Global. He is also a member of IEEE, IET, and SAE, and is a Charted Engineer.

Emmanouil Pateromichelakis received his Diploma in Information and Communication Systems Engineering from University of the Aegean in Greece. He received his MSc and Ph.D. degrees in Mobile Communications from the University of Surrey, UK, in 2009 and 2013, respectively. He is a senior wireless researcher at Lenovo, focusing on 5G and beyond solutions. Previously, he worked as Senior Research and Standardization Engineer at Huawei Technologies, at the Europe Standardization & Industry Development Department. He has co-authored numerous publications, has filed multiple patents, and has actively contributed to 3GPP 5G standards (RAN and SA).

End-to-End Mobile Communications: Evolution to 5G

Syed S. Husain

Athul Prasad

Andreas Kunz

JaeSeung Song

Adrian Buckley

Emmanouil Pateromichelakis

New York Chicago San Francisco
Athens London Madrid
Mexico City Milan New Delhi
Singapore Sydney Toronto

Library of Congress Cataloging-in-Publication Data

Names: Husain, Syed S., editor.
Title: End-to-end mobile communications : evolution to 5G / Syed S. Husain, editor.
Description: New York : McGraw Hill, [2021] | Includes bibliographical references and index. | Summary: "This book presents the foundations of mobile communications, from architecture to function, leading up to a special focus on 5G services, networks, and applications. The book also presents a primer on the vast topic of mobile technology security as well as a peek at the future evolution of mobile systems"—Provided by publisher.
Identifiers: LCCN 2020016135 | ISBN 9781260460254 (hardcover ; acid-free paper) | ISBN 9781260460261 (ebook)
Subjects: LCSH: 5G mobile communication systems.
Classification: LCC TK5103.25 .E54 2021 | DDC 621.3845/6—dc23
LC record available at https://lccn.loc.gov/2020016135

McGraw Hill books are available at special quantity discounts to use as premiums and sales promotions, or for use in corporate training programs. To contact a representative please visit the Contact Us page at www.mhprofessional.com.

End-to-End Mobile Communications: Evolution to 5G

1 2 3 4 5 6 7 8 9   LCR   25 24 23 22 21 20

ISBN 978-1-260-46025-4
MHID 1-260-46025-8

The pages within this book were printed on acid-free paper.

Sponsoring Editor	**Proofreader**
Lara Zoble	Alekha C. Jena
Editorial Supervisor	**Indexer**
Donna M. Martone	Edwin Durbin
Acquisitions Coordinator	**Production Supervisor**
Elizabeth Houde	Lynn M. Messina
Project Manager	**Composition**
Rishabh Gupta	MPS
Copy Editor	**Art Director, Cover**
MPS	Jeff Weeks

Contents

Acknowledgments

The authors would like to acknowledge the following individuals for helping in promoting this textbook idea and critiquing its different parts, which has helped immensely both in its realization and in making necessary improvements:

Farooq Bari, PhD, Standards & Industry Alliances, AT&T, USA

Sheeba Backia Mary Baskaran, PhD, Researcher, Lenovo, Germany

Aliya N. Husain, MD, Department of Pathology, University of Chicago, USA

JaeYoung Hwang, MSC, Sejong University and Korea Electronics Technology Institute, South Korea

Ravi Kuchibhotla, PhD, Executive Director, Motorola Mobility, USA

Robert Love, MSC (EE), Executive Director, Motorola Mobility, USA

Prateek Basu Mallick, MSC (EE), Researcher, Lenovo, Germany

Serge Manning, PhD, Technology Standards, T-Mobile, USA

Srisakul Thakolsri, PhD, Senior Researcher, NTT DOCOMO Euro Labs, Germany

Preface

The idea of writing a textbook on the fifth-generation (5G) mobile communications was discussed in early 2018 among the authors, with the goal of educating engineering students on the key concepts behind the mobile system, leading up to 5G and beyond. Currently, there are many textbooks that provide a good deal of information on the fundamentals of telecommunications and mobile system standards, but there are a limited number of textbooks that provide a holistic background on the mobile system evolution from 0G to 5G and details on the evolution of mobile system functions. This background is helpful in understanding the concepts and how the mobile technology has gradually evolved, taking into account the technological advancements in both radio access and core network domains. The other available channels to learn about mobile communications is to sort through a plethora of standards documents or by reading various academic papers, which is a cumbersome task without knowing where to start.

The goal of this textbook is to help in preparing students as they embark on their professional careers in mobile communications and its related fields. In addition, it provides engineering students the ins and outs of mobile technology not found in other textbooks, e.g., the services that are offered today and those that are currently under development. Another discipline that can also benefit from this textbook is security (a branch of mathematics and computer science) for development of data encryption techniques used in secure transmission of data over the radio interface. Also, this textbook is a useful reference for system engineers working in the field of mobile communications in further enhancing their knowledge on specific topics of interest related to 5G communications.

This textbook provides an extensive review on how mobile technology has developed and evolved up to 5G, what are its key components, and how it is likely to further evolve in the future. Topics covered will include the technology behind the mobile system, which is made up of radio access network, core network, and mobile terminals like smartphones. In addition, it will cover aspects of different types of services offered by the mobile system for the benefit of users.

The wide area of expertise of a team of six authors, who have worked in the mobile telecommunications field for over 30 years and directly engaged in the 3GPP and other industry standardization bodies' standards development process, will provide a breadth of knowledge in mobile telecommunications that is almost unrivaled.

1

Introduction

The internet has had a major influence on the evolution of the mobile system due to its growing use in every aspect of our daily lives. The mobile system has embraced the internet technology in order to ease the delivery of internet services on the mobile devices. The mobile systems have moved away from using traditional signaling protocols, like Signaling System 7 (SS7) for out-of-band signaling and asynchronous transfer mode (ATM) for connection-oriented in-band packet switching, to using internet protocols, such as session initiation protocol (SIP), hypertext transfer protocol (HTTP), and transmission control protocol/internet protocol (TCP/IP). Also, the mobile systems of the past, which were mainly used for human-to-human communications, are now being used more and more for human-to-machine and machine-to-machine communications.

The worldwide communication enterprise is quite remarkable, since it enables us to communicate seamlessly with each other from/to anywhere whether stationary or on the move. How is this possible? One of the major ways is through the development of rigorous standards by various standardization bodies, both regional and worldwide, making it possible for all telephonic devices (fixed telephone, mobile phone, etc.) to communicate over standard signaling protocols (i.e., same language between two end-points). The communication enterprise consists of many players (or companies) whose primary business is the transmission of data using standard signaling techniques from point A to point B in a secure manner. These companies consist of fixed network operators, mobile network operators, satellite network operators, internet operators, undersea cable operators, etc. It is not possible for a single company to be the sole provider of data/voice services worldwide. Therefore, all of the players in the communication enterprise are interconnected using standardized interfaces, making it possible to transmit data/voice signals across the world seamlessly and without actual know-how of the different players involved in between. In order to enable the interworking of these players in a seamless manner, standardized interfaces are required for the transmission of data from point A to point B, so that both ends understand each other. This is why the standards are necessary and play a major role in the seamless operation of the worldwide communication enterprise. The global standards making bodies, such as ITU, IETF, IEEE, NGMN, oneM2M, and 3GPP, are considered key in the development of industrywide communication standards. The network operators and product manufacturers are heavily engaged in these standards bodies, along with government agencies. A brief description of each of the major global standards bodies mentioned follows.

- The International Telecommunication Union (ITU) is an agency of the United Nations (UN) created in 1865 primarily to coordinate global telecommunication operations and services. It is headquartered in Geneva, Switzerland. It consists of three sectors: ITU-radiocommunication (ITU-R) sector, ITU-telecommunication (ITU-T) standardization sector, and ITU-telecommunication development (ITU-D) sector. Each ITU sector performs a different role. ITU-R's role is to coordinate the growing range of radiocommunication services, as well as the international management of the radio-frequency spectrum and satellite orbits. This is to ensure harmonization in the use of radio spectrum. ITU-T's role is to develop standards for core network functionality and myriad of other standards related to information and communication technology (ICT). ITU-D's role is to promote ICT leadership in emerging markets, i.e., learning how to put good *policy* into practice. The outputs of ITU are recommendations and guidelines that are used by other global or regional standards bodies in their standards development process.

- The Internet Engineering Task Force (IETF) started out as an activity (ARPANET—Advanced Research Projects Agency Network) supported by the US Federal Government, but since 1993 it has operated to perform as a standards-development organization under the auspices of the Internet Society, an international membership-based nonprofit organization. IETF is an open standards-development organization with the mission to make the internet work efficiently by producing high-quality, relevant technical documents that influence the way people design, use, and manage the internet. The outputs of IETF are the famous RFCs (Request for Comments) documents that drive the development of internet technology.

- The Institute of Electrical & Electronics Engineers (IEEE) is a leading developer of industry standards in a broad range of technologies that drive the functionality, capabilities, and interoperability of a wide range of products and services, transforming how people live, work, and communicate. One of the more notable standards are the IEEE 802 local area network/wireless local area network/metropolitan area network (LAN/WLAN/MAN) group of standards, with the widely used computer networking standards for both wired (Ethernet, aka IEEE 802.3) and wireless (Wi-Fi, aka IEEE 802.11 and IEEE 802.16) networks.

- The Next Generation Mobile Network (NGMN) alliance is an open forum founded by world's leading mobile network operators, in which telecommunication vendors, software companies, and research institutes are contributing members. Its goal is to ensure that the standards for next-generation network infrastructure service platforms and devices will meet the requirements of operators and, ultimately, will satisfy end-user demands and expectations. Currently, there is special focus in the telecom industry on how 5G will meet the NGMN requirements.

- The oneM2M organization is a global partnership project established in 2012. Eight leading regional standards development organizations are members of oneM2M: ARIB, ATIS, CCSA, ETSI, TIA, TSDSI, TTA, and TTC. The goal of oneM2M is to create global technical standards for interoperability concerning the efficient deployment of machine-to-machine (M2M) communications systems. oneM2M specifications provide a network agnostic framework to support applications and services such as the smart grid, connected car, home automation, public safety, and health. The oneM2M specifications are internationally recognized and transposed by ITU-T under the Y.4500 series recommendations.

- The Third Generation Partnership Project (3GPP) is an organization formed in 1998 to develop complete end-to-end technical specifications for mobile systems. It has seven regional standards development organizations (i.e., ARIB, ATIS, CCSA, ETSI, TSDSI, TTA, and TTC) as its organizational partners. The output of 3GPP is a set of technical documents called technical reports (TRs) or technical specifications (TSs). The TRs are used to capture the results of technical studies which are then used to write TSs. The TSs are the actual normative standards documents[1] that manufacturers of mobile system products have to comply with. This textbook is primarily based on the technology behind the development of 3GPP standards specifications. More than 1500 standards experts are involved in the various working groups of 3GPP to advance the standards development work on every aspect of the mobile system. These standards are continuously being enhanced through the efforts of a large group of companies and are managed by the project management group resident in ETSI headquarters, Sophia Antipolis, France.

In addition to global standards organizations, there are several regional standards bodies, as already mentioned before, that develop standards specific to the operation of in-country or regional telecommunication networks. The regions or countries that do not have their own standards organizations utilize ITU guidelines or adopt standards developed by other leading regional or global standards organizations. The leading regional standards development organizations (SDOs) are as follows:

- Association of Radio Industries and Businesses (ARIB), Japan (www.arib.or.jp)
- Alliance for Telecommunications Industry Solutions (ATIS), North America (www.atis.org)
- China Communications Standards Association (CCSA) (www.ccsa.org.cn)
- European Telecommunications Standards Institute (ETSI) (www.etsi.org)
- Telecommunications Industry Association (TIA), North America (www.tiaonline.org)
- Telecommunications Standards Development Society of India (TSDSI) (www.tsdsi.in)
- Telecommunication Technology Association (TTA), South Korea (www.tta.or.kr)
- Telecommunication Technology Committee (TTC), Japan (www.ttc.or.jp)

---

[1]It must be stressed that 3GPP develops technical specifications, not standards, even though many consider these as actual standards documents. The TSs developed by 3GPP are transposed into standards by its seven regional organization partners to be used in their respective regions. These TSs might be slightly modified by the regional standards bodies to comply with regional regulatory requirements. The regional standards bodies are also responsible for enforcing Intellectual Property Rights (IPR) policy.

All standards organizations focus on creation of guidelines or technical specifications in accordance with their charter and in so doing interact with other standards organizations to leverage their expertise wherever necessary. This is required to ensure development of unified standards, which can be used by different vendors in building standards-compliant products, so that when deployed in the communication enterprise they work seamlessly. The worldwide communication enterprise is built using products that comply with standards from the above-mentioned standards organizations.

In the past, mobile systems were based on regional standards. For example, TIA developed AMPs & cdma2000® radio standards for use in the United States, ETSI developed GSM radio standards for Europe, and TTA developed JTACS & W-CDMA radio standards for use in Japan. Mobile systems based on these radio standards were deployed in other countries as well, with GSM-based mobile systems taking a stronger foothold worldwide. The mobile systems based on different regional standards resulted in fragmented markets around the world, making global roaming a challenge and building mobile devices complex. To alleviate this situation, ITU took on the task to develop international mobile standard (IMT-2000) in 1996. Several countries welcomed this effort, especially Japan and the United States. However, the ETSI standards body felt a big danger that GSM standard, which was already widely deployed would be obsolete with the development of a new ITU IMT-2000 standard, making GSM mobile system lose its market superiority. To circumvent this situation, ETSI, through participation from leading GSM equipment manufacturers and mobile operators, decided to create 3GPP global standards organization as a competing organization to ITU, inviting regional standards bodies to become members (organizational partners). The formation of 3GPP was thus realized in 1998.[2]

The main goal of 3GPP was to drive the evolution of GSM standard. NTT DOCOMO, a mobile operator from Japan who was a big proponent of ITU IMT-2000 in 1996, with strong participation, decided to leave ITU to join 3GPP and adopt GSM-based standard for their mobile market. This led to ITU IMT-2000 losing its importance, with reduced participation from other leading companies. Moreover, ITU was considered by many as too slow to meet growing demands of the mobile market. Hence, ITU's efforts to develop a global mobile system standard diminished and ITU was left with just developing high-level guidelines for future mobile systems. During the same timeframe when 3GPP formation was under consideration, the American National Standards Institute (ANSI)-accredited TIA organization was developing cdma2000 standard and wanted 3GPP to adopt cdma2000 as another alternative to the GSM evolution. This was not agreed by proponents of ETSI GSM standard, i.e., the major players in the formation of 3GPP. The proponents of US-based cdma2000 standard under the auspices of ANSI then decided to form a mirror global standards body to 3GPP, called 3GPP2, to drive the evolution of US-based cdma2000 standard. Hence, in 1998 two global standards bodies were created: one to drive GSM evolution, i.e., 3GPP, and the other to drive cdma2000 evolution, i.e., 3GPP2. Regional standards bodies joined both organizations as members since both standards had a market presence. The two standards progressed in parallel and were deployed in various countries by different mobile operators.

---

[2]3GPP was formed in 1998 to advance the work on third-generation mobile systems. The name 3GPP remains today even though the work in 3GPP has moved on to fifth-generation mobile systems. The reason is that 3GPP is a registered organization well known globally as the leading body to write TSs for the mobile systems used around the world. Hence, 3GPP name will not change.

The two mobile standards instead of multiple regional standards in the past was a much better situation, and a step in the right direction, but still did not end the fragmented mobile market. Around 2014–2015, the 3GPP-driven standard became prominent with a much larger share of the global mobile market, leaving the 3GPP2-driven standard with limited market share. The work in 3GPP2 started to decelerate along with participation in the working meetings. 3GPP2 eventually stopped its operations as all mobile operators adopted 3GPP-driven standard. Today, 3GPP is the primary standards organization on which products are built. We have finally reached a single global standard for mobile systems, improving the economies of scale. This is a win-win situation for customers, equipment manufacturers, and operators. It has been financially beneficial for equipment manufacturers to develop products for one type of mobile system rather than two competing ones in the past and avoid sending their company delegates to two different standards organizations.

Since the focus of this textbook is on the use of 3GPP standards, it is befitting to briefly explain the 3GPP organizational structure and its standards development process. The organizational structure of 3GPP is shown in Fig. 1.1.

Within the 3GPP organizational structure, there are three technical specifications groups (TSGs) where all the standards development work is conducted, i.e., TSG Radio Access Network (RAN), TSG Service and Systems Aspects (SA), and TSG Core Network and Terminals (CT). The TSGs are under the overall management of the Project Coordination Group (PCG), which outlines the technical work to be handled under each TSG and its timeframe. Each TSG is responsible for its work program under the terms of reference laid out by the PCG. As shown in Fig. 1.1, each TSG consists of

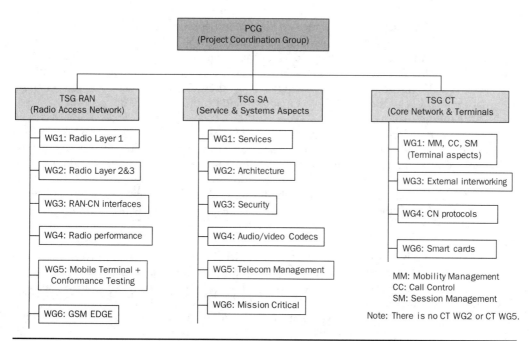

Figure 1.1   3GPP organizational structure

several working groups (WGs) to handle standards development for different technical areas. The outputs of the working groups are TRs that capture the results of technical studies, and TSs that are actual standards documents. The 3GPP standards development process follows the ITU-T-defined three-stage methodology outlined in ITU-T Recommendation I.130 [1]. The three stages are as follows:

- Stage 1: In this stage, the overall service description from the user's point of view is studied, followed by specifying the service requirements and the expected system behavior in delivering the required service. These are high-level service requirements to be used by Stage 2.

- Stage 2: In this stage, the functional capabilities needed in the system are studied and how these interact (information flows, call flows, or procedures) to support the service requirements as described in Stage 1. The Stage 2 description includes the system architecture showing how the different functions (or entities) in the system interact via signaling interfaces. The information flows specify message exchange between the functions.

- Stage 3: In this stage, the information flow and SDL diagrams from the Stage 2 output form the basis for producing the signaling system protocol recommendations and the switching recommendations. The actual protocols are developed during this stage. The Stage 3 specifications are the key documents used by the vendors in developing standardized products for the mobile system market.

The work on technical reports (TRs, studies, informative standards) or technical specifications (TSs, normative standards) is a lengthy process and is carried over several 3GPP WG meetings. The work on these documents is progressed via proposals submitted by member companies to the WG meetings. These proposals are then discussed, and if agreed are included in the draft TR or TS by the editor. When TRs or TSs are completed and agreed by the WGs, they go for final approval to the TSG plenary held quarterly. This is the final stop, after which TRs and TSs are approved becoming official 3GPP documents. Any future changes to them are managed through the change control process by the 3GPP project management group. All 3GPP documents are open to the public.

In the development of this textbook, it was felt that before delving into the 5G technology, it would be beneficial to provide background on how the mobile technology has evolved over the years up to 5G. This is to give readers a better understanding of the underlying concepts in the evolution of technology.

Chapter 2 summarizes some of the basic services of the mobile system and then delves into describing enhanced mobile system services that will be enabled in 5G. This chapter also provides a few use cases to help in understanding how the enhanced services are used in different applications.

Chapter 3 is devoted to providing a historical background of radio access network, core network, and terminals, along with some example services. This chapter is a welcome refresher on the initial concepts used in data transmission and shows the evolution toward 5G.

Chapter 4 talks about how 5G technology will be used for new services, especially for human-to-machine and machine-to-machine communications. The 5G new radio (NR) and core network (5GC) technology are superior to 4G in terms of reliability in handling of data transmission to meet the data delivery demands of superior 5G devices. This chapter will provide an overview of recent 5G commercial deployments around the world and some prominent 5G-based services, e.g., mission critical services, smart cities, and smart health.

Chapter 5 provides details of 5G Radio Access Network (RAN), mainly from an architecture and functional perspective, while explicitly describing the key technology enablers which distinguish NR from previous technologies. The key requirements in the design of the 5G radio are provided. Finally, this chapter provides a brief overview of some of the key 5G use cases where artificial intelligence (AI) technology is leveraged. Since radio spectrum is an expensive commodity (or pipe) which links the user to the outside world, its proper use is of utmost importance. The radio technologists are always on the quest to manage this expensive commodity in the most efficient manner. From mobile operator perspective, this is where the rubber meets the road.

Chapter 6 focuses on providing an in-depth technical analysis of the core network which is another important part of the overall mobile system. The core network has evolved along with the evolution of radio technologies from 1G to 5G. This chapter provides details on this evolution, followed by describing key aspects of the 5G core network.

Chapter 7 discusses a very important aspect of the mobile system, which is security. The radio link between the device and the cell tower carries user data in the air, which can be spoofed and must be secured. This chapter provides details on what security techniques are employed in securing the radio link (also called radio interface). Not only the radio data link must be secured from spoofing but also the unauthorized use of the radio link must be ensured, i.e., stop the entry into the mobile system by unauthorized users. A lot of effort goes into providing adequate security measures for the radio and link and other mobile system interfaces as well that could be hacked. For this reason, 3GPP has a separate security working group that addresses security requirements for the whole mobile system.

Chapter 8 provides an overview of the currently ongoing work on the new features for Release 17 and also gives an outlook on what is under discussion for the next generation, 6G. Yes, research is already under way on studying new capabilities for 6G.

Reference

1. ITU-T Recommendation I.130, "Method for the Characterization of Telecommunications Services Supported by an ISDN and Network Capabilities of an ISDN," 1988.

2

Background on Mobile Network Services (Pre-5G)

2.0   Smoke to 5G Services

The mobile systems' existence is relatively new compared to fixed wireline systems. The early mobile systems were based on the same principles as the fixed wireline systems with the addition of mobility support. They provided the same service offerings as the fixed wireline systems, and as time went on, the service offerings in mobile systems got richer and enhanced due to their wider use. This chapter provides background on mobile service offerings in pre-5G mobile systems to lay the groundwork for discussions that will follow in later chapters.

A mobile network is basically a means to provide some form of information to its subscribing entity (e.g., the human user), which can be visual, audible, or another sensory system. The mobile network delivers either one of these or a collection of them.

Before looking at what type of information can be provided, one needs to look at the history of information transfer. Initial types of information transfer between humans have been expressing oneself with sounds, movements, or drawings/paintings. The basic form of communication is speech, which is a direct communication medium that requires a person to make a sound and another to listen. This can only be achieved over short distances and cannot be stored. People have always wanted to be able to communicate over longer distances where speech was not possible. In such circumstances, other forms of communication mechanisms were created. A few of them are described as follows:

Paintings/Drawings: People would draw, e.g., on a wall or rockface a series of pictures or glyphs that would communicate/describe a story. The communication was stored, but it could not be transported/communicated over longer distance without creating a copy that could be moved.

Mail: This is the process of a written document being transported from point A to point B directly or indirectly. Its origins can stretch back to the "before Christ (BC)" period. It allowed communications to exist between people where speech was not possible, e.g., over long distances.

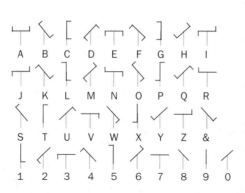

FIGURE 2.1 Semaphore example 18th century (Telegraph signals invented by French technician, Claude Chappe)

Smoke signals: A form of visual communications that allowed communications to take place over long distances. The only limitation to this form of communication was that it was limited to visual line of sight and weather.

Semaphore: A form of visual communication for conveying information by means of moving instruments. See Fig. 2.1.

Pigeons: These birds were used to carry written messages from point A to point B. There was no need for a visual line of sight as with semaphore; however, the speed of transferring the message was limited to how fast the pigeon could fly. The size of the message was limited, and there was no guarantee the pigeon would reach its final destination.

Telegraph: Very similar to smoke signals, telegraph was an electronic way of sending a signal using a code that represented each letter of the English alphabet. It enabled communications to take place between two parties, which was not limited by visual limitations or the weather. Its origins go back to two researchers in England and two in the United States. The code used consisted of a series of long and short signals or marks that would be known as "dots" and "dashes," which was standardized by ITU-R and called International Morse code as specified in ITU-R M.1677-1 [1]. See Fig. 2.2.

Telephone: The communication methods listed above were slow, required knowledge by a person to decode, e.g., Morse code or semaphore, or were subject to other limitations such as line of sight or weather. The telephone was the first communication method that made it possible for people to transport speech over long distances in order to communicate with other people afar. It was first developed in the 1870s. It allowed a person in one location to talk to another at a distant location using a telephone network, when otherwise those two persons would not have been able to communicate. Hence, the voice service came into existence.

2.1   Basic Mobile Network Services

2.1.1   General

Basic mobile network services started by taking what had been developed in the past for the fixed wireline telephone network and applying them to work over the mobile

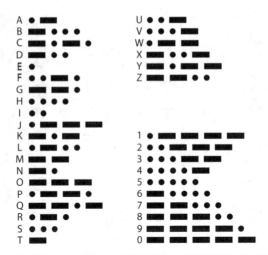

FIGURE 2.2   International Morse codes

communications system. Initial systems supported voice services and some basic call-forwarding services.

In the early European-based Global System for Mobile Communications (GSM) mobile systems (the so-called second-generation mobile system, 2G), the concepts of bearer service and teleservice were adopted from ITU-T Recommendation I.210 [2]. The Integrated Services Digital Network (ISDN) was already developed for the fixed wireline networks in the 1980s, and the idea was to adopt the concepts of ISDN for the GSM mobile system. A bearer service is the capability to transfer data between two end points. An example would be 9.6-kbit synchronous data. A teleservice is akin to an enhancement of a bearer service, and it provides an end-to-end voice service which consists of data encoding/decoding of voice signals using codecs at each end of the bearer.

A bearer service or teleservice provides a service to the end user, which can be further supplemented by additional capabilities called supplementary services (SS). SS cannot exist without a bearer service or teleservice but adds capabilities onto the bearer service or teleservice.

In later wireless systems (e.g., 2.5G[1], 3G, and then 4G), the concept of quality of service (QoS) was introduced, where a service could request a set of characteristics from the network to render acceptable level of user service expectations. The services would use the Internet Protocol (IP) as their transport, and the QoS would be applied to the IP packets.

2.1.2   Bearer Services

2.1.2.1   General

Before describing bearer services, one first needs to understand why the characteristics exist. Initial communication networks used analog technology with no digitization. The system would take the tones generated by phone or fax machines and/or voice and send them down the line with no alteration. When digitalization started, these

---

[1]Another name for the GSM system when GPRS was added.

voices/tones were used as the basis to create the digitalization methods, meaning that the system still natively supported the services that were initially using the telephone network. Tones, voice, and fax used 3.1-kHz bandwidth that voice uses (300–3400 Hz). However, as the usage of the telephone network increased, compression methods were introduced to increase capacity. Compared to pulse code modulation (PCM) that used 64 kbits to code a voice circuit, adaptive differential pulse code modulation (ADPCM) could use 16–40 kbits and digital circuit multiplication equipment (DCME)[2] 16 or 24 kbits. These compression methods affected the working of some of the services that used 3.1 kHz, e.g., carrying of tones used for a fax machine, where fax machine used an analog access line that connected to the digital network. With the advent of moving the digital line all the way to the user and the ability to use compression methods end-to-end, it became necessary to include information transfer capabilities to state what service or capability was required from the network to avoid unnecessary corruption of the end-to-end service. Thus, a bearer service has a set of characteristics associated with it. They define how a bearer is to be set up within the network. These characteristics are described in Secs. 2.1.2.2 to 2.1.2.8. They were initially proposed as part of the ISDN standards.

2.1.2.2 Information Transfer Capability

Information transfer capability defines capabilities associated with the transfer of different types of information through the network.

1. Unrestricted digital information (UDI): It requests the network to provide a bearer where the data, the bits, are not altered. It instructs the network not to perform any compression of the data. A number of circuit-switched (CS) data services were defined in the GSM system, initially at 9.6 kbits/s but later higher.

2. Speech: It requests the network that the bearer will be used for speech as defined within a set of specifications. Those specifications define the codec(s) to be used and how tones are to be transported in the network. It allows compression methods to be used that are tailored for voice service.

3. 3.1 kHz: It tells the network that a service that uses the legacy analog voice method is to be used, e.g., fax. The network then knows what types of compression, if any, can be used.

2.1.2.3 Transfer Mode

Transfer mode describes how transportation and switching occurs. The only value is:

1. Circuit switched (CS)

2.1.2.4 Transfer Rate

Transfer rate describes bit rate between two reference points:

1. Appropriate bit rate
2. Throughput rate

---

[2]PCM ITU-T Recommendation G.711 [3], ADPCM ITU-T Recommendation G.726 [4], and DCME ITU-T Recommendation G.763 [5], G.767 [6], G768 [7] are forms of digital transport mechanisms. They take analog signal and digitize it.

2.1.2.5 Establishment of Communication

Establishment of communication describes the mode used to establish a given communication.

1. Demand mobile originated (MO) only
2. Demand mobile terminated (MT) only
3. Demand mobile originated or terminated (MO, MT)

2.1.2.6 Communication Configuration

Communication configuration describes the spatial arrangement for transferring information between two or more access points.

1. Point-to-point communication: this applies when there are only two access points.
2. Multipoint communication: this applies when more than two access points are provided by the service. The exact characteristics of the information flows must be specified separately, based on functions provided by the public land mobile network (PLMN).
3. Broadcast communication: this applies when more than two access points are provided by the service. The information flows are from a unique point (source) to another (destination) in only one direction, i.e., one participant sends the data while the receiving participants only receive and do not respond back.

2.1.2.7 Symmetry

Symmetry describes the relationship of information flow between two (or more) access points or reference points involved in a communication.

1. Unidirectional: this applies when the information flow is provided in only one direction.
2. Bidirectional symmetric: this applies when the information flow characteristics provided by the service are the same between two (or more) access points or reference points in the forward and backward directions.
3. Bidirectional asymmetric: this applies when the information flow characteristics provided by the service are different in the two directions.

2.1.2.8 Data Compression

Data compression indicates whether the use of a data compression function is desired (and accepted) between a mobile terminal (MT), which is basically the mobile phone, and an inter-working function (IWF), which could be collocated with a mobile switching center (the MSC is described in Chap. 6).

1. Use of data compression requested/not requested
2. Use of data compression accepted/not accepted

Dominant attribute	Category of teleservices		Individual teleservices	
Type of user information	No.	Name	No.	Name
Speech	1	Speech transmission	11	Telephony
			12	Emergency Calls
Short message	2	Short message service	21	Short message MT/PP
			22	Short message MO/PP
			23	Cell Broadcast Service
Facsimile	6	Facsimile transmission	61	Alternate speech and facsimile group 3
			62	Automatic Facsimile group 3
Speech	9	Voice Group service	91	Voice Group Call Service (VGCS)
			92	Voice Broadcast Service (VBS)

TABLE 2.1   Teleservice categories and teleservices from 3GPP TS 22.003 [8]

2.1.3   Teleservices

Teleservices use bearer services to set up different types of service. They are a predefined set of services that a user can use (see Table 2.1). They set the characteristics (see Secs. 2.1.2.2– 2.1.2.8) to specific values for those teleservices.

There are two groups of speech services each having two individual teleservices. For example, speech has telephone TS11 and emergency calls TS12. Short message service has mobile terminated, mobile originated, and cell broadcast service.

2.1.4   Supplementary Services (SS)

2.1.4.1   *General*

Supplementary services are "add-ons" to other existing services, which cannot work unless a user has another service, e.g., voice, CS data. These services were first introduced in the wireline world with ISDN. Presently, users of devices don't have much visibility of how these services work or how they are controlled; at best, a user will select a menu option on the phone. However, before smartphones, users would have to dial a star, hash (*#) sequence to activate or provision one of the SSs. The *# sequence had a structure that identified the service, and any data to be appended to the service. The *# sequence would then trigger the device to send a message or sequence of messages to the network. The codes would allow the service to be activated, cancelled, and interrogated. Some services might even allow data to be registered, where registering would not mean the service was activated, e.g., a call-forwarding service requires data, i.e., the number the call should be forwarded to, to be registered so that the service can work.

2.1.4.2   *SS Categories*

2.1.4.2.1   Call Forwarding   Call forwarding is part of mobile terminated services. They were initially deployed in Europe to increase the completion rate of calls and thus guarantee revenue generation for the cellular operator. Different features of call-forwarding service are as follows:

- Call forward unconditional: The call is forwarded to another number. There is no condition that has to be met before the service is invoked.
- Call forward busy: The call is only forwarded if the subscriber is busy. It means that either the user is unable to be alerted when there is another mobile terminated call, or the user is already engaged in a call.
- Call forward no reply: The call is only forwarded if the user has been alerted for a period and the call has not been answered by the user.
- Call forward not reachable: The call is only forwarded if the mobile device cannot respond to a paging attempt.

2.1.4.2.2 Call Barring   Call-barring services are offered to allow control over where and when calls can be made or received by the user. Types of call-barring services are as follows:

- Barring of all outgoing calls (BAOC): As the name suggests, no outgoing calls can be made.
- Barring of outgoing international calls (BOIC): No international calls can be made.
- Barring of outgoing international calls EXCEPT those directed to the home PLMN country (BOIC-exHC): When the device is in another country, only calls to the home country can be made. No local calls or other international calls can be made.
- Barring of all incoming calls (BAIC): As the name suggests, no incoming calls are allowed.
- Barring of incoming calls when roaming outside the home PLMN country (BIC-Roam): Incoming calls are not allowed when the device is roaming outside the home country.

2.1.4.2.3 Call Completion Services   These services would provide capabilities to increase the opportunity for a call to be answered that otherwise may not. Types of call completion services are as follows:

- Call completion to busy subscriber: If a device is busy, the originating party could select to be notified when the terminating device is no longer busy, and the call would be attempted again.
- Call waiting: If the device was already engaged in one call and another mobile terminated call arrived, call waiting would provide an indication for a period of time before the call would be assumed to have failed. When the call is considered to be failed, either the call attempt is terminated or a call-forwarding service is invoked.
- Call hold: It allows the device to put the call on hold, effectively muting the mic and the earpiece. The service also allows the user to switch between calls.

2.1.5   Display Supplementary Services

The display services provide relevant information to the user device based on the current active supplementary service. They provide information that can be displayed on the screen to help users determine if they should answer an incoming call request, e.g., who is calling and what is the phone number of the caller. In addition, while making a call, the user can control the information to be released to the called party, e.g., they can decide not to have their phone number presented to the called party.

2.1.5.1   *Calling Line Identification*

This service allows the mobile terminated party to be presented with the identity of the originating party.

2.1.5.2   *Calling Line Restriction*

This service allows an originating party to withhold their calling line identification (CLI) from the terminating party unless the terminating party is allowed to override that option. The overriding option could be something emergency services have; however, some countries allow anonymous calling to emergency services as well.

2.1.5.3   *Calling Name Display*

This service is similar to the CLI service; however, it presents the name of the originating party.

2.1.6   Unstructured Supplementary Service Data (USSD)

It was foreseen that the GSM system would evolve over time and devices would not always support new services. USSD was created so that if a user entered a *# sequence into the device and the device did not recognize it, the device would send a USSD message to the network to process it. This allowed for services that were executed in the network to be supported by all devices.

USSD codes or strings consisted of a *# sequence with numbers. The numbers would identify the service and also provide additional data. For example, service X might also require a telephone number to be registered with it. In principle, the data was limited in scope to a limited character set. There was mobile-originated and mobile-terminated USSD to allow for services to be initiated either in the device or the network.

Later, USSD was expanded to allow it to be a generic data transport (e.g., binary data, alphanumeric data). Instead of just characters, applications could be loaded onto the device and they could use USSD as a way to send data from the device to, e.g., a server in the network. A very common use for USSD is the pre-pay services where a user types in a command and can retrieve their balance, etc. The pre-pay service is usually either a menu set on the phone or a set of entries stored in the phone textbook of the universal subscriber identity module (USIM). The phone textbook entry, e.g., "Check balance" would have a *# sequence, so when the user selects it the device would send this sequence to the network.

2.1.7   Management of Services

In order to provide services to a device belonging to a user, there needs to be a way how to manage what the device is allowed to do. This management can extend to, if a device is allowed to go to a particular location, what services the user equipment (UE) is allowed to access. In a GSM network, this user management was done via the Home

Location Register (HLR). In later generations, it was renamed to Home Subscriber Server (HSS). It is basically a database that contains information specific to the user (i.e., user subscription information). This information includes location of the user, authentication vectors to authenticate a user, and list of features and services a user has access to or not allowed to use at the user's present location. The latter case is more common to barring of features and services, to limit the subscriber's activity, e.g., disallowing user making phone calls while roaming.

2.1.8 Service Descriptions

2.1.8.1 *Voice Service*

Voice service allows an individual to do voice communication with another. The service is split into two services in the mobile network.

Normal voice calls: These are calls where a user (a user has a public and private identity) enters a public identity of a destination they want to call and resources are allocated in the network, and then the call is routed to the destination (subject to any restrictions imposed by operator, regulations, etc.). In order to make a phone call, the mobile device needs to be authenticated and authorized. Authentication is the process in the network to ensure that the identity is what it claims to be (see Chap. 7 for more information). Once an identity has been confirmed, it needs to be authorized for that service, in this case voice. This service in GSM and UMTS is known as TS11. If the service has not been authorized, it will not be allowed. In GSM and UMTS, the voice service is provided over CS bearers, where a dedicated set of resources are allocated. In LTE and 5G, the voice service is exclusively provided over IP (voice-over IP, VoIP) within the IP multimedia subsystem (IMS; see Chap. 6 for details); in this situation specific QoS is assigned to a set of resources to guarantee acceptable voice quality. Voice may be provided by CS in GSM or UMTS and via IMS in LTE or 5G, but it means that the mobile device will have to decide to use either of those two technologies for the voice service. When voice-over LTE (VoLTE, basically VoIP but via IMS) was defined, there was also a set of functionalities defined that aided the device to choose how to make the voice call, either over CS or IP, i.e., LTE. This is called voice domain selection.

Emergency call: This is a call with a higher priority in the network that allows the originating party to identify a mitigation situation. The user usually dials a number that is known to be associated to an emergency service. The mobile phone detects that the dialed number is an emergency call and sets up an emergency teleservice. The network will then assign higher priority resources to that call and route the call. This service in GSM and UMTS is known as TS12. The destinations that can be called are limited and usually handled by Public Safety Answering Point (PSAP). The PSAP may then further route the call, e.g., some countries have a generic emergency number like 999 and then the user will request for fire, police, and ambulance once the call is answered by the PSAP. Some other services may have dedicated numbers, such as poison control, animal rescue, and mountain rescue. Each country has its own specific numbers. In order to aid the caller sometimes emergency numbers can be downloaded to the mobile device and then the mobile device might have a menu that allows the user to select the type of emergency, e.g., coast guard. The mobile device will then send the local emergency number for that service or an indication of the type of emergency, e.g., coast guard.

In some situations, it is not always possible to download every type of emergency number, e.g., police may have different numbers for tourist police, local city police, or national police. In such situations, the mobile device will not detect the number dialed by the user as an emergency call, but the network will detect it and inform the mobile device of the type of call. This initial type of call is called "an undetected emergency call," meaning that the device did not originally request resources for an emergency; however, the call was identified by the network as an emergency call.

In some situations, a mobile device does not have to be registered with the network to make an emergency call. It may be mandated by law that unregistered mobile devices should be able to originate emergency calls.

PSAPs or emergency answering services may also be able to originate calls to mobile users that originated an emergency call; these types of calls are called PSAP callback. In some situations, it is possible to identify the call as a PSAP callback and inform the mobile device as such.

2.1.8.2 *Short Message Service (SMS)*

2.1.8.2.1 General   SMS is a teleservice defined as part of the GSM set of specifications, with the first SMS message tested in the early 1990s and its commercial deployment done within the next couple of years. In the 3GPP specifications, it is split into two services: one is mobile originated and the other mobile terminated. Mobile originated is where the user or the device sends the message to a destination and mobile terminated where a user or device receives a message. It is more traditionally known as allowing a device to send or receive text messages. The service is a store and forward one, in that messages can be sent to the recipient or stored for a period of time if the recipient could not receive the message, e.g., if the recipient was out of coverage. In such a circumstance, the system reattempts to deliver the message based on how the operator has configured the system to reattempt delivery. Given this nature of the SMS service, it cannot be relied upon to send real-time information.

The SMS protocol allows for 140 bytes of data/payload. Text messages could be coded in different ways, allowing up to a maximum of 160 characters (7-bit encoding), 140 characters (8-bit encoding), or 70 characters (16-bit encoding[3]) in an SMS. With the development of concatenation, longer messages could be sent. Concatenation was achieved by using some of the 140 bytes to carry a header field to extend SMS in a number of ways. In addition, SMS can be used by operators to send subscription data to the SIM card to configure it. These messages would be known as over-the-air (OTA) messages. The SMS was later enhanced with the ability to send pictures.

2.1.8.2.2 Uses for SMS   Initially, SMS was used to send short messages between devices. In some regions of the world, it is known as "texting"; however, SMS has been used for many new services, where the SMS service has become a transport for other over-the-top (OTT) services. Some examples are listed below:

- User verification: Web or phone-based applications use SMS as a mechanism to verify a user. Some mechanisms send a code via SMS to a device and then ask the user to type that code into a webpage or phone-based application.

- Voting: TV programs can instruct users to send a text message to a specific number to cast a vote.

---

[3]This type of encoding is used for non-Roman characters such as Chinese, Japanese, etc.

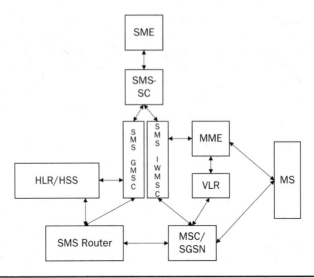

Figure 2.3   SMS architecture

- Notifications: Airlines, travel companies, etc., use SMS to send notifications to inform the user of an event.

Note: Most of the applications listed above are user-to-machine in some form. There are examples where the user can be removed and then it can be purely a machine-to-machine transaction. This is typically seen on phone-based apps, where a user has given permission to the application to read their SMSs, hence avoiding the need for user intervention.

2.1.8.2.3   SMS Architecture
2.1.8.2.3.1   General   SMS is specified in 3GPP TS 23.040 [9]. A copy of the basic SMS architecture is shown in Fig. 2.3.

A brief description of the functions in the SMS architecture is provided below.

2.1.8.2.3.2   SMS-Service Center (SC)   The SMS-SC is a function in the network that supports delivery of an SMS message from the mobile phone, also called mobile station (MS) in GSM/UMTS or user equipment (UE) in LTE and 5G. It is the entity that receives the SMSs and forwards them to the recipients. It basically acts as a store-and-forward function. If the destination is unable to receive the message, it will be stored for a period of time and delivery will be reattempted. It will also generate reports that can be sent to a short message (SM) end point, e.g., the entity that originated the short message. When messages cannot be delivered, it will set a flag in the HLR/HSS indicating the status why the message could not be sent.

2.1.8.2.3.3   HLR/HSS   This is the database function that is described in Chap. 6. It maintains a number of additional data items per subscriber related to SMS independent of whether 3G or 4G is used, as follows:

   a. Messages Waiting Data (MWD) flag, which contains at least one service center address

    b. Last known MS location as reported by the serving MSC, SGSN, or Mobile Multihop Relay (MMR)

    c. Mobile Station Not Reachable Flag (MNRF)

    d. Mobile Station Not Reachable for GPRS (MNRG)

    e. Mobile Station Not Reachable via the MSC Reason (MNRR-MSC)

    f. Mobile Station Not Reachable via the SGSN Reason (MNRR-SGSN)

    g. Mobile Station Memory Capacity Exceeded Flag (MCEF)

The items (b)–(g) are flags that are stored in the HLR/HSS. These flags indicate to the system a reason why the SMS could not be delivered to the end subscriber (see Sec. 2.1.8.2.1 regarding SMS is a store and forward service). The system is then able to take appropriate action based on the flag (e.g., if flag MNRR-SGSN is set, the SMS-SC can attempt to send the SMS via a different access mechanism, and if flag MCEF is set, it means that the UE has no more memory for SMSs and the system should not send any more messages until it has been informed about availability of memory). Item (a) contains one or more SMS-SC addresses, so when the HSS is aware that the MS can be reached again for the SMS it will alert the SMS-SCs that are stored in item (a).

2.1.8.2.3.4   ***MSC/SGSN***   The MSC is an entity in the 2G and 3G networks that support CS services such as voice, data, and SMS. It performs mobility management and subscriber management functions (see Chap. 6). For the SMS service, it is responsible for determining where to page the MS/UE and the paging strategy to reach the MS/UE in case there is no response to initial paging attempt. It also executes a number of supplementary services that are described in Sec. 2.1.4.

    The serving general packet radio service (GPRS) support node (SGSN) is a 2G and 3G entity that is responsible for packet-based data transport. It performs mobility management and subscriber management functions but supports no supplementary services. For SMS, it performs the same functions as the MSC.

2.1.8.2.3.5   ***MME***   The mobility management entity (MME) is the 4G equivalent of the SGSN; a more detailed description can be found in Chap. 6.

2.1.8.2.4 Support of SMS via IMS Architecture

SMS is a short data delivery mechanism that can be subject to delay. Its architecture, see Sec. 2.1.8.2.3.1, is tightly integrated with the underlying cellular network, making the service dependent on the cellular transport methods. In order to make the SMS service transport agnostic, the IMS was used as the transport mechanism over IP. IMS (see Chap. 6, Sec. 6.9) supports a similar mechanism to SMS whereby the IMS sends data in a single message through the system. As the use of a common service delivery mechanism shows better performance, IMS was seen as the next logical way to transport SMS. This was achieved using the SIP MESSAGE method and encapsulating the SMS payload inside the SIP MESSAGE, making for an easy migration from the legacy 2G/3G mechanisms to using SIP.

    SMS can be also supported/delivered using IMS. IMS is defined in more detail in Chap. 6; however, for the purpose of this section, it is a system that uses SIP as a control protocol. The SIP MESSAGE method is used to transport SMS messages. It effectively provides a common framework for supporting services. Figure 2.4 shows the architecture to support SMS delivery over IP, called SMSoIP. A new gateway (GW) entity called

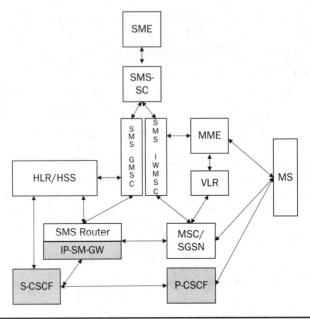

FIGURE 2.4   SMS over IMS

an IP-SM-GW is introduced that allows the legacy SMS architecture to interwork with the IMS architecture. The new parts of the architecture are shown in light gray. The S-CSCF and P-CSCF are discussed more in Chap. 6.

2.1.8.3 *Text Broadcast*
SMS is a point-to-point service, in that it goes from one user to another. There is also a broadcast version of SMS called cell broadcast service (CBS). This first originated in GSM system and is commonly used today in public warning systems (PWS).

2.1.8.3.1 Cell Broadcast   Cell broadcast is a service that allows a single or group of cells to send out information periodically. Devices can, in most situations (see Sec. 2.1.8.2.4), decide if they wish to receive such information.

In GSM and UMTS implementations, the CBS interfaces to the RAN; however, in 4G and 5G it interfaces to the core network. These later architectures are shown in Fig. 2.5. The MME or AMF is responsible to deliver the message, created by the cell broadcast entity (CBE) to the UE.

Cell broadcast center function (CBCF): It takes the message sent by the CBE and ensures that it is sent over the cellular system. It allocates a serial number, and determines the time to start and stop sending the message, which cells the message should be sent to, and the time period at which they are broadcast.

CBE: This is the entity that creates the SMS message to be broadcast by the cell(s). It could be a computer with a screen where someone types in the message. It could also represent the PWS as a single functional entity.

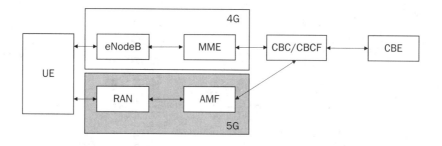

FIGURE 2.5 Cell broadcast architecture

2.1.8.4.2 Public Warning System (PWS)   PWS is a service that uses CBS as a transport and is provided to the general population. It is used to inform users of a pending situation. In the United States, this is called Commercial Mobile Alert System (CMAS). In Canada, the system is called Wireless Public Alerting System (WPAS), whereas in Europe it is called Early Warning and Response System (EWRS).

A PWS may have different levels of information that can be transmitted—some are of a more informative nature where a user may opt to "opt out" to receiving, whereas others are mandatory to be received. Messages are created by authorized entities such as national or local governments and then directed to geographical areas where the incident notification would be appropriate.

The CMAS has a number of different levels of warnings called:

- Amber Alert—used to indicate child abduction
- Extreme/Severe threats—for weather and other situations that may affect life
- Presidential Alert—used by the president to issue a warning when there is a national emergency or threat

Another well-known PWS is deployed in Japan and is called Earthquake Tsunami Warning System (ETWS). As the name suggests it is usually used to inform the public of earthquake, tsunami, etc.

Figure 2.6 shows a simple architecture for PWS, where there are three agencies that can provide messages to the public. An alert aggregator takes those and provides them to the entity that communicates to the wireless network.

2.2   5G Enhanced Mobile Network Services

2G and 3G services were mostly fixed in nature, meaning that all operators basically offered the same set of services. In order to launch a new service, excessive standardization was required to specify that service and it did not change the status quo; operators again would all be on equal ground.

The 4G network was designed to be a packet-based (All-IP) network that provided a set of capabilities that could be adapted to provide differentiated services. The key enabler to providing the services would be IMS (see Chap. 6). IMS was a session-controlled framework that would allow operators to develop services independent

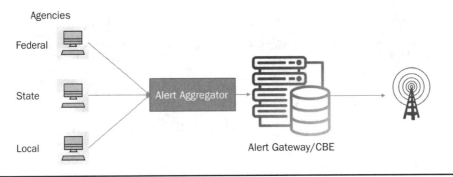

FIGURE 2.6   Simple PWS architecture

of each other; the first services to be launched would be voice, SMS, and video calls. However, with the advent of 4G packet-based delivery system, a tidal wave of "OTT" providers, who used the basic IP connectivity, which even though was best effort, provided a complete suite of new and exciting services.

IMS was also extended in 4G to allow a migration path from CS voice to IMS-based voice. The capability was called IMS centralized services. IMS would be used to provide supplementary services that in 2G/3G where executed in the cellular core network would now be executed in the IMS network. In addition, as IMS allowed a user to share their telephone number across multiple devices IMS was extended to allow a user to move a voice call, video call, etc., from one device to another or even split components of a call. An example scenario could be that a user would be having a video call outside and enter into the house. The video component of the video call could be transferred to a TV while the voice component stays on the phone. The video call would now be hosted on two devices—TV and phone.

5G aims to expand on the opportunities for operators/service providers by having a system that allows vertical industries to be able to use the mobile network. The vertical industries can concentrate on what they are best at and the operator can provide managed mobile services to those verticals. Vertical industries are, for example, the car industry with vehicle to anything (V2X) communication or manufacturing industry where the production machines are connected via 5G.

2.3   Service Evolution Toward 5G (Historical Background)

The following sections provide an overview of some services that were developed for pre-5G networks or may have also used noncellular technologies. They are now drivers for new capabilities from 4G and 5G networks.

2.3.1   Internet of Things (IoT)

Initial mobile networks were about providing the people with the ability to communicate with each other. 2G system provided voice, texting, and very basic data connectivity; 3G provided higher rate data that was packet orientated; 4G and 5G provide very high speed data capabilities. In addition, as the mobile system has evolved, the number of types of devices has increased to include the ability of machines to communicate with machines, or an "internet of things." The mobile network can now aid in this new paradigm. Some examples of services are:

- *Utility meters*: Traditionally, the utility company would send a person out to read the meter once a month, so a bill could be generated for the customer. The customer would have no easy way to determine how much they had consumed, and if they did, creating an estimate of the bill could be challenging. An IoT-capable utility meter can now report back on usage, traditionally a daily figure, but if the IoT device has a long-term power source, it can report back more often. The utility company can see how much of the utility is being used and act accordingly. If there is a high usage, e.g., water and its unusual consumption for the time of day and time of year, the utility could take action. Consumers can also see how much they are using and how much it is costing, and take appropriate actions.

- *Pressure sensors on suburban railway coaches*: The pressure sensors can report the weight in the coach, which gives an indication how full that coach is. Users can use an application on their smartphone to determine what train and even which coach they should use for the best comfort.

Given the wide potential for the number of different use cases IoT devices can be used for, the cellular system supports many features. Some include aiding battery-constrained devices to perform a lot longer on their battery, e.g., the device informs the network that the device will go into a hibernation state. In the case of utility meters, where there are potentially thousands, the system instructs the meters to report their data over a period of time so as to not congest the network and allow the meter to successfully report its data.

2.3.2 Vehicle to Anything (V2X)

2.3.2.1 General
V2X enables the vehicles to communicate to other vehicles (V2V), pedestrians (V2P), business (V2B), home (V2H), and roadside infrastructure (V2I), as shown in Fig. 2.7.
Some typical services are described in the following sections.

2.3.2.2 Weather/Road Condition Reporting
This service helps the car in reporting to the network about the criteria that are relevant to road conditions, such as windscreen wiper speed, vibration characteristics, outside and inside temperature, visibility, lights on the car, humidity, dew point, etc. The purpose is to aid other road users and road authorities to act upon this information. They can change the speed limit, perform road maintenance, such as snow clearing, etc.

2.3.2.3 Basic Safety
This service reports basic characteristics of a moving vehicle, such as location, speed, heading, vehicle type, position of steering wheel, if brakes are being engaged, etc. This type of information can be used for cooperative adaptive cruise control, speed harmonization, pedestrian warnings, etc.

2.3.2.4 Platooning
Platooning is a formation of vehicles that have an interest in going in a common direction in the same lane, as shown in Fig. 2.8. There is usually a leader of the platoon that

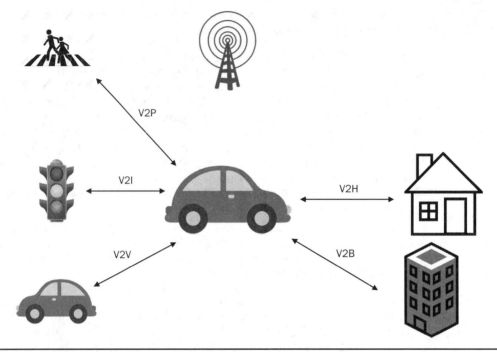

Figure 2.7 V2X communications

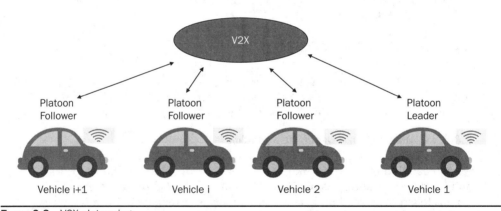

Figure 2.8 V2X platooning

is responsible for the direction of the platoon and its speed. It signals to the other cars the direction to take, its speed, etc. The other vehicles will signal things like their position in the platoon, maintain a safe distance from each other, and follow closely behind. Vehicles can leave and join the platoon. A platoon can increase road safety, reduce congestion, and improve energy efficiency due to the flow dynamics of the air across the platoon.

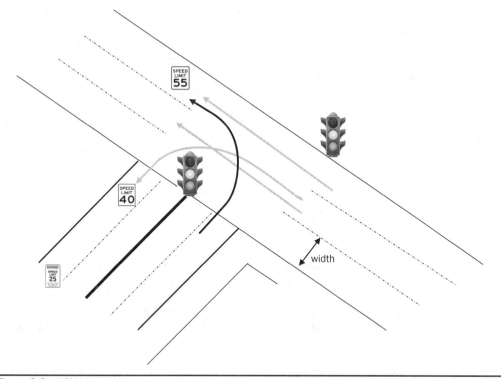

2.3.2.5 Traffic Light Timing and Intersection Maps

When cars and buses approach intersections, they do not know the layout of the intersection nor the timing of the traffic lights. In such circumstances, vehicles can be in wrong lanes and larger length vehicles can end up blocking intersections or causing congestion due to their less aggressive acceleration capabilities. In addition, roadworks can also cause problems as the road layout can change, lanes can get narrower, and they can move the orientation totally. Providing an intersection, as shown in Fig. 2.9, to the approaching vehicle is useful information for the driver and can greatly improve the flow of traffic through the intersection.

Traffic light timing can also be used to help improve traffic flow. If a bus or long vehicle is wanting to progress through the intersection, information on traffic timing could aid the vehicle to move at the right speed to either stop or go through. Emergency vehicles could also provide their position, so lights can be set to not hinder their progress to an incident.

2.3.3   Mission Critical Communication (MCC)

2.3.3.1   General

Mission critical communication (MCC) is about providing services to users where the services themselves are imperative to users performing their tasks/jobs. Examples are firefighters, police, ambulance, etc., where the operations involve high risk to life and property. However, the services are equally applicable for utility companies, sports arena staff, transportation systems, etc. This latter group might be known as business critical communications. The main difference between the two is the amount of fault tolerance or failure that the system can sustain before all communication ends.

MCC systems are usually push-to-talk (PTT) systems based on land mobile radio (LMR) system. Later, digital Project 25 (P25)[4] [10] suite of standards were developed in North America and TETRA[5] [11] standard was developed in Europe; both are adopted by numerous other countries. An MCC system is required to deliver the service in a time critical fashion. The MCC system should be able to provide:

- Groups management
- Broadcast
- Low latency
- Fault tolerance

With the requirement to support the above characteristics and more feature-rich data services over PTT systems, there was an interest to use cellular technologies as they could provide all of these capabilities to the consumer market.

Now, there is a move to use LTE and NR for MCC leveraging the functionality and capabilities that are available to consumer devices. In addition to MCC voice, MCC data and video was also specified, bringing consumer-based functionality to the MCC community.

2.3.4   Healthcare (Remote Operation by Doctors from Different Locations)

As depicted in Fig. 2.10, healthcare is another area that telecommunication services are being used in. Some examples include remote monitoring of patients. Here, a patient is provided with one or a group of devices that monitor the patients' health, e.g., watch, ring, and body sensors. These sensors then collect vital statistics of the patient and send them to a remote monitoring facility. The network should be able to support low latency as medical situations are critical as well as support devices that are battery constrained. In addition, sensors that are small may make use of another device, e.g., a smartphone or a wireless gateway in the home to relay the information.

A similar use case to the one above is an ambulance that has a number of medical equipment devices including a blood pressure monitoring, ultrasound, and vital statics monitoring devices. In addition, if the ambulance is at an incident, paramedics will have to work remotely from the ambulance, and the paramedics will have devices to treat people as well as there may be body cameras and devices monitoring the paramedics.

---

[4]P25 [10] is a set of standards produced through the joint efforts of the Association of Public Safety Communications Officials International (APCO), the National Association of State Telecommunications Directors (NASTD), selected Federal Agencies and the National Communications System (NCS), and standardized under the Telecommunications Industry Association (TIA). P25 is an open architecture, user-driven suite of system standards that define digital radio communications system architectures capable of serving the needs of Public Safety and Government organizations.
[5]TETRA [11] is an open standard developed by the ETSI to define a series of open interfaces, as well as services and facilities, to enable independent manufacturers to develop infrastructure and terminal products to provide interoperability with each other.

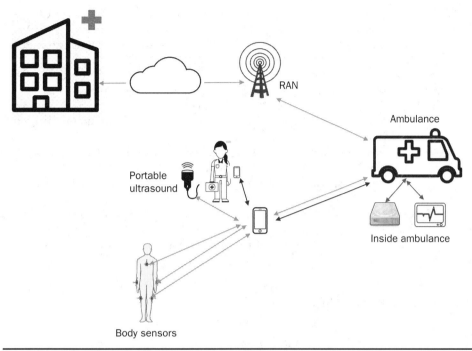

FIGURE 2.10   Healthcare in 5G

The network should be able to cope with a remote incident center that is reporting high bandwidth delay sensitive applications as well as low bandwidth, low data.

One final use case for medical applications is remote operation, whereby a surgeon may be involved in an operation remotely, either because of an incident where a patient requires immediate attention, or a patient is in a hospital and the surgeon is situated either in another room or location. In this situation, there is a need for high-quality imaging (e.g., 4K+) where there is almost zero data loss. Data loss is not acceptable if a surgeon is performing a remote surgery that requires precise working without any data loss.

2.3.5   Manufacturing (Vertical LAN, URLLC)

Manufacturing industries are another use case for vertical services that can utilize 5G for connecting the production machines. Key requirements are ultra-reliable low latency communication (URLLC) and redundant transmissions in order to send control messages to the machines without interrupting the production process.

References

1. ITU-R M.1677-1, "International Morse code."
2. ITU-T Recommendation I.210, "Principles of Telecommunication Services Supported by an ISDN and the Means to Describe Them," March 1993.

3. ITU-T Recommendation G.711, "Pulse Code Modulation (PCM) of Voice Frequencies," March 1993.

4. ITU-T Recommendation G.726, "40, 32, 24, 16 Adaptive Differential Pulse Code Modulation (ADPCM)," March 1993.

5. ITU-T Recommendation G.763, "Digital Circuit Multiplication Equipment using G.726 ADPCM and Digital Speech Interpolation," March 1993.

6. ITU-T Recommendation G.767, "Digital Circuit Multiplication Equipment using 16 kbit/s LD-CELP, Digital Speech Interpolation and Facsimile Demodulation/ Remodulation."

7. ITU-T Recommendation G.768, "Digital Circuit Multiplication Equipment using 8 kbit/s CS-ACELP."

8. 3GPP TS 22.003, "Circuit Teleservices Supported by a Public Land Mobile Network (PLMN)."

9. 3GPP TS 23.040, "Technical Realization of the Short Message Service," Release 4.

10. TIA TSB-102-B, "Project 25 TIA-102 Document Suite Overview."

11. ETSI EN 300 392-1, "Terrestrial Trucked Radio (TETRA); Voice plus Data (V+D); Part 1: General Network Design."

Problems/Exercise Questions

1. Name two early forms of communications.

2. Name two types of digital voice encoding.

3. What is teleservice T12?

4. Describe the characteristics of a bearer service.

5. What is a supplementary service? List four of them and what they are used for.

6. How many bytes of data (payload) are there in a short message?

7. How can you transport more characters in a short message than the available payload?

8. Draw the SMS architecture and briefly explain each function's purpose in the architecture that you draw.

9. What happens if an SMS message cannot be delivered?

10. Name the three types of text delivery services.

11. What is V2X? Describe two use cases for it.

3

Evolution of Mobile Technologies

3.0   Introduction

The mobile system consists of two subsystems, as shown in Fig. 3.1: the radio access network (RAN) subsystem covered in Chap. 5 and the core network (CN) subsystem covered in Chap. 6. One of the main functions of RAN subsystem is to provide ingress for wireless devices (e.g., mobile phone users) into the mobile system via the radio access technology mounted on cell towers, followed by the CN subsystem enabling mobile phone users to reach their desired destination addresses (e.g., dialed phone number they want to reach) anywhere in the world. So, essentially the mobile CN subsystem provides an egress for the mobile phone users to the public network, called public-switched telephone network (PSTN), making it possible for mobile phone users to reach anyone anywhere, just like the fixed landline phone users (i.e., PSTN users). The backhaul transport network shown in Fig. 3.1 allows RAN and CN to be interconnected as they can be located far away from each other.

When a cellular service is initiated by an active (powered up) mobile device, the RAN subsystem gets involved first to determine the actions required, which then involves the CN subsystem to take further actions, as necessary, in the handling of the requested cellular service toward its final completion toward the PSTN.

An example of a typical mobile (cellular) service: All active (i.e., powered up) mobile devices even while idle, i.e., not involved in an ongoing phone call, are constantly in radio communication with the RAN to ensure that their location is known to the mobile system at all times. This is required to ensure that active mobile devices are reachable by the mobile system for delivery of incoming service request, and also the mobile system is instantly ready to accept an outgoing service request, as initiated by active mobile devices.

As a black box, the CN subsystem shown in Fig. 3.2 responds to input commands and produces outputs based on these commands. Inputs and outputs can come from the RAN subsystem or the PSTN. A simple input command from the RAN would be to complete a voice call initiated by the mobile phone user (subscriber). The subscriber dials the digits, which are sent to the RAN subsystem and forwarded to the CN subsystem. The CN subsystem takes appropriate actions on the dialed digits by analyzing the dialed digits to determine the intended destination of the voice call, and forwards to the PSTN. The PSTN locates the destined called party and sends a response back to the CN subsystem. The CN subsystem in turn sends the accepted response to the RAN subsystem to notify the subscriber device to initiate connection of the voice call between

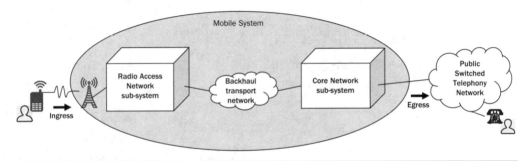

FIGURE 3.1   Mobile system

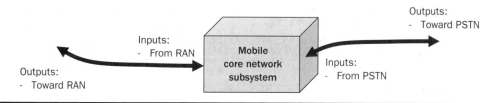

FIGURE 3.2   CN subsystem inputs and outputs

Generation	Year introduced	RAN technology	Speeds	Usage
1G	1980	AMPS, NMT, TACS	14.4 Kbps	For voice only
2G	1990	GSM, PCS	14.4 Kbps	For voice and data
2.5G	2001	GPRS, EDGE	56–256 Kbps	For internet services
3G	2004	UMTS	500–700 Kbps	For multimedia services
3.5G	2006	HSPA	14.4 Mbps	For high-speed data needs
4G	2011	LTE	100–300 Mbps	For video streaming
5G	2020	5G NR	10 Gbps	For massive data communications & IoT

TABLE 3.1   Radio technology evolution and its usage

source party and destination party on the communication path reserved by the CN subsystem and the PSTN. All actions taken by the CN subsystem are done through input commands to arrive at desired output commands. The input and output commands can become complex based on the type of cellular service requested by the subscriber. Further description of the cellular services is provided in more detail in other chapters.

The mobile system is continuously evolving with enhanced capabilities due to the introduction of advanced radio access technologies to meet the growing demands of the data traffic. Initially, the mobile system was mainly used for voice communications. However, as the use of data services (text messaging, video streaming, internet browsing, etc.) has grown significantly over time, the mobile system evolved from just supporting voice communications to supporting data communications as well. The following chart in Table 3.1 shows how the radio technology and its usage have evolved over the years from 1G to 5G. It should be mentioned that start of the mobile system was motivated by the pre-cellular wireless telephony service, referred to as 0G, made available as a commercial service via the PSTN, using a transmitter/receiver mounted in a small suitcase and requiring operator intervention to connect the call. The pre-cellular wireless telephony service was meant for voice calls only with no handover feature; hence, there was no service continuity, i.e., handover support, if a user was moving out of coverage of a cell resulting in an interrupted call.

The focus of this chapter is on the evolution of global switch mobile (GSM)-based system originally standardized by European Telecommunications Standards Institute (ETSI; *https://www.etsi.org*) and deployed in many parts of the world. In December 1994, when the Federal Communications Commission began the auction of 1900 MHz frequency band for offering broadband personal communication services (PCS) in the United States, the GSM standard took foothold in the United States. The GSM-based

PCS standard was standardized under the auspices of American National Standards Institute (ANSI) around 1996/1997 timeframe. It was deployed by several US operators, e.g., Pacbell Mobile Services, Voicestream, Powertel, etc. It can be noted that international roaming in different countries was not possible using 1G, because of the variety of incompatible 1G standards around the world. However, international roaming was facilitated in 2G using multimode cellular devices.

When 3rd Generation Partnership Project (3GPP; *https://www.3gpp.org*) was formed in 1998 as an international mobile standardization body to primarily promote GSM-based mobile standard as a global standard, it was joined by the major international GSM mobile operators participating in regional standards development organizations. The current organizational partners of 3GPP consist of seven regional standards development organizations from Asia, Europe, and North America who determine the general policy and strategy of 3GPP. These regional standards bodies are ARIB, TTC, TTA, CCSA, TSDSI, ETSI, and ATIS (see Chap. 1). Since 1998, the standardization of GSM-based system in 3GPP has continuously evolved from Release 1999 to current 5G Releases 15/16. When GSM-based standards were adopted by 3GPP, they were managed and progressed using the existing ETSI standards development process until Release 1999. Then, 3GPP adopted a new process to manage these standards starting with UMTS in Release 4. The progression of GSM releases under 3GPP is as follows:

- GSM Phase 1 used ETSI standards development process, considered as 3GPP Release 1.

- GSM Phase 2 used ETSI standards development process, considered as 3GPP Release 2.

- Release 1999 used ETSI standards development process, considered as 3GPP Release 3.

- Release 4 onward started using the new 3GPP standards development process.

3.1   The Gs

The evolution of mobile system has continued over the years starting from 1G around 1970 to 5G of today with the evolution of radio technology. This evolution will continue in the future as the usage of mobile technology expands starting initially from human-to-human (voice calling, SMS, etc.) to human-to-machine (web services) and now toward machine-to-machine/IoT communications used to support a plethora of vertical services such as smart cars, factory automation, healthcare, etc. With each generation, there has been an improvement over the previous one. The pre-cellular wireless telephony service (referred to as 0G) was basically a fixed wireless service provided on the fixed wireline network, with no mobility support. The first-generation (1G) mobile wireless communication network was analog used mainly for voice calls followed by 2G which used digital technology providing voice and text messaging and later packet data services. Then, with the arrival of 3G, higher data transmission rate was provided for enabling multimedia support. With the introduction of 4G the wireless mobile internet came into existence; hence, making it possible for users to use internet services with similar ease as possible on the fixed wireline network, but not with similar speed due to inherent limitations of the wireless technology. The 5G mobile system will bring about a new revolution in wireless communications, mainly geared for the realization of internet of things (IoT), where all objects (i.e., devices, appliances, etc.) will become smart (i.e., equipped with computer chips that can collect and transmit data via sensors) and interconnected via internet using WLAN or IEEE 802.11 or the ultra-reliable low-latency 5G radio technology. The sensor technology will be a huge enabler in this revolution.

The mobile system generations (0G, 1G, 2G, etc.) from a technology perspective are mainly associated with the evolution of radio access technologies (from a mobile device to the cell tower) and not so much on the mobile CN. However, the mobile CN does evolve (or is enhanced) with the introduction of each new generation of radio access technology. The mobile CN is a centralized system providing interconnectivity for the radio access network to the outside world. The mobile CN technology has evolved as well, but more to support enhanced services, such as when moving from an analog-to-digital signaling, or from supporting voice services only to voice/data services, or enhancing the network architecture from a point-to-point-based architecture to a service-based architecture (as done in the 5G system). From a customer perspective, the evolutions of the core network are not that evident to the general user compared to radio technology evolution due to the changes in mobile devices which are more visible with enhanced data speeds.

3.1.1   Zero Generation (0G)

Wireless telephone service started with 0G (zero generation) which was deployed sporadically after World War II. It was called pre-cellular telephony service made available on some automobiles before the advent of cell phones in 1970, triggering the start of cellular service. In 0G, there was no mobility or handover capability; therefore, it was similar to fixed wireless service, used primarily by construction foreman, realtors, and celebrities. The push-to-talk feature was one of the main features supported in 0G and could also be known as land mobile radio (LMR) networks. The core network used was the PSTN. This pre-cellular wireless service (referred to as 0G) was only mentioned here briefly as background information and will not be discussed further.

3.1.2   First Generation (1G)

The actual advent of wireless cellular systems came about in the mid-1980s offering mobility and handover capabilities. These systems were referred to as 1G (First Generation) used for wireless mobile telecommunications. Several countries in North America and Australia used the ANSI-developed Advanced Mobile Phone System (AMPS) standards, but there were several other analog standards developed by other countries, which were not compatible to each other, e.g., Total Access Communications System (TACS) in the United Kingdom, JTACS in Japan, Nordic Mobile Telephone (NMT) in the Nordic region, and others. Multiple analog standards were used worldwide, making it difficult to support global roaming for users traveling to other countries using a cell phone supporting a different analog standard. The main difference between 1G and 2G is that 1G networks used analog signaling, while 2G networks used digital signaling. The circuit-switching technology was used in the mobile core network for voice communication. However, communication with data networks was made possible by connecting a circuit-switched voice connection to a modem on the far end.

3.1.3 Second Generation (2G)

The 2G cellular networks were deployed in 1991 in Europe using the GSM standard for voice and messaging service and later GPRS and Enhanced Data Rates for both GSM Evolution (EDGE) standards for limited data services. The SMS and MMS features (see Chap. 2 for details) were introduced in 2G.

The introduction of data services started with 2G mobile core network, initially low-bandwidth circuit-switched data services which were enhanced later when a packet-switched network and radio access technology enhancements were added.

3.1.4 Third Generation (3G)

The achievable data rates with GPRS and EDGE as enhancements to 2G were very limited and only improved the data performance when user was close to the base station. For this reason, 3G radio technology (WCDMA) was developed and was first commercially deployed in Japan in 2001, providing faster internet speed, making it possible for the deployment of broadband services.

When 3GPP (the GSM-led worldwide mobile standards initiative) was formed in 1998, it adopted the GSM core network architecture (standardized in ETSI) as its first release for 3G—Release 1999 or Rel-99 or R99. The technology was called Universal Mobile Telecommunications Service (UMTS). The main focus for UMTS, from its predecessor GSM, was the redesigned radio interface technology based on WCDMA instead of TDMA/FDMA in 2G (see Chap. 5 for details).

3.1.5 Fourth Generation (4G)

In 2008, the 4G radio technology (LTE) was introduced for providing capabilities, as defined by ITU in International Mobile Telecommunications (IMT) Advanced, making it possible for services such as IP telephony, video conferencing, and gaming. In 2008, the mobile CN went through a major change when the network evolved from using a combination of circuit-switched digital technology and packet technology to an All-IP core network in 4G eliminating circuit switching all together.

3.1.6 Fifth Generation (5G)

In 2019, 5G new radio (NR) technology came about which is geared to provide support for enhanced mobile broadband (eMBB), enhanced vehicle to everything (eV2X), critical communication, massive connectivity, and ultra-reliable low-latency communication needed for machine-to-machine communications (massive MTC) (see Chap. 6). The focus of this textbook is on technology evolution toward 5G. Therefore, prior to getting into details of 5G in subsequent chapters, it is important to provide the motivations behind the development of 5G. Chapter 4 provides a description of the market trends and business motivations for new 5G services.

3.2 Radio Network Architecture

3.2.1 Radio Access Network Evolution

This section presents an overview of the key features, architecture, and roadmap evolution of radio access technologies from the first to fourth generation of mobile networks evolution, as highlighted in Fig. 3.3. A detailed description of the evolution of the various generations of radio access networks from 1G to 5G and requirements for 6G has been presented in [1].

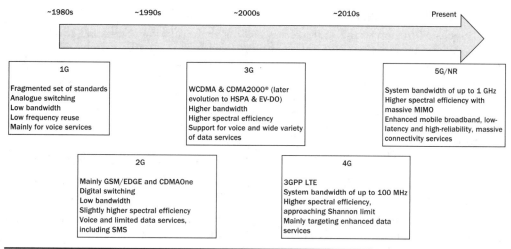

FIGURE 3.3   Roadmap of cellular technologies

The first generation (1G) of radio access networks utilized an analog transmission technology which was prevalent at that point of time and was mainly designed to provide basic voice service. Due to the fragmented nature of the industry, there were a significantly large number of standards that were available in different parts of the globe, which provided limited mobility between the parts that adopted different standards. The second generation (2G) was able to provide a significantly improved voice quality and gave birth to the first data service offering in the evolution of mobile networks. It also provides a platform that was adapted by most of the world's operators to allow global roaming. South Korea and Japan adapted the CDMA technology that was being deployed in the United States. For Global System for Mobile Communications (GSM)-based networks (standardized by European ETSI body), the enhancements for providing data services was called Enhanced Data Rates for both GSM Evolution (EDGE) [2] and for code division multiple access (CDMA)-based networks CDMAOne (standardized by North American ANSI body). The third generation (3G) networks provided dedicated digital networks used to deliver broadband/multimedia services. The support for an enhanced data rate (throughput speed) and quality of service (QoS), together with services such as global roaming—with the consolidation of cellular standards—and enhanced voice quality within the architecture could be considered as some of the key technology enablers for 3G networks. There were two key global standards development organizations (SDOs) created in 1998 to initially promote 3G—the third-generation partnership project (3GPP) providing enhancements based on 2G/GSM, and 3GPP2 which provided CDMA2000 mainly based on the enhancements of the CDMA technology. With the advent of smartphones, the 3G standards were further enhanced into what is broadly classified as high-speed packet access (HSPA) and evolution-data optimized (EV-DO), due to the significantly increasing data demand from the end users. The fourth generation (4G) represented the generation of mobile cellular communication technology, which was widely anticipated to productively deliver the demands for broadband data transmission and broadcasting, in addition to supporting significantly high volume of voice users. The 4G LTE also consolidated the two different SDOs, 3GPP and 3GPP2, with network operators across the globe adopting 3GPP-based standards for deploying the mobile networks.

3.2.2 Key Functions of the Radio Access Network

The basic function of the radio access network is to provide access and connectivity to the end user device. With the evolution of various generations, the key functions of the RAN have also evolved, depending on the evolution of various hardware and software capabilities. The connectivity provided by the RAN includes user plane connectivity toward the data network, and control plane connectivity which ensures efficient radio resource control (RRC) through nodes within the core network. The RAN consists of a multitude of user and control plane interfaces between base stations and core network.

Some of the functions traditionally hosted within the RAN are as follows [3,4]:

- Radio admission control: Ensuring appropriate initial access and mobility for the UEs, and ensuring the setup and release of RRC connections.
- Radio resource management (RRM) and scheduling: Allocating appropriate amount of resources to the UE, mitigating interference, and load balancing.
- Some of the other functionalities include:
  - Internet protocol (IP) header compression, encryption, and integrity protection.
  - Appropriate transport of control and user plane information between core network and UEs.
  - Delivering paging messages to the UE from the core network and messages originating within the RAN.
  - QoS flow management and policy enforcement ensuring the application of quality of service constraints on the air interface.

An overview of some of the key RAN functions is shown in Fig. 3.4. RAN also supports some other key features as follows:

- Network sharing which enables the same RAN infrastructure to be shared between a multitude of network operators

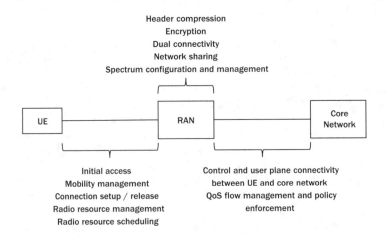

FIGURE 3.4 Key functions of RAN

- Selection of appropriate core network control and user plane entities for connectivity, and managing an appropriate session between the core network and the UEs
- Providing single or dual connectivity between 4G/5G RAN and core network in an interchangeable manner

3.2.2.1   Network Selection

The UE has been described in Chap. 2 in some detail. In this chapter, the UE will be described in a number of processes of how it acquires service on a network. Before obtaining service, a UE needs to select a network.

3.2.2.1.1   Network Acquisition   When a UE is fist powered on, it needs to register on a network. However, before the UE can register on a network, it needs to find the RAN serving the area where the UE is located. The RAN information (network ID, etc.) is continuously broadcast on many different radio bands, so the UE first needs to find its most preferred network. Having discovered the most preferred network, it will then select it and attempt to register on it.

3.2.2.1.2   Preferred Network   The most preferred network for a UE can be different over time and geography. The simplistic way of looking at how this works is the following:

> Home network always has the preference. The UE determines its home network by comparing its IMSI mobile country code (MCC) and mobile network code (MNC) with the network code broadcast by the network. If the UE discovers that the MCC and MNC broadcast by a network matches that of the IMSI, the UE will stop looking and register on that network.

Now, if the UE cannot find its home network, a UE could contain two lists of networks to help it choose a network. These are stored in the USIM.

- User-controlled list: This list contains a list of networks that the user prefers to be the most preferred. This list takes priority over the operator-controlled list. Networks in the list are stored in order and are identified by MCC and MNC.
- Operator-controlled list: This list is similar to the user-controlled list, but this list is provided by the home network.

The UE will scan the radio bands it supports; as it does it will find MCC and MNC being broadcast. If it finds highest priority network in the user-controlled list, it will stop scanning, else it will continue scanning until it has searched every frequency/radio band it supports. Note if the user-controlled list is not present or present but empty, the highest priority network will be the first entry in the operator-controlled list. After scanning all the bands and frequencies the UE supports, it will choose the highest available network. If the user-/operator-controlled lists are not present, or present but empty, or no network is found on the user-/operator-controlled lists, the UE will choose any network. To complicate matters further an entry in the user-/operator-controlled list could be qualified by a technology indicator, so that you could have:

1. Network A using Technology 1
2. Network B using Technology A
3. Network A using Technology 2

So, Network B could be classified as higher than Network A because supports preferred it Technology 1.

If a UE has not selected its home network or the highest preferred visited network (e.g., first entry in either user-controlled or operator-controlled lists), then the UE will periodically perform a scan and look for the next highest preferred available network.

3.2.2.1.3 Steering of Roaming   When a UE registers on a network, the home network might prefer the UE to move to another network. In this instance, the home network sends a message to the UE to:

- replace the first entry in the operator-controlled list, and
- cause the UE to perform a scan again.

An operator might do this because of commercial reasons, as they will get better roaming rates if they increase the number of roamers on a visited network.

3.2.2.1.4 Other Considerations   The initial way a UE selects a network dates back to the GSM system. However, over the years the commercial landscape has changed. In some countries operators only had specific geographical regions they could operate in. Over time that limitation was removed as operators in different regions merged and consolidated. In the United States, the larger GSM operators consisted of OmniPoint, Voicestream, Aerial, Powertell, Pacbell Mobile, and BellSouth. The latter two operators merged and created Cingular and eventually merged with AT&T. The other four operators are now part of T-Mobile USA, which has also merged with Sprint making its foot print much larger. As such, a visited network now would become a home network. To handle this situation, the concept of "Equivalent HPLMN" was created. It allowed the operator to create another list in the USIM that contained network identities that could be considered as home networks.

3.2.3 3G UMTS RAN Architecture

The 3GPP-based 3G network, called the UMTS Terrestrial Radio Access Network (UTRAN), consists of base stations and radio network controllers. A detailed overview of the 3G UMTS architecture can be found in [5]. The base stations are called Node B. A Node B can support frequency division duplex (FDD) mode, time division duplex (TDD) mode, or dual mode operation. Several base stations are managed by a radio network controller (RNC). The RNC is responsible for the handover decisions that require signaling to the user equipment (UE). A handover is a procedure when an ongoing communication between a UE and a base station is moved to a new base station, e.g., due to the mobility of the UE and the loss of coverage to the old base station. A logical view of the network is shown in Fig. 3.5. Iub is the interface between an RNC and a Node B. Iur is the logical interface between two RNCs. The interface between the RNC and the UMTS circuit-switched core network (CS-CN) is called Iu-CS, and one between the RNC and the UMTS packet-switched core network is called Iu-PS.

The 3G RAN architecture consists of two main parts: the user equipment (UE, also called mobile station/MS) and the UMTS UTRAN. The UE is the end user mobile terminal, and the UTRAN is the base station where the network intelligence is located. Both the UE and the UTRAN are composed of different protocol layers. The four lowest layers are the physical layer (PHY), the medium access layer (MAC), the radio link layer

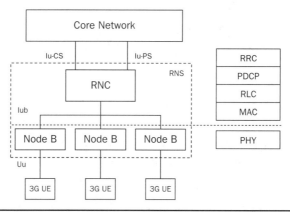

Figure 3.5 3G UMTS RAN architecture

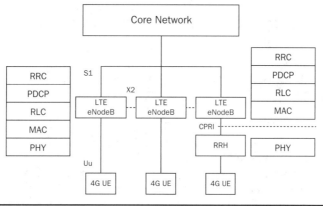

Figure 3.6 4G LTE RAN architecture

(RLC), and the radio resource layer (RRC). This radio protocol stack could be considered as the fundamental basis for the design of 4G LTE networks (Fig. 3.6), and later on 5G NR, with significant enhancements.

3.2.4   4G LTE and LTE-Advanced RAN Architecture

The LTE RAN, usually denoted as enhanced-UTRAN (E-UTRAN), handles the radio communications between the mobile terminal (or UE) and the evolved packet core and has only one component, the evolved base stations/NodeB, called eNodeB or eNB. The evolution from UTRAN to E-UTRAN was also essentially a transition from centralized network architecture toward fully distributed networks. Each eNB is a base station that controls the mobile terminals in one or more cells. The LTE mobile terminal communicates with just one base station and one cell at a time. Each eNB connects with the 4G core network (aka EPC) by means of the S1 interface and it can also be connected to nearby base stations by the X2 interface, which is mainly used for signaling and packet forwarding during handover. The functionalities of the protocol structure are shown purely for illustration and described in detail in Chap. 5.

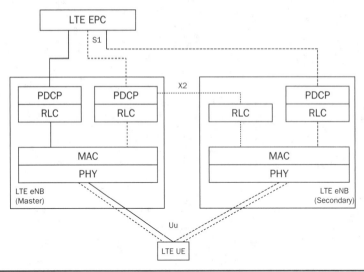

FIGURE 3.7 Simplified RAN architecture of LTE-advanced with dual connectivity [6,7]

As particular scenario, LTE RAN supports the functional split in lower layers, more specifically between the PHY ad MAC layers, and allows for a centralized-RAN (C-RAN) deployment where Layer 2 (L2, i.e., MAC, RLC, and PDCP layers) processing of the base station is split between the PHY and RF components, which can be seen as Layer 1 (L1, PHY layer) relays, aka, Radio Remote Heads (RRHs).

An overview of the 4G LTE RAN architecture is shown in Fig. 3.7 [6,7]. It enables a device to connect to multiple eNBs (called master and secondary). The solid line in the figure indicates the data flow from the EPC directly to the UE and the dotted lines indicate the data flow that has been split between the master and secondary eNB. This formed the basis for later enhancements in 5G standards to enable tight connectivity between 4G and 5G RAN for non-standalone (NSA) deployments (see Chap. 5 for more details). A detailed overview of LTE and LTE-Advanced RAN architecture and functionalities has been presented in [8].

3.3 Core Network Architecture[1]

Mobile core network technology has evolved over the years, in conjunction with the evolution of radio technologies from 1G to 5G (and is expected to further develop in the future); however, these domains have progressed independently. The mobile service providers do not readily change their mobile core networks to accommodate next-generation radio technologies. In general, the mobile CN subsystem is used for a much longer period with continued enhancements to accommodate new features offered by introduction of the next-generation radio technology.

Initially, the mobile core network followed the traditional landline core network technology with one of the big differences, which was the addition of mobility

---

[1]The mobile core network technology discussed in this textbook is based on the European Global System for Mobile Communications (GSM) standard.

functionality. The mobility functionality provides a means of keeping track of the location of devices that are in motion (i.e., mobile devices/handsets). Other than that, there was not much difference between the traditional landline core network and the mobile core network. However, since cellular service grew at a fast pace in the 1990s, the mobile core network continually enhanced over the years making it much different and feature rich than the fixed landline network. The mobile core network switching was also enhanced from circuit switching using traditional landline Signaling System No. 7 (SS7) to an All-IP packet-switching network using signaling based on internet protocols. The previous core network architectures were based on the point-to-point signaling between the different nodes of the core network, but 5G mobile core network adopted the service-based architecture going forward. The service-based architecture for mobile CN was adopted in 5G to make future evolutions easier. More details of point-to-point versus service-based architectures are described in Chap. 6.

3.3.1   Historical Background[2]

In the beginning, the mobile system was used as a replacement for fixed wireline PSTN, also known as traditional circuit-switched telephone network. It was primarily used for voice service, which included voice-based fax service, using a circuit-switched core network. In addition, short message service and circuit-switched data calls were supported over the circuit-switched core network. It took several years to make the mobile service commercially available. Initially, the operator's revenue generation was primarily based on the amount of voice calls (measured in Erlangs—a unit of traffic density in a telecommunications system).

As the use of internet services grew, the mobile systems were enhanced in the late 1980s and into the 1990s to provide users access to data services, such as short message service (SMS), multimedia messaging service (MMS), and internet access using wireless application protocol (WAP), via the addition of a separate packet data network. For several years, the mobile system continued with two separate parts—one for the voice services (using circuit switching) and the other for the data services (using a packet data network GPRS). Today, the entire mobile system is packet based as data services have outgrown the voice services and voice services are realized over IP. The mobile system of today is packet based using internet-based protocols, enabling internet services to be fully integrated. In fact, the voice service is basically offered as one of the data services.

3.3.2   Circuit Switching

Circuit switching is a type of communication in which a dedicated channel (or circuit or path) is established between sender and receiver for the duration of a transmission. The dedicated channel provides a guaranteed data rate to ensure that data can be transmitted without any delays once the circuit-switched path (channel) is established. The establishment of the dedicated channel is done using control signaling as each intermediate node communicates with the next one in the path to request reservation of each dedicated intermediate channel, until the full path is established. Only then the communication between the end parties can begin. The dedicated channel is reserved for the entire duration of the communication and then torn down after either party hangs up.

---

[2]Motorola was the first company to produce a handheld mobile phone and make the first mobile telephone call in April 1973. Mobile service was commercially launched in 1985.

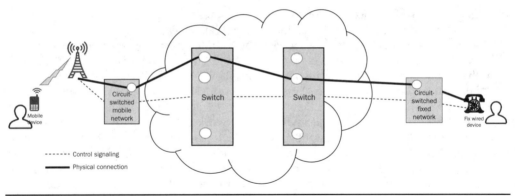

Figure 3.8 Circuit switching

Circuit switching is a connection-oriented network-switching technique. Here, a dedicated route is established between the source and the destination and the entire message is transferred through it. The main advantage of circuit switching is that it is highly reliable for its purpose, i.e., voice communication, and also that once the circuit is set up, the communication is fast and mostly error-free. Ordinary voice phone service is circuit switched. The bottleneck in circuit-switching communication is that the communication channels can get exhausted quickly if the calling load is heavy which limits the number of users that can communicate at the same time. Therefore, packet switching is a preferred method and more cost effective as, e.g., in voice communications there are periods of time people do not talk or there is sometimes the ability to code the voice differently.

Figure 3.8 shows the voice call establishment between a mobile telephone and a landline telephone connected via circuit-switched connections. All the intermediate boxes represent the switching offices/nodes. The dotted black lines represent the control signaling between the nodes used for sending requests to the next node for reservation of a communication link. As each link in the path is reserved, the request goes back to the source node that a communication path is established and communication between source and destination parties can begin. The links remain established as long as communication continues.

3.3.3 Packet Switching

The packet switching is another method used in mobile systems to connect multiple communicating nodes in the path with one another. While circuit switching is connection oriented, packet switching is connectionless. Packet switching is more efficient, as far as utilization of network resources is concerned, because it does not deal with limited number of connections, as in circuit switching. In packet-switched networks, the data stream is broken into smaller units called packets, which are then routed through the network based on destination address contained in each packet. These packets consist of a header containing destination address, and a payload containing user data. The header is decoded at each node (server) in the chosen path to direct the packet to its next destination in the path. At the final destination, the packets may arrive at different times which are assembled by the application software in the originally ordered sequence. A packet-switched network uses transmission control protocol/internet protocol (TCP/IP) suite or open systems interconnection (OSI) layer.

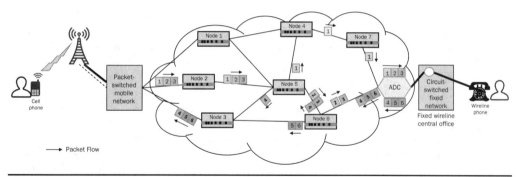

FIGURE 3.9   Packet switching

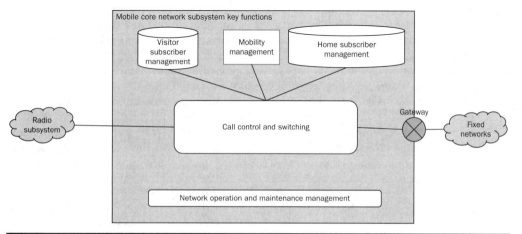

FIGURE 3.10   Mobile 3G/4G CN architecture

Figure 3.9 shows the voice call establishment between a mobile telephone and a landline telephone using packet switching. All the intermediate nodes represent server nodes for routing packets between the end offices (or central offices). Outgoing packets from the packet-switched mobile network take different routes to their destination, while incoming packets from the destination fixed line circuit-switched office are routed via different servers. The decision on how the packets are routed is taken by the individual servers on which routes are least congested to reach the next server in the hop. This method of routing packets describes the basic packet-routing technique. There are other more sophisticated packet-routing techniques, for example, software defined network (SDN)-based routing technique, which is described in Chap. 6. The arrival of packets can be random, and it is the job of the originating and terminating nodes to reorder the packets in the correct sequence and deliver to the device.

3.3.4   Key Functions of Mobile Core Network

The mobile core network consists of several functional blocks that are essential for its operation. These functional blocks make up the mobile core network architecture, as shown in Fig. 3.10.

The functions performed by each of the CN functional blocks are described below.

Call Control and Switching: It determines the next steps to be taken for an incoming call request received from the RAN subsystem, based on some subscriber action (e.g., dialed digits). In the PSTN, the subscriber lines directly communicate with the call control and switching functional block when service is requested, while in cellular system the RAN subsystem is the first point of contact for the subscriber line (in this case, cell phone) using radio signaling, which then forwards the call to the CN call control and switching functional block. The RAN is one of major differences to the PSTN, in addition to the mobility provided by the mobile system.

Home Subscriber Management: This is the master user database repository containing pertinent information about all the home subscribers registered on the operator's network (i.e., all paying customers). This database contains information about the subscriber's unique identity, any public identities the subscriber is known by, and the list of services signed up for by the subscriber. The subscriber's unique identity is needed for authenticating the subscriber and in assisting the network to route the call to the proper destination. The master user database repository is permanent and under operator control and, so, cannot be changed by without operator permission. The home subscriber management function also contains subscriber location information, e.g., MSC location to determine where the subscriber is currently registered.

Visitor Subscriber Management: This is the temporary user database repository containing information about the roaming subscribers currently being served in a visited operator's network, which might be a roaming network. This database is populated when a roaming subscriber enters a visiting network, and the visiting network then queries the home network of the visiting subscriber to authenticate the visiting subscriber, using received subscriber credentials, followed by giving the subscriber desired service. This database is changing all the time as visiting subscribers come and leave, while the home subscriber database has permanent subscriber-related data stored. The database also assigns local temporary identities to the user. The purpose of these temporary identities is that when the user moves around the same network and needs to make the network aware of his/her location, it uses local temporary identity instead of its unique identity for authentication to avoid disclosing the unique identity all the time and compromising its disclosure to rogue intruders (eavesdropping).

Mobility Management (MM): This provides the unique capability in a mobile system to be able to track, at all times, where the active (powered up) mobile devices are located so that mobile phone services can be delivered to them, based on their current known location. If the mobile device is in a dead zone (with no radio signal available), then the mobile device cannot be tracked and will not be able to receive service.

Session Management (SM): Session as its name implies is a voice or data connection (or a session) between two communicating entities. There can be multiple voice or data sessions started by a user. A session is a temporary and interactive information interchange between two or more communicating entities. Each session is identified by a session ID in the network. These session IDs last only for the duration of an active session. Therefore, session management is used to manage sessions in the mobile network, i.e., make/break, assignment of QoS level, etc.

Network Operation and Maintenance Management: This provides the necessary processes, activities, and tools, required in operating, administering, managing, and maintaining the mobile CN. This functional block is important for operators as it collects different types of data regarding the health of the network. The collected data allows operators to take corrective actions, as necessary, if there are problems or bottlenecks identified during the operation of the network.

Gateway (GW): This provides access to external fixed networks. The functions performed by the GWs are dependent on the type of fixed network to which they are connected (e.g., PSTN, enterprise NW, satellite NW, etc.).

3.4 Terminals

3.4.1 General

The terminal is the end device that communicates with the radio subsystem. It is also known as mobile station (MS) or user equipment (UE). Figure 3.11 shows an overview of the UE.

UICC is the piece of plastic that goes into the phone. In layman's terms, it is known as a subscriber identity module (SIM) or smart card and contains the unique identity of the subscriber and the necessary algorithm to validate that subscriber. It may also contain storage for messages that have been sent to the subscriber and any necessary configuration information that is specific to the user. In 2G the data was stored on a SIM; however, in 3G and beyond the SIM card was split into the physical platform called the UICC and the application on the UICC called the USIM for accessing the cellular network. Another application was later developed to access the service network responsible for voice, etc.; this was called the IMS SIM or ISIM. A UICC could also have the SIM application.

Mobile equipment (ME) is the physical terminal that contains the modem to communicate with the radio network. It is further split into two entities: mobile termination (MT) and terminal equipment (TE). These are described more in Sec. 3.4.3.

3.4.2 UICC and Applications

The UICC, sometimes known as a smart card, is basically a small computer on a piece of plastic. There have been four different sizes of smart cards over the years. The physical

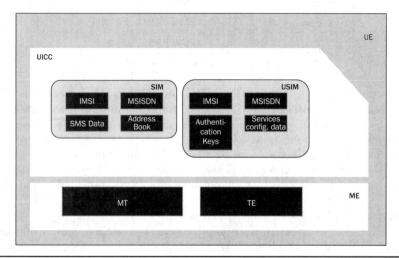

FIGURE 3.11 UE overview

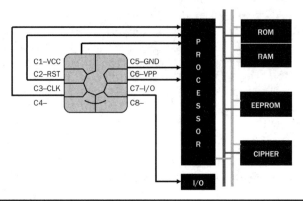

FIGURE 3.12   UICC overview

and electrical aspects of the card are based on ISO7810 (physical characteristics) [9] and ISO7816-2 (circuit card layout) [10].

Figure 3.12 shows an overview of a UICC, which has a processor, EEPROM, ROM, and RAM. There is also a ciphering application that is used to perform the necessary security computations required for the mobile system. Newer UICCs may combine the ROM/EEPROM. The Pin layout is based on ISO/IEC 7816-2 [10].

ISO/IEC 7816-2 pinout [10]		
Pin #	**Pin name**	**Pin description**
1	VCC	+5v or 3.3v DC
2	Reset	Card Reset (Optional)
3	CLOCK	Card Clock
4	AS	Application Specific
5	GND	Ground
6	VPP	+21v DC [Programming], or NC
7	I/O	In/Out [Data]
8	AS	Application Specific

Communication between the UICC and the ME takes place over Pin#7 using a serial interface (9600 bps) and over Pin#3 for synchronization with the clock provided by the ME. Other pins provide the electrical aspects and ability to reset the card. The two application-specific pins—4 and 8—have also been assigned to supporting USB 2.0 interface allowing the dataflow to and from the card to be a lot faster.

The SIM card over years has also changed in size. The original SIM card was the same size as a credit card, and then three other sizes appeared in the market as UEs became smaller and more functional, as shown in Fig. 3.13. The current smallest SIM is the eSIM (see Sec. 3.4.4).

On the UICC numerous applications can be installed, supporting cellular-based applications such as mobile banking, mobile ticketing, presence-based messaging and other services.

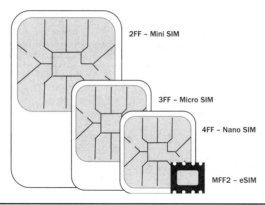

FIGURE 3.13 SIM sizes

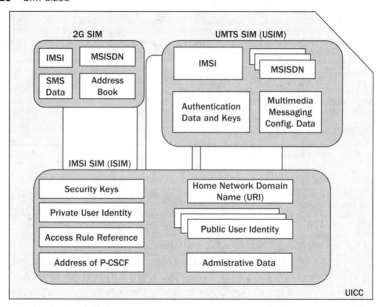

FIGURE 3.14 UICC applications overview

Figure 3.14 shows the type of information stored on applications on the UICC. There are three applications: the 2G SIM specified in GSM 11.11 [11], the 3G+ USIM specified in 3GPP 31.102 [12], and the ISIM specified in 3GPP 31.103 [13]. Each contains a private identity and optionally a public identity. Each application can also contain storage for data as well as authentication information.

The USIM is the application that is used in all current mobile systems. It contains:

Private user identity: This is the IMSI, which is used by the network to locate the subscribers' profile and their authentication information.

Administration data: This can be a set of data used for many things. Typically, one set of data that is stored is a set of networks that the subscriber ME should choose when it roams. It may also contain network names that accompany these "preferred roaming" networks.

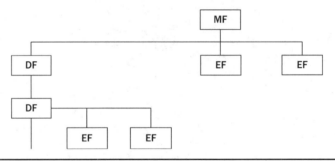

FIGURE 3.15   Partial USIM file structure 3GPP TS 31.102 [12]

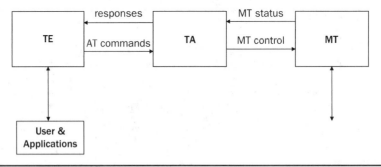

FIGURE 3.16   ME overview

Optionally, the USIM may contain the subscribers' public identity, which the subscriber will give out to be contacted, e.g., telephone number or the so-called MSISDN.

The data is organized into a file structure, as shown in Fig. 3.15.

There is a master file (MF) that contains the core files to be used by the application. Under this are directories in the dedicated files (DF) followed by the content or documents within the directory called elementary files (EF). When an operator procures a USIM application, they need to decide what files will be available on the USIM.

A UICC/USIM can last many years, so there needs to be a way to update the data on the application to ensure it is up to date. A mechanism called over-the-air (OTA) has been developed that allows data to be sent to the card so that specific files can be modified.

3.4.3   Mobile Equipment

Figure 3.16 shows an overview of the subsystem components of the ME.

Mobile Termination (MT): This provides the necessary functionality to work with the RAN and CN.

Terminal Equipment (TE): This is an external interface to the Terminal Adaptor. In some contexts, the interface is over existing serial (ITU-T Recommendation V.24) cables, infrared link, and all link types with similar behavior. The TE could also be the operating system of the ME. It communicates with the TA using AT commands specified in

3GPP TS 27.007 [14]. AT commands are a set of commands with a specific syntax that can read or write information to the MT via the TA.

Terminal Adapter (*TA*): This provides the necessary interface between the TE and MT. It takes AT commands and performs the necessary procedures with the MT to provide a response back.

3.4.4   eSIM

Embedded SIM (eSIM) or embedded UICC (eUICC) is an evolution of the UICC and applications that can be supported on it. Whereas the SIM card today is a physical piece of plastic, eSIM provides an electronic way to deliver the SIM card. It provides the user with the ability to download some secure software into a secure platform in the ME and has also the full capability of the SIM card. An operator can now provide the SIM card via electronic means, e.g., internet to the ME, all that is required is an internet connection. Operators are no longer tied to stores; they don't need to manage the logistics of SIM cards. Subscribers can pick and choose their operators even if the operator is halfway around the world.

References

1. K. David, and H. Berndt, "6G Vision and Requirements: Is There Any Need for Beyond 5G?" IEEE Vehicular Technology Magazine, vol. 13(3), pp. 72–80, 2018.
2. A. Furuskar, S. Mazur, F. Muller, and H. Olofsson, "EDGE: Enhanced Data Rates for GSM and TDMA/136 Evolution," IEEE Personal Communications, vol. 6(3), pp. 56–66, 1999.
3. 3GPP TS 36.300, "Evolved Universal Terrestrial Radio Access (E-UTRA) and Evolved Universal Terrestrial Radio Access Network (E-UTRAN); Overall Description; Stage 2," v15.8.0, Jan. 2020.
4. 3GPP TS 38.300, "NR; Overall description," v15.6.0, June 2019.
5. X. Li, "Radio Access Network Dimensioning for 3G UMTS," Vieweg+ Teubner Verlag, 2011. (Available Online: https://link.springer.com/content/pdf/10.1007/978-3-8348-8111-3.pdf)
6. S. C. Jha, K. Sivanesan, R. Vannithamby, and A. T. Koc, "Dual Connectivity in LTE Small Cell Networks," IEEE Globecom Workshops (GC Wkshps), Austin, TX, 2014, pp. 1205–1210.
7. C. Rosa, et al., "Dual Connectivity for LTE Small Cell Evolution: Functionality and Performance Aspects," IEEE Communications Magazine, vol. 54, no. 6, pp. 137–143, June 2016.
8. E. Dahlman, S. Parkvall, and J. Sköld, "4G—LTE-Advanced Pro and the Road to 5G," Academic Press, 2016.
9. International Standard—ISO/IEC 7810, "Identification Cards—Physical Characteristics," 2003.
10. International Standard—ISO/IEC 7816-2, "Identification Cards—Integrated Circuit Cards—Part 2: Cards with Contacts—Dimensions and Location of the Contacts," Second edition, 2007-10-15.
11. GSM 11.11, "Digital Cellular Telecommunications System (Phase 2+); Specification of the Subscriber Identity Module–Mobile Equipment (SIM–ME) Interface."
12. 3GPP TS 31.102, "Characteristics of the Universal Subscriber Identity Module (USIM) Application," v16.2.0, December 2019.

13. 3GPP TS 31.103, "IP Multimedia Services Identity Module (ISIM) Application," v15.5.1, January 2020.

14. 3GPP TS 27.007, "AT Command Set for User Equipment (UE)," v16.4.0, March 2020.

Problems/Exercise Questions

1. What are the three major components of a mobile system?

2. From the roadmap diagram of the cellular technologies, what are the key characteristics of 5G?

3. Provide a list of key functions of the radio access network.

4. Prior to UE getting service from the mobile system, what is the first step it has to perform to get access to the mobile system?

5. Is circuit switching still used in mobile core networks?

6. What is the biggest benefit of using packet switching?

7. What constitutes the overall mobile equipment (ME)?

8. What is the UICC?

9. Name 2 applications that can be on the UICC.

4

5G Services and Applications

4.0   Introduction

The commercialization of 5G technology is currently underway by the telecom operators in order to accommodate the ever-increasing data generated by devices equipped on smart vehicles, factory robots, medical remote patient monitoring equipment, etc., all part of Internet-of-Things (IoT) ecosystem. The 5G new radio (NR) and core network (5GC) technology is superior to 4G (i.e., LTE/EPC) in terms of reliability in handling of data transmission to meet the data delivery demands of superior 5G devices. More importantly, it is necessary to have a look at the business aspects of 5G technology on top of its technical specifications.

First, the 5G technology (IMT-2020), unlike the previous generations of mobile technologies, cannot be defined by any single form of network technology. As 5G is often referred to as "the network of networks," it binds various existing network technologies such as IoT, Edge Computing, and softwarization of network functions, including the current advanced LTE networks. This will be of great help in early commercialization of 5G. In addition, as the convergence between wired and wireless communications becomes more advanced and strengthened, various services and new business models are expected to emerge in the 5G era. From the users' perspective, which is a more general level, 5G will provide mobile users with far higher and reliable data throughput than ever experienced before. The on-the-go internet experience will be more enjoyable with much faster response times. The increase in data usage is evident in our daily lives as mobile devices are becoming our personal assistants on which we rely upon for everything. The data usage is going up ten-fold each year with the increase in web browsing, social networking, video down streaming, file sharing, online games, and future advanced services to come.

IoT applications using 5G networks can manage various assets such as roadway sensors, actuators, and vehicles, or manage entire complex physical infrastructures such as smart buildings and cities. These applications rely on different tools to process data, generate insights, and automate different business- and society-related processes, which require data exchange over 5G networks. The following sections provide an overview of recent 5G commercial deployments and some prominent 5G-based services.

4.1   Market Trends

Since the introduction of smartphones, high-speed internet, and LTE, data services have exploded overtaking voice services. In fact, voice service is now included as a part of the data services. In particular, data services associated with video content is expected to continue to grow, increasing by more than 34% by 2024 based on market studies [1]. Video-on-demand (VoD) streaming services such as YouTube™ and Netflix™, as well as smart devices' high-resolution screens, are driving the explosion in data traffic. In addition, with the emergence of numerous IoT devices/sensors mounted on autonomous cars and drones, the need for mobile networks to efficiently accommodate and handle data traffic generated from these devices/sensors is growing exponentially. To cope with the changing environment, major carriers around the world are scrambling to commercialize 5G network technology, which is much superior to LTE technology in terms of transmission speed, latency, handset capacity, and data processing.

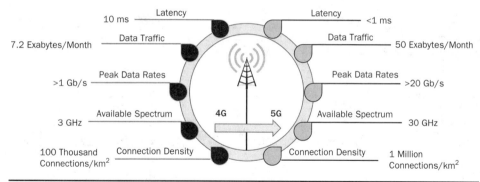

FIGURE 4.1 Comparing 5G and LTE-A (4G)

Figure 4.1 shows a comparison of key performances between 5G and LTE-Advanced (LTE-A). Generally, mobile network technologies can compare performances in the following five areas:

- Connections (terminals/km^2)
- Latency (ms)
- Data Traffic (Exabytes/Month)
- Peak Data Rates (Gb/s)
- Available Spectrum (GHz)

In addition, 5G performs better than LTE-A in terms of performance comparison indicators, such as consumer sentiment speed (Mbps), frequency efficiency, mobile speed (km/h), network energy efficiency, and capacity per area (Mbps/m^2). Table 4.1 shows the differences regarding the performance of 4G and 5G.

In the same manner, 5G has an advantage over LTE-A in all comparison areas, which shows performance that is incomparable to LTE-A, especially in terms of latency and handset capacity. For example, in terms of speed, LTE can provide 1 Gbps during a standby condition, whereas 5G can provide up to 20 Gbps, which is 20 times faster than LTE. Due to this rapid advantage, not only video contents, but also various services that generate large data traffic such as virtual reality (VR)/augmented reality (AR), and holograms used for remote education can be used to provide seamless service using 5G technology.

For 5G network technology, the processing delay time that occurs during the data transmission and reception process is 1 ms, which in effect provides an environment in which real-time services are available. This shows that the 5G technology is suitable for services that require high reliability such as self-driving, remote driving, smart factories, and remote surgery.

5G is also showing better performance than LTE in terms of connected devices. Although LTE can connect up to 100,000 terminals per square kilometer, 5G can accept up to 1 million terminals, which theoretically allows 10 times more terminals. 5G will be able to provide optimal network technology to connect and manage numerous IoT devices and sensors that are connecting everything around our lives to the internet.

Mobile multimedia services (e.g., video conferencing) have been discussed as killer services of 3G technology in the past, but these practical services had inadequate

	4G (LTE-A)	5G
Latency	10 ms	Less than 1 ms
Peak data rates	1 Gbps	20 Gbps
No. of mobile connections	8 billion (2016)	11 billion (2021)
Channel bandwidth	20 MHz 200 kHz (for Cat-NB1 IoT)	100 MHz below 6 GHz 400 MHz above 6 GHz
Frequency band	600 MHz to 5.925 GHz	600 MHz–mm Wave (e.g., 28 GHz, 39 GHz, and onward to 80 GHz)
Uplink waveform	Single-carrier frequency division multiple access (SC-FDMA)	Option for cyclic prefix orthogonal frequency-division multiplexing (CP-OFDM)
User equipment (UE) transmitted power	+23 decibel-milliwatts (dBm) except 2.5 GHz time-division duplexing (TDD) Band 41 where +26 dBm, HPUE is allowed IoT has a lower power-class option at +20 dBm	+26 dBm for less than 6 GHz 5G bands at and above 2.5 GHz

TABLE 4.1   Major differences between 4G and 5G technology

performance due to inherent technical limitations of the 3G technology. With the separated circuit-switched (CS) and packet-switched (PS) core networks (CNs) in the 3G environment, it was not fundamentally easy to provide the desired level of quality for multimedia services to consumers, and the fees were also recognized as a barrier to entry into using these services.

In a similar way, the LTE technology was considered by telecom carriers as a means to integrate the wired and wireless services. But this was not realized, again due to limitations of the LTE technology. Going forward, many operators have a plan to make an investment to integrate wired/wireless networks using 5G technology, facilitated by the evolution of both wired and wireless networks toward becoming All-IP networks. In the 5G era, there is a strong desire for having hyper-speed internet via wired or wireless network utilizing 5G access technologies.

Verizon® and AT&T®, the two US carriers, plan to use 5G wireless networks as an alternative to wired networks. (Note that we did not use a wireless network to watch TV in the living room until now, but this is about to be changed.) It is expected that high-speed internet and on-demand service with 5G technology will be implemented; and this will save a huge amount of money and time for building existing wired networks through a full-scale implementation of fixed wireless access (FWA). FWA is expected to be actively used in mountainous regions and countries where there is insufficient wired infrastructure or terrains where wired infrastructure is difficult to build.

For the general use of 5G in wired networks, customer premise equipment (CPE), a handset that can receive signals from 5G base stations, should be installed in the home. Figure 4.2 shows the configuration of "Verizon 5G Home" service presented by Verizon® [2]. This shows the CPE installed within each home communicating with the external base station, the Access Unit, and providing various services to the home. CPE will also be replaceable using a 5G smartphone with the same function implemented, indicating that wired and wireless services can be provided simultaneously by a single network.

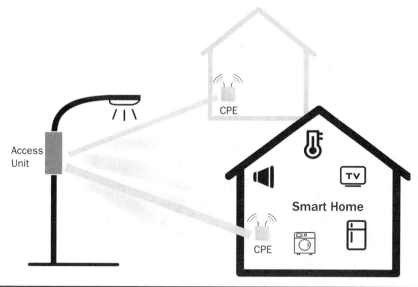

FIGURE 4.2   Prototype example of using CPE

On the contrary, wired network operators are also planning to provide services using 5G technology. If only the most problematic 5G frequency license can be secured, it is possible to configure and deliver services similar to the FWA led by the wireless carrier. Naturally, it is an efficient choice for existing cable operators to provide services using 5G rather than building new wired cables. In fact, the United States is conducting 5G tests centered on the Charter Communications®. As a result, cable operators are also entering the wireless market [3]. In the United States, satellite broadcasting operator Dish® has secured 5G frequencies and is speeding up the development of technologies and services with the aim of launching commercial services in 2020–2021 [4]. As such, 5G technology provides new opportunities for service providers in diverse fields as well as existing mobile network operators, as it drives toward integration of wired and wireless.

The increase in participating businesses is expected to ultimately be a big boost to the expansion of 5G markets. As the number of operators increases, the number of new services will also increase, resulting in the launch of various 5G terminals/devices. This will lead to fierce competition among the various operators, which will naturally result in price cuts for 5G terminals and services. As 5G has a wide range of applications, it is expected that there will be a variety of services and new business models that have not been explored thus far.

The network slicing feature developed by 3GPP specifically for 5G system will provide a unique management tool for operators to offer dedicated services based on customer needs. This will be achieved through utilization of two technologies, namely software defined network (SDN) and network functions virtualization (NFV). It will provide a way to slice an operator's network to behave like a collection of number of dedicated, smaller networks to handle specific tasks based on customer needs. Chapter 6 provides technical details on network slicing.

As shown in Fig. 4.3, network slicing can occur across multiple parts of the network (mobile terminal, access network, core network, and transport network). One network slice

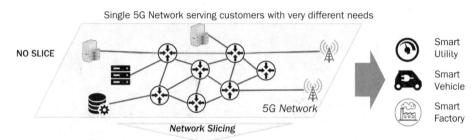

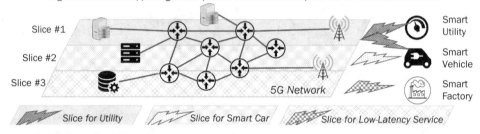

FIGURE 4.3 Network slicing in 5G

is composed of independent and shared resources for storage, processing power, and bandwidth according to the requirements, and operates separately from other network slices.

By using a single network as if it were multiple independent networks, users (personal user or specific service) get the same effect as using their own private networks. Today, in order for an operator to provide a dedicated network (virtual private network) to a particular user or business, many resources must be invested and the cost of maintaining various functions such as billing, quality of service (QoS), user information, and security had to be separately configured and reflected in the requirements of the respective enterprise. However, network slicing technology provided by 5G provides creation and management of dedicated network services in a much easier way. This suggests that more business users or service users can use dedicated private networks using 5G system.

Figure 4.3 shows vertical industries where network slicing technology can be used.

- Slice for Smart City: Network slicing of 5G can be used to provide necessary services for the federal government or local governments in smart cities where many verticals coexist.

- Slice for Massive IoT: IoT devices show different network usage patterns depending on environment and service used. Operators will be able to use a number of network slices to effectively support different IoT users. Each network slice may have a different billing policy and network traffic limit depending on the service.

- Slice for Automotive: Autonomous cars and connected vehicles have considerable versatile requirements. This includes the high-definition in-car input for providing a high-definition in-car entertainment for passengers to watch while driving and the ultra-reliability and low latency (URLLC) for use in autonomous driving by receiving and processing real-time information from various sensors. These various requirements can be implemented and provided on the network slice.

- Slice for Industry Automation: In the industrial sector, URLLC type slices can be used to control precision robots used in product lines.

- Slice for AR/VR: High-density computing power may be required for video processing for AR/VR services requiring high-resolution video transmission, which may result in the use of a network slice. For one-to-many downlink services, a network slice that implements an enhanced mobile broadband (eMBB) can be used.

4.2   Deployment in Asia and the United States

In this section, we show 5G preparations and commercialization of each country at the time of writing this textbook (see Fig. 4.4).

South Korea: South Korea, which started commercializing business-to-business (B2B) 5G service in December 2018 for the first time in the world and commercializing business-to-consumer (B2C) smartphone 5G for the public in April 2019, has made many preparations for 5G.

For instance, at the 2018 PyeongChang Winter Olympics watched by people around the world, Korea Telecom (KT) has successfully introduced various 5G-based real-life services such as autonomous 5G buses, pigeon performances through 1200 LED controls and Sync View, showing a video of the games from the perspective of players [5].

Since the first commercialization of 5G in the world, South Korea's three mobile carriers (LGU+, Korea Telecom, and SK Telecom) have competitively set up 5G base stations across the country to dominate the 5G market. As of September 2019, KT has set up 34,557, LGU+ has set up 30,433, and SKT has set up 22,448 base stations across the country. As of August 2019, 49,266 in SKT, 48,564 in KT, and 40,165 in LGU+ devices have been operational, totaling 137,995 devices [6].

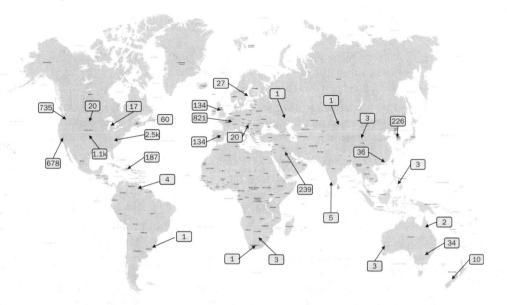

FIGURE 4.4   The world 5G deployments map as of February 2020 (The number in a box represents the number of deployed 5G base stations)

KT selected communication and game media as the core service areas for 5G and released communication services such as "Real 360" and "Narle," and "eSportsLive" services in the gaming sector [7] as follows:

- Real 360: A service that allows users to exchange 360° videos with multiple people in ultra-high definition.
- Narle: A new concept communication service that supports 3D avatar/augmented reality (AR) emitters.
- eSportsLive: A service that divides game images into up to five screens and relays them in ultra-high definition.

In the case of SKT, it has created eight 5G slice clusters across the country: 5G factory, smart hospital, smart logistics, smart city, media, public safety, smart office, and national defense, providing 5G services through convergence/composition between industries. For example, SKT is building a 5G network at its semiconductor plant and building a 5G smart factory based on mobile edge computing (MEC) and plans to spread these 5G MEC technologies nationwide to provide 5G edge cloud services to industries such as manufacturing, media, and finance.

For LGU+, the company selected various services based on high-definition video for B2C and remote control and image recognition for B2B and is planning to provide various services based on the following [8]:

- 5G Live Broadcast: Viewing various live videos from desired point of view.
- Ultra-High-Definition Virtual Reality (8K VR): A service that provides 360° video of 8K high-definition movies and performances.
- Mixed Reality Game: A service that combines real and virtual space to enable you to enjoy realistic games.
- Intelligent CCTV: A service that transmits and analyzes a filmed video in real time.
- 5G Smart Drone: A service that develops smart drones with cloud drone control systems automatically flies when the destination is set and allows viewing of the flight footage on IPTV.
- Driver Monitoring: A service that analyzes the driver's facial movements, blinks in real time through an internal camera, and provides the driver with various information.

The United States: The US-based mobile network operator has different 5G response strategies depending on the number of subscribers. Verizon® and AT&T®, two US operators, are pushing ahead with the installation and service provision of 5G infrastructure based on fixed wireless access at home or in buildings. Verizon® from the United States is approaching the 5G market with the aim of securing leadership in all areas of the United States, including wired communications networks through 5G. Verizon® mainly dealt with home-oriented products and services until 2018, but also provided mobile services since 2019 [9].

AT&T® is building 5G infrastructure with fixed wireless access at the center [10].

On the other hand, new T-Mobile USA (merged with Sprint) is continuing aggressively with its mobile-focused 5G strategy [9]. It is deploying 5G infrastructure centered on mobile rather than wired services that require massive investment for infrastructure to deliver services smoothly, thus providing various 5G services with a focus on smartphones.

The United States is also working on global initiatives for 5G technologies, including national policies such as creating more than four times the spectrum of high-bandwidth so that 5G technology can be used flexibly than before in the Spectrum Frontier Order of FCC (Federal Communications Commission).

Japan: Japan finished defining 5G technologies by mid-2018 and allocated low- and high-band 5G frequencies to four mobile network operators on April 10, 2019.

NTT DOCOMO launched the 5G pre-service on September 20, 2019 and plans to provide hands-on content first. NTT DOCOMO5G broadcasting and hands-on contents are planned to be provided at the Rugby World Cup to be held in Japan in September 2019 [11]. NTT DOCOMO considers providing vivid on-the-spot experiences and other services aimed at launching commercial 5G services at various large events such as Olympics and Paralympics.

After providing high-speed telecommunication services in 2019, SoftBank plans to provide VR/AR, multiple device connection services, ultra-low-delay services, traffic congestion mitigation, and construction site remote operation services from 2020. SoftBank established a research center in Odaiba, Japan, in May 2018 to build 5G technologies and services, and is preparing for a demonstration of low-latency remote operation combining 5G with a robot arm.

Rakuten aims to provide high-speed services in 2020 and plans to introduce smartphone/mobile broadband services, IoT services, and solution services that utilize multiple device connection and ultra-low-latency features sequentially later.

The Ministry of Internal Affairs and Communications of Japan plans to allocate local 5G frequencies to establish 5G networks and operate its own services like Wi-Fi and commercialize them in 2019. As a result, traditional mobile communication system vendors such as NEC, Panasonic, Toshiba, Fujitsu, and non-mobile telecommunication businesses such as NTT East and Optage are declaring and reviewing their entry into the local 5G market.

China: Since the standardization of 3G, China has actively participated in mobile telecommunication technologies and expanded its influence through 4G standardization. In the 5G era, China has been making many efforts to lead the technology. China was also selected by the United States Telecom Association as the leading country in the 5G sector.

In this direction, a demonstration test was held in 2018 based on a stand-alone (SA) standard that connects handset and base station only with 5G technology. SA is considered to be a more advanced technology standard than non-stand-alone (NSA), which mixes 4G and 5G. In the process, vendors of equipment and mobile telecommunication with the support of the Chinese government started testing in major cities in early 2018 and completed demonstration tests in 16 cities. These demonstration tests linked indoors and outdoors within the 5G frequency 3.5 GHz and 4.9 GHz bands, and identified links between core networks and base stations, which means they are ready to perform commercial services from the infrastructure perspective of 5G.

In 2019, China started conducting compatibility tests between network systems and 5G terminal chipsets. The test is generally the final step in preparing for commercialization, and a successful completion of this step means that the 5G service could be provided to customers as the 5G chipset could fit to a number of 5G terminals, including smartphones.

The 5G strategies of China's major telecommunications companies are as follows:

- China Mobile defined the 5G strategy as a "Big Connectivity Strategy" that will "change society as a whole" and started working on the project with the aim of having a total of 300,000 5G sites across China by the end of 2020. To this end, China Mobile has set up more than 100 5G base stations in Hangzhou, Shanghai, Guangzhou, Suzhou, and Wuhan and is coordinating the timing of commercialization.

- China Unicom has also started 5G pilot projects in Chinese cities to commercialize the new communication technology in 2020 on a similar schedule to China Mobile [12].

- China Telecom is expected to proceed with its 5G strategy for the time being under the NSA structure where 4G and 5G coexist to provide services [13].

4.3   Smart Vehicles

Conventional vehicle communication was first introduced and spread by General Motors (GM) in the mid-1990s for emergency and vehicle management purposes, such as accident reporting and failure alarms [14]. This vehicle communication was conducted by mounting a camera inside the vehicle and ensuring safety by informing the roadside unit (RSU) installed on the surrounding vehicle or on the road, such as a forward road situation, or an approaching vehicle in a blind spot, obstacles in the range of vision, and various accidents while driving. Recently, with the addition of the concept of space for entertainment or relaxation in simple mobility, the 5G technology killer service is gaining attention for sending and receiving large amounts of data at high speeds.

Autonomous cars self-assess and perform recognition, judgment, and control procedures previously handled by the driver through various sensors mounted on the vehicle. In order to achieve this, cameras, lidar, radar, ultrasonic waves, and geometrical sensors are installed inside the vehicle to collect various data on the driving environment, with the route determined by compiling various recognition technologies and map information and performing autonomous functions by controlling steering and acceleration.

- In the perception phase, data on driving environment above the dozens of Mbyte to some Gbyte units per second are collected from various sensors such as cameras, radar, and lidar, which identify the driving environment such as obstacles around the vehicle or road conditions.

- In the determination phase, the route for driving is determined in real time through the results from the perception phase and current vehicle conditions, maps, and weather information.

- For autonomous, it can be divided into six levels, as shown in Fig. 4.5, depending on the degree of autonomy. In order to be level 4 or higher, vehicle communication technology that can provide traffic information in real time beyond simple measurements of vehicle sensors attached inside for intelligent driving judgment and support dynamic precision guidance will be required. Vehicle sensors such as cameras and radars are difficult to detect for long distances of 200 meters or longer, and non-visible areas hidden in front cars or buildings are difficult to detect with internal sensors of vehicles, and 5G communication technology can solve such problems.

	SAE level	Name	Steering and acceleration / deceleration	Monitoring of driving environment	Fallback when automation fails	Automated system is in control
Human driver monitors the driving environment	0	NO AUTOMATION	Human driver	Human driver	Human driver	N/A
	1	DRIVER ASSISTANCE	Human driver and system	Human driver	Human driver	System driving modes
	2	PARTIAL AUTOMATION	System	Human driver	Human driver	Some driving modes
Automated driving system monitors the driving environment	3	CONDITIONAL AUTOMATION	System	System	Human driver	Some driving modes
	4	HIGH AUTOMATION	System	System	System	Some driving modes
	5	FULL AUTOMATION	System	System	System	All driving modes

FIGURE 4.5   SAE J3016 levels of automation [53]

Driving status data should be collected through sensors attached to the vehicle in order to make more intelligent driving decisions in different situations that may occur while the vehicle is driven, and this should be transmitted live to the surrounding vehicle and road infrastructure and to the cloud control center. To do this, 5G can be a good solution for managing communication characteristics such as delay, latency, and a number of vehicle support, which require assistance from the infrastructure.

5G technology has 10 times shorter latency than LTE. And the capacity and transfer rates of data are also expected to increase by up to 1000 times, enabling the transfer of large amounts of data from multiple vehicles to be efficiently analyzed by cloud servers, creating an optimal environment for autonomous use.

One of the major areas of vertical services mentioned with the introduction of 5G is automotive service, with various types of composite services being proposed, such as platooning, self-driving, and car-sharing to support simultaneous vehicle movements beyond simple safe driving. In 3GPP Release 15, various use cases, e.g., platooning, limited or full self-driving, sensor and map data sharing of autonomous vehicles, and dynamic ride sharing were proposed. In order to support this self-driving, the 5G communication characteristics that provide broadband, low latency, and high reliability can meet service requirements because a large number of image or laser sensor data must be processed in real time.

Use cases and service requirements applicable to vehicle-to-everything (V2X) completed development in 3GPP System Aspects (SA) Working Group 1 and published as a 3GPP TS 22.186 standard document [15]. 5G V2X service is largely divided into services that do not involve safety related to driving.

Examples of safety-related services include autonomous driving and platooning. Safety-related services are primarily intended to improve driving convenience, including large-capacity transmission for advanced driving features (e.g., lane merging, see-through driving), mobile hot-spot and digital map updates. Service requirements for some typical use cases are shown below.

- Vehicle platooning: It allows a number of vehicles to travel in a close line, as if the train were going. To maintain the distance between adjacent vehicles, information such as speed, direction, and acceleration is shared among the vehicles belonging to the platooning. It allows the vehicle except the lead vehicle to be driven automatically without driver intervention, thereby improving fuel efficiency and driving safety. Composition and dismantling of clusters is dynamic and such information shall be shared whenever individual vehicles enter or leave the cluster. For this purpose, cluster internal message transmission should be done quickly and accurately.

- Extended sensor: It allows each vehicle to have a comprehensive situational awareness by allowing it to share raw data or processed data between vehicles. This can help to improve the safety of intersections by predicting/preventing various accidents between platooning and by sharing information on pedestrians/emergency vehicles. The characteristics of mission critical services make achieving high reliability and low latency essential, and high-data transmission is generally required depending on the amount of sensor information being transmitted.

- Remote driving: It enables people to drive remotely using communications technology from a distance. This is achieved by transferring live driving images taken by a camera mounted on the vehicle to the driver or cloud server located in a remote location and by receiving control commands generated. Therefore, it must support at least 20 Mbps of Uplink transmission capacity and achieve a reliability of at least 99.999% and a delay rate of less than 5 ms.

- Advanced driving: Autonomous technical phase includes full autonomous driving (Level 4/5) and half-autonomous driving (Level 2/3), and the technical phase determines the information sharing requirements (see Fig. 4.5). This supports the cooperative perception of the object commonly detected and the information sharing function of the cooperative maneuver for the purpose of driving, such as changing lanes and moving/stop/parking.

- Tethering: When the vehicle is connected to the network, it enables access to the network in the form of a relay to passengers inside the vehicle or to pedestrians around it. This can satisfy the demand for internal entertainment of the vehicle and reduce the communication distance, thereby reducing the power consumption of the user terminal.

Table 4.2 summarizes the requirements for each service classification presented in 3GPP, which may vary depending on the detailed services within each classification and the autonomous level.

As seen in Fig. 4.6, 3GPP has standardized communication with wireless networks arising from vehicle to person (V2P), vehicle to infrastructure (V2I), vehicle to network

	Transmission rate	Delay	Reliability
Vehicle platooning	70 Kbps–65 Mbps	10–25 ms	90–99.99%
Advanced driving	60 Kbps–53 Mbps	3–100 ms	90–99.999%
Extended sensor	120 Kbps–1000 Mbps	3–100 ms	90–99.999%
Remote driving	Uplink: 25 Mbps Downlink: 1 Mbps	5 ms	99.999%

TABLE 4.2 Summary of eV2X requirements defined by 3GPP

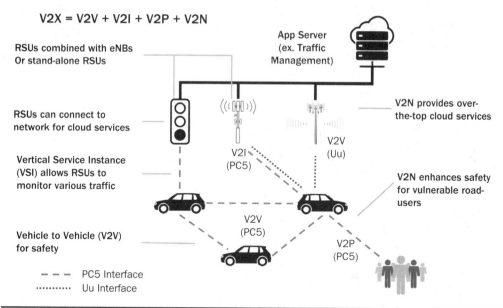

FIGURE 4.6   V2X services using 5G

(V2N), and vehicle to vehicle (V2V) with developed standards that are mainly limited to network standardization. V2V and V2P are also popular concepts in the existing intelligent transport system (ITS), and V2N refers to communication with vehicles and mobile networks [16]. V2I refers to communication with the roadside unit (RSU) in the ITS concept, which assumes that the RSU may be a stand-alone base station in the same form as the on-board unit (OBU), or it may be in the form of a mobile network base station.

As a first step in V2X standardization, 3GPP presented V2X use cases and derived performance requirements from them. In the past, there were many research and pilot services in the European Union (EU) and North America related to safety based on wireless LAN and dedicated short range communications (DSRC),[1] so 25 similar safety-oriented use cases were first standardized and reflected in the requirements of LTE V2X [17]. Table 4.3 shows 25 use cases related to V2X as defined in Release 14 LTE.

Use cases for more complex high-bandwidth scenarios, such as automatic driving and cooperative driving, were proposed, and these requirements were reflected in the LTE V2X enhancement (LTE-eV2X) Release 15. The requirements for use cases in eV2X include those related to latency, message size, frequency, vehicle speed, and security, with different requirements for each use case. Table 4.4 shows the major use cases for eV2X as defined in 3GPP Release 15 divided into four groups: automotive driving, cooperative driving, multi network, and service safety.

V2V requires 20 ms of latency, 100 ms for other V2I/V2N, and within 1000 ms for application servers. Vehicle platooning is a service that allows vehicles to travel together

---

[1]DSRC is a term used to encompass a set of V2X standards based on IEEE 801.11p and SAE specifications J2945.

Category	Use case name
Safety	Forward collision warning
	Control loss warning
	V2V emergency stop
	V2I emergency stop
	Queue warning
	Wrong way driving warning
	Pre-crash sensing warning
	Curve speed warning
	Warning to pedestrian against pedestrian collision
	Vulnerable road user safety
	Pedestrian road safety via V2P awareness message
Network Operation	V2X message transfer under MNO control
	V2X in areas outside network coverage
	V2X by UE-type roadside unit (RSU)
	V2X minimum QoS
	V2X access when roaming
	Enhancing position precision for traffic participants
	V2N to provide overview to road traffic participants and interested parties
Information	V2V for emergency vehicle warning
	Cooperative adaptive cruise control
	Road safety service
	V2X road safety service via infrastructure
	Traffic flow optimization
	Mixed use traffic management
Service	Automated parking system
	Privacy in the V2V communication environment
	Remote diagnosis and just in time repair notification

TABLE 4.3   Summary of eV2X requirements defined by 3GPP

in groups dynamically. Vehicles in the group must periodically receive a series of data from the lead vehicle in order for the vehicle to function normally. These data are used to keep the distance between vehicles in the platooning group to a minimum. Table 4.5 illustrates the requirements for each scenario in the use case of vehicle platooning, as defined in 3GPP TS 22.186 [15], for payload, Tx rate, Max end-to-end latency, reliability, data rate, and minimum required communication range.

4.4   Mission Critical Service Using URLLC

5G is expected to support various mission critical services using ultra-reliable low-latency communication (URLLC), which is one of the 5G New Radio features supported

Category	Use case name
Automotive Driving	eV2X support for Remote Driving
	Information sharing for limited automated driving
	Information sharing for full automated driving
	Information sharing for limited automated platooning
	Information sharing for full automated platooning
	Video data sharing for assisted and improved automated driving
	Changing driving-mode
	Cooperative lane change (CLC) of automated vehicles
Cooperative Driving	eV2X support for vehicle platooning
	Information exchange within platoon
	Automotive sensor and state map sharing
	Automated cooperative driving for short distance grouping
	Collective perception of environment
	Cooperative collision avoidance (CoCA) of connected automated vehicles
	3D video composition for V2X scenario
Multi Network	Communication between vehicles of different 3GPP RATs
	Multi-PLMN environment
	Use case on multi-RAT
	Use case out of 5G coverage
Service and Safety	Dynamic ride sharing
	Tethering via Vehicle
	Emergency trajectory alignment
	Teleoperated support (TeSo)
	Intersection safety information provisioning for urban driving
	Proposal for secure software update for electronic control unit

TABLE 4.4   Release 15 eV2X use cases from 3GPP TR 22.886 [16]

by the 5GC. URLLC use cases include public safety, smart factory, smart healthcare, and communication for intelligent transport services.

4.4.1   Public Safety and Disaster

First of all, we should note that the most common type of service in the field of disaster is group communications. Private or emergency calls, which are also included in commercial services, are important in the field of disasters, but group communications are the main services because more than three people usually perform a specific mission critical task in a disaster environment. Basically, all services including group communications are required to be provided smoothly and reliably in various situations. Also, noticeable differences here from conventional commercial services are in terms of survival and reliability. In addition to the factitious situation such as traffic congestion and network infrastructure collapse due to natural disasters, wireless communication

Communication scenario description		Payload (Bytes)	Tx rate (Message/Sec)	Max end-to-end latency (mess)	Reliability (%)	Data rate (Mbps)	Min required communication range (meters)
Scenario	Degree						
Cooperative driving for vehicle platooning	Lowest degree of automation	300–400	30	25	90		
Information exchange between a group of UE-supporting V2X application.	Low degree of automation	6500	50	20			350
	Highest degree of automation	50–1200	30	10	99.99		80
	High degree of automation			20		65	180
Reporting needed for platooning between UE-supporting V2X application and between a UE-supporting V2X application and RSU.	N/A	50–1200	2	500			
Information sharing for platooning between UE-supporting V2X application and RSU.	Lower degree of automation	6000	50	20			350
	Higher degree of automation			20		50	180

TABLE 4.5 Performance requirements for vehicles platooning from 3GPP TS 22.186 [15]

service should be provided by using technology such as device-to-device (D2D) communication or relay technology in 5G.

The most crucial mission-critical feature of disasters is the fault-tolerant survivability and high-speed of group communication services. In group communication, several people participate in a single emergency voice call, and one person acquires a voice through the control of the voice and the other people listen to the voice. We can easily imagine this situation when several firefighters are working together to rescue people. Therefore, in order for the group call service to support mission-critical characteristics, the group call setup, the control of the right to speak, and the speech transmission process must always succeed normally in various wireless signal environments. In addition, these processes need to be done more quickly than a certain standard in order to support mission-critical services in various situations of disaster. For example, even when the voice call is delivered normally, the group communication service is difficult to perform smoothly between people performing disaster mission through the group call service.

One example of mission critical services is mission critical push-to-talk (MCPTT) (see Fig. 4.7), which was initially defined by the Open Mobile Alliance (OMA) and then subsequently redefined and adapted by 3GPP for LTE. A push-to-talk (PTT) service provides an arbitrated mechanism by which multiple users may engage in communication. PTT users can request permission to talk to transfer their voice. The MCPTT service supports an enhanced PTT service, suitable for mission critical scenarios. In particular, 3GPP TS 22.179 [18] quantitatively defines the performance requirements criteria for these mission-critical features. In particular, the following three Key Performance Indicators (KPIs) are important to support MC service:

- The MCPTT access time (KPI 1): The time between when an MCPTT user request to speak (normally by pressing the MCPTT control on the MCPTT UE) and when this user gets a signal to start speaking.

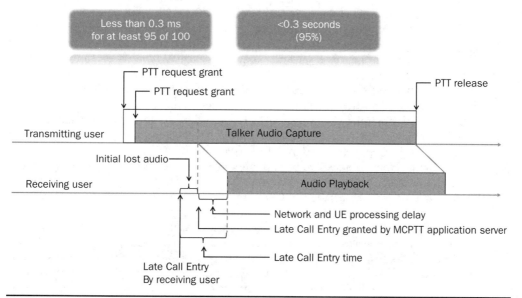

FIGURE 4.7   3GPP MCPTT performance indicator in 3GPP TS 22.179 [18]

- The end-to-end MCPTT access time (KPI 2): The time between when an MCPTT user requests to speak (normally by pressing the MCPTT control on the MCPTT UE) and when this user gets a signal to start speaking, including MCPTT call establishment (if applicable) and possibly acknowledgment from first receiving user before voice can be transmitted.

- The mouth-to-ear latency (KPI 3): The time between an utterance by the transmitting user and the playback of the utterance at the receiving user's speaker.

KPIs 1 and 3 should be provided less than 300 ms for 95% of all MCPTT request and voice bursts, respectively. In 3GPP, the mission-critical characteristic is defined around 5% level, and even in the worst case, the delay time for the two KPIs is 300 ms or less. Here, 300 ms corresponds to the minimum time that a person can feel a response time when a person performs a specific action. That is, when a response to an event takes more than 300 ms, a person considers the response as a late response, and when less than 300 ms, a response to the event is considered as an immediate response.

4.4.2 Smart Factory Including Robotics and Automation

The most notable country in this field is Germany. Germany, a leading manufacturing country, is planning to allocate 3.7–3.8 GHz as a frequency band that can be used by 5G convergence industries [19]. The 3.4–3.7 GHz band, under auction, can be seen as a strategic band in which mobile operators provide various 5G applications nationwide through large investments. This frequency band can be seen as an additional frequency to rapidly spread 5G. Although this gives a 10-year grace period without undergoing through another government auction, however, unused frequency for more than one year will be immediately recalled and applicants will be screened for how to utilize it.

The United Kingdom supplied an additional 3.8–4.2 GHz band in 2019 [20]. This band is dedicated for the applicability of 5G standards, and its purpose is to promote the development of 5G ecosystem. Unlike the 3.4–3.8 GHz band, there is a limit that the band is already being used as satellite earth stations, fixed links, and FWA depending on the region. Therefore, the United Kingdom has a focus on low-power licenses with narrow base station coverage and expects to be used to provide 5G applications in urban or industrial complexes.

Japan is making similar moves. The 4.6–4.8 GHz band, 28.2–29.1 GHz band under consideration for additional supplies [1], is expected to be utilized as 5G smart factory and remote control–related applications [21]. China has yet to make noticeable moves. However, Huawei has been making efforts to spread 5G in industrial internet for a long time and is leading standardization in this area.

Meanwhile, there is a need to pay attention to the 5G-ACIA (Alliance for Connected Industries and Automation) [22] formed in June 2018. 5G-ACIA was launched by the Electronic Industries Association of Germany (ZVEI) to utilize 5G technology in automating manufacturing processes and to convey industry needs in 5G standardization. Based on the membership status as of December 2018, Germany consists of 23 members, the rest of Europe consists of 9, China consists of 2, Japan consists of 2, and South Korea consists of 1. Because its influence is significant as leading companies in process automation industries such as Siemens, Yokogawa, and ABB participate, the 5G diffusion in smart factory field will gain significant momentum. Germany is moving fast in terms of frequency and global ecosystem leadership. Germany has long pushed for smart factory and focused on automated production through robots. The reason why Germany is trying to quickly incorporate 5G into manufacturing industries is because of the high sense of crisis that it is lagging behind the United States in software that can be applied to manufacturing sectors such as artificial intelligence (AI), Big Data, and Cloud.

4.4.3 Smart Healthcare

The paradigm of the medical industry is evolving from a treatment center to a disease prevention, diagnosis, and health-monitoring center. According to McKinsey & Company [23], it is estimated that US healthcare spending is expected to top $5.34 trillion by 2025. In addition, they identified a $284-billion to $550-billion opportunity for value creation from the healthcare applications. 5G will play an important role in creating a service model that can be used for disease prevention and health monitoring based on the internet of medical things (IoMT). The IoMT is a concept encompassing medical devices, wearable devices, remote sensors, and wireless patches that transmit biometric information such as a person's electrocardiogram, brain waves, blood pressure, blood sugar, and body temperature. Although 5G can still send and receive only basic biometric information to individual smartphones and medical institutions, the establishment of a 5G communication environment will enable real-time interconnection between various IoMT devices, creating an environment where individuals can accumulate information reliably and continuously. According to a report [24] by Ericsson Consumer Lab, 5G mobile communications will move the healthcare delivery hub from the hospital to the home and is predicted that healthcare services using biometric information will be possible anytime, anywhere. When big data and AI technologies are incorporated into this, users will be able to actively manage their health through AI's recommended healthcare service. For healthcare providers, targeted, personalized care will be possible based on patient biometric information, and precision medical services will also be available to reduce side effects and increase treatment effectiveness.

To perform surgery remotely that requires super-precision or to connect hospitals and accident sites in an emergency situation, a 5G communication environment without delay is needed. Communication speed and stability of 5G, which is close to real time, can greatly reduce the risk of remote surgery. Residents in areas where it is difficult to find healthcare providers will also be able to receive medical services remotely through 5G communications technology. In fact, in January 2019, a medical school in Fujian, China, conducted a remote robot operation of surgically removing the liver of an experimental pig 50 km away using Huawei's 5G technology with a successful completion. According to market research firm Statista [25], the size of the global telemedicine market is forecasted to grow from $26.5 billion in 2018 to $41.2 billion in 2021.

5G is expected to be especially useful in emergency situations. While only basic emergency measures are still being taken in emergency vehicles to transport injured or seriously sick patients to hospitals, patients and doctors are expected to be remotely connected within the emergency vehicles that can communicate with 5G and be treated on the spot. In addition, data such as patients' past history, camera images, 3D, and VR images will be able to be received in real time by connecting with data servers in ambulances. There is a fierce competition to dominate this market in anticipation of the opening-up of the dream healthcare market that overcame time and space constraints in the 5G era. Global IT companies such as Google, IBM, Apple, and Amazon are developing digital healthcare-related business models linked to their devices or AI/OS/distribution platforms. 5G technology in Korea is ahead of other countries, but innovation in the healthcare industry is relatively slow due to security issues on personal information and distributed medical data. 5G is expected to pave the way for Korea's healthcare industry in taking a big step forward, which will be a model for other countries.

4.5   Massive Internet-of-Things (IoT)

Internet-of-things (IoT) refers to a technology that allows us to exchange data between objects and provide new services by attaching communication functions to various objects around our lives and connecting them to the internet. IoT is expected to connect objects, houses, offices, and all urban infrastructure (referred to as smart things or smart devices) and share and utilize information in the upcoming hyper-connected society. A 5G environment is essential for connectivity and data collection, control, and data transmission from various smart devices. Especially for objects in environments that require mobility, such as vehicles and drones, 5G technology is essential to connect to the internet and exchange information. Although there are some differences between research institutes, International Data Corporation (IDC) [26], a global ICT market research firm, predicts that 50 billion devices will be connected to IoT by 2030, and super connectivity of 5G that will be supported like IoT is expected to cause major changes in smart home, office, smart city, and smart energy fields.

For smart homes and offices, distributed ecosystems are expected to be created in the form of direct communication with individual objects that are equipped with IoT and AI platforms. Smart city provides hyper-connectivity based on 5G and various network technologies, through which various objects are connected to every corner of the city like a human neural network, data is collected through IoT platform, and data collected is analyzed in real time using AI technology, and various responses are possible for the efficiency of the city. The 5G can be used for the integration of the city's operations by fusing urban infrastructure such as roads, power grids, gas pipes, and water supply with ICT. In the smart energy sector, digitalization of the energy industry is underway as fourth industrial revolution technologies such as IoT, AI, and Cloud, including 5G are applied to the energy industry. This is expected to bring about major changes in the overall ecosystem of the energy industry, including production, storage, distribution, and consumption of energy. It is shown that variety of smart energy–related services can be provided through power grids tightly connected to massive IoT.

4.5.1   Smart Home/Office/Building

4.5.1.1   Smart Home

Until now, smart homes have remained mostly in smartphones or separate remote controls, such as remotely connecting various devices to a network to operate air conditioners or to turn a reserved washing machine, which has remained at the level of control of electronic devices. It is predicted that smart home in IoT and 5G era will develop into an intelligent smart home that is more personalized and more user convenience by embedding AI into IoT devices itself and utilizing data that is provided on IoT platforms.

AI is already penetrating our daily lives. Traditional internet service providers are scrambling to introduce AI services [27]. Google's "Google Assistant™," Amazon's "Alexa™," and Apple's "Siri™" are available in various household appliances and AI speakers, enabling users to control a wide range of devices in the home even with voice.

They are working to create a more extended smart home ecosystem by working with external data as well as data on users.

Traditional home appliance manufacturers are also rushing to provide services that incorporate their own AI platforms and technologies [28]. Samsung Electronics plans to install an AI platform called "Bixby" on all its products by 2020, connecting homes, offices, and cars. LG Electronics is also establishing a smart home ecosystem by linking home appliances with its AI brand "ThinQ."

4.5.1.2 Smart Office Using Hologram

Smart offices based on 5G are expected to improve business efficiency and revolutionize business processes. Because smart office requires multiple members to use multiple services with different requirements, supporting 5G's high-capacity network, which enables many to support simultaneous and reliable access, is essential.

The stability of the network and the speed of data processing and transmission will have an absolute impact to support hologram-based communications like "Holodeck" in the American sci-fi movie Star Trek, or services where agents located in different locations in the movie Kingsman meet wearing special glasses. Until now, video conferencing has been limited in quality because the video conference was not guaranteed speed to provide stable service. However, video conferencing and hologram conference services, which are combined with new technologies, are being attempted recently based on the high-speed transmission provided by 5G network, so holographic technologies that have been seen in movies can be commercialized in the near future.

Smart offices also acquire information through IoT sensors like Smart Home and provide various services to office users based on analysis of such aggregated data. You can automatically locate users, control lighting in a building, restrict access to a secure office, or automatically prepare screens and resources for a meeting. Smart offices have a significant number of connected devices and services that need to be real time compared to smart homes, so super-low power, super-connectivity, and reliability features of 5G are essential factors.

4.5.1.3 Building Management

Smart building refers to a high-tech building that incorporates information and communications technology (ICT) in a building. Building management system, in particular, refers to a system that provides integrated monitoring and control of lighting, heating ventilation, air conditioning (HVAC), and security of a building with building automation system (BAS), energy management system (EMS), and facility management system (FMS) to efficiently manage and operate the building.

Building management using BMS is also an area where 5G can provide evolved services. By integrated management of data collected from sensors in buildings, such as big data analysis or AI, various facilities in buildings can be efficiently managed.

For example, a smart lighting system that works with or without a person with sensors installed in the office will reduce energy and operating costs, while remote control and monitoring of multiple buildings will enable rapid response in an emergency to ensure efficient management and safety.

4.5.2 Smart City

Smart city refers to one that uses ICT to efficiently operate its infrastructure and to solve such problems as traffic congestion, parking difficulties, crime, waste disposal, floods, fires, and energy overconsumption to enhance citizens' convenience in life. Although data needs to be collected through IoT sensors that are installed in every corner of the city to implement smart city, there have been limitations in establishing a massive IoT environment that connects large-scale IoT devices to telecommunication network.

It is predicted that 5G network technology will connect urban, ICT, and spatial information infrastructures to IoT and make it possible for entire cities to maintain optimal conditions on their own like organic bodies. If the previously developed ubiquitous city focused on the installation and monitoring of sensors, smart city is being developed based on urban data. Only when the various information that constitutes a city is digitized, collected, and analyzed in a standardized manner, it can be used to solve problems caused by urbanization.

An example of smart city is the "Virtual Singapore" project that the Singaporean government has been pushing since 2014 along with the Singapore National Research Foundation (NRF) [29]. The project, based on the 3D platform to create a sustainable city, is a virtual reality for all of Singapore's land, including buildings, roads, power, wind direction, noise, population, and even solar power devices installed on building rooftops. By converging predictive analysis and Digital Twin technology using cutting-edge technologies such as Big Data, IoT, and AI, the company can identify all moving objects in Singapore city and make various predictions. For example, an accurate simulation of solar power can be made by analyzing the solar generation of each building, and a simulation of how air flows in the city when the wind blows. By adjusting the layout of the buildings, the air quality as well as the overall ventilation of the city was improved.

South Korea is also pushing to build a data-driven smart city, moving away from a smart city that is focused on installing simple sensors. Smart city is a space where complex and diverse services converge, and the role of a standard-based integrated platform for this is becoming important. The Korean government has established a sustainable IoT-based smart city as oneM2M, an international standard for IoT, in three cities, Busan, Daegu, and Goyang, for three years from 2014. This ensured compatibility for various services and data provided by smart city. Recognizing the importance of data produced by smart cities, the company has since invested a total of KRW 115.9 billion ($14 million) from 2018 to 2022 to develop a common smart city platform and implemented a "CityHub" project [30] and a demonstration project based on it. In particular, the smart city will be linked to autonomous cars built on 5G and utilize 5G technology in various services.

4.5.3   Smart Grid and Energy Management

4.5.3.1   *The Role of 5G in Smart Grid*

Smart grids are considered a key element of future energy infrastructures by applying ICT technologies to existing power grids and enhancing efficiency in energy use. Smart grids require numerous power devices and systems such as energy storage system (ESS), advanced metering infrastructure (AMI), intelligent transmission and distribution system, and energy management system (EMS). There are various types of power plants, such as hydropower/firepower/wind power, that produce electrical energy, and smart grids minimize the energy loss caused to move power through an intelligent power distribution system. Utilizing the power network and AMI, the use of electricity will be planned and operated systematically according to the accurate measurement of the usage of electricity and its results.

In order for smart grids to operate reliably, speed and reliability must be ensured in the transmission of measured data from each device, and it is expected that 5G's hyper-connectivity and reliability will play a major role in this process. Smart meter, which monitors power from AMI infrastructure and transmits it to servers, is expected to have more than 10 million devices in the future, requiring a large network connection.

South Korea's SK Telecom [31] is launching a smart energy meter for homes, while KT [32] is continuing attempts by telecom operators to enter related markets by building network solutions that use it for monitoring and diagnosing solar power. Because 5G network has 10–100 times more energy efficiency than current 4G network, it is predicted that introduction of low power sensor will connect more energy consumption devices to network and enable smart power management.

5G is expected to play an important role in intelligent power management through massive IoT, which is connected to power plants, transmission/distribution facilities, energy storage systems, energy management systems, and devices from power consumers in the future. This can change the grid, which consists of power generation, power transmission/distribution, and sales of smart grid product phases, and contribute to new business innovations. Consumers can monitor power usage and charges in real time, and suppliers can save on facility investment and power generation costs while keeping power reserve rates low through sophisticated forecasting and coping with power demand.

In a 5G-related report [33], Britain's mobile carrier O2 said a smart grid through the 5G connection could reduce energy consumption by 12% for all households in the United Kingdom. The annual value generated by the 5G smart meter is expected to reach GBP 6.47 billion by 2025 and GBP 7.37 billion by 2030, according to data estimated by the EU's board .

4.5.3.2 Energy Management Services

It seems that 5G diffusion will create many new services in smart energy field. Smart meters, which are connected power detectors that measure electricity consumption and transmit data in real time, are also considered promising areas.

Korea, for example, has more than 20 million analog power meter, and Korea Electric Power Corporation (KEPCO) plans to invest a total of KRW 1.2981 trillion by 2020 to distribute 22.5 million smart meters. 5G can be used to provide hyper-connectivity and reliability as a means of transferring data from the smart meter [34]. Smart meter using the wireless network is also preferred in Europe, and UK mobile network provider O2 and others are considering using 5G in smart meter market. Huawei is also considering use in the global market, citing the use case of 5G network in smart meter and energy management systems [35]. In the future, devices utilizing 5G technology may be actively used in the smart meter market as smart grid systems evolve.

In fact, many countries are testing systems that can effectively distribute power by deploying pilot services using smart meter. For example, let us consider that as a smart meter provider, you distribute 5000 kWh of electricity to all households. The advantage of these services is that, if the average electricity use is 3000 kWh and only few households use 7000 kWh, it is possible to provide 3000 kWh of electricity to households using 3000 kWh of electricity, and 7000 kWh to households using 7000 kWh of electricity. Also, if you are away for a month for a vacation in the middle of summer, you may not supply electricity at all. In addition to these advantages, the deployment of 5G smart meters will eliminate the discomfort of the meter reading method and the errors

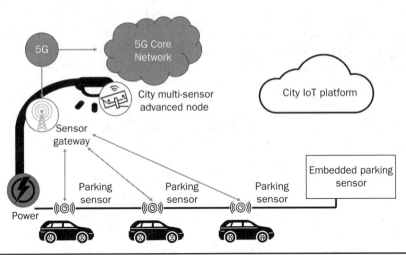

FIGURE **4.8** Smart LED light using 5G

caused by the human meter reading. More efficient meter reading of various electrical appliances, especially in mountainous areas or remote areas, can be supported by 5G smart meter.

As Fig. 4.8 shows, smart electricity management device is a device that intelligently measures and controls power consumption to reduce power. For example, smart streetlamps that incorporate 5G network modules into streetlights for LED lighting automatically produce energy and adjust brightness depending on the environment and weather around them. Here are some examples of how smart street lighting works.

- Telefonica, a Spanish mobile carrier, changed 70% of its downtown streetlights in the Spanish city of Malaga to smart streetlamps, saving more than 2.9 million USD a year in electricity bills.

- Taipei installed 2,600 streetlamps in five regions with smart LED streetlamps in 2019 and plans to install additional 110,000 of these by 2021.

- At Los Angeles, 110,000 LED street lights and 100 LED smart poles were installed to reduce energy use by 70%, reducing administrative costs and accidents at night.

Smart street lighting installed in smart city is also actively researched and developed, especially as a 5G base station. (See Fig. 4.8.) In this case, there are many advantages over regular street lighting as follows:

- Power supply to the street lighting can be shared to the 5G base station.

- 5G base station managed with sensors installed in street lighting at lower cost.

- 5G coverage expanded.

- Space rental costs for 5G base station installations reduced.

- Various 5G services connected to street lighting (e.g., RSU for Cloud-based Autonomous Driving).

	Smart Meter	Smart Energy Control	Energy Management System (EMS)
Concept	Automatic Meter Reading (AMR), two-way communication system between a meter and a control unit, smart electric meter allowing various measurement options such as time-based	An intelligent control system dramatically reducing power consumption for various devices connected to the network	A smart energy management system to monitor, control, and optimize the performance of the generation and/or transmission system
Use of 5G	Support reliable data transmission and connectivity more than tens of millions smart meter units via 5G technology	Get real-time device energy usage level insight and analytics using 5G	Real-time power monitoring and power generation control
Deployment Scenarios	Smart meter using 5G and power line communications	Smart streetlights automatically adjusting lighting levels on the fly based on circumstance A smart plug supporting remote control and switch on/off based on schedule	Connecting power generation, transmission and distribution systems using 5G and supporting real-time management system

TABLE 4.6 Smart energy-related services using 5G

5G can also be utilized for EMS, one of the underlying infrastructures for smart grids. In particular, high stability along with super-connectivity of 5G is needed to monitor power volume in real time and control power generation precisely. It is also likely that 5G-based network solutions will be installed in distributed renewable energy generators such as solar and wind power along with electric vehicle charging stations and ESS to create more intelligent EMS by integrating and managing data.

Table 4.6 shows some of the smart energy-related services using 5G technology. In smart meters, tens of millions of smart meter terminals can be connected over a 5G network, and various network technologies and sensors can be connected through 5G network to provide real-time control in smart energy and monitoring and control of real-time power volumes can be reliably achieved through 5G in EMS.

4.6   Enhanced Mobile Broadband

4.6.1   On-Demand Media

Network bandwidth is considered the biggest obstacle to ultra-high-definition services such as 4K on mobile communication platforms. The capacity of video clips increases rapidly as quality levels of full high definition (FHD), 4K, and 8K increase. Because UHD 4K (3840 × 2160) video has four times more pixels than FHD (1920 × 1080), it also increases capacity by four times. Netflix™ recommends more than 25 Mbps network bandwidth for 4K video streaming. Therefore, 4K video streaming is not possible on the 4G LTE network, which currently provides a 10 Mbps level of speed, and is only available on 5G which supports above 100 Mbps. Ultra-high-definition video in 4G causes bottlenecks in mobile data traffic due to limited bandwidth.

According to telecom equipment maker Ericsson [36], video traffic accounts for 60% of all mobile traffic as of 2018, and is expected to grow 35% annually until 2024, accounting for 74% of the total mobile traffic. Accordingly, the average monthly traffic of smartphone video streaming data worldwide is 3.4 GB in 2018, compared with 16.3 GB in 2024. The increase in video mobile traffic stems from increased playback of high-definition videos along with users viewing more videos on their smartphones.

The 5G network provides a seamless high-definition video service environment, while reducing the burden on data traffic along with increased speed. However, because ultra-high-definition video services that are introduced with 5G are still slow to supply 4K smartphones and it is difficult to feel 4K quality in mobile devices, it is predicted that they will spread gradually. It is predicted that gradual upgrades such as video streaming speed, which is mostly viewed on mobile networks at 720p picture quality, will be made first, while 4K streaming will be fully commercialized and expanded in the market following the supply of supporting terminal, service price, and efficiency that consumers can feel. In line with the multi-screen trend of expanding screens available to consumers in the future, demand for the seamless service, which follows images from one unit to another, is also expected to increase.

While 5G allows real-time transmission of 4K and 8K ultra-high-definition images, it is also causing huge changes in contents production. YouTube started supporting 4K video replay in 2014 and has been providing real-time 4K broadcasting in 2016 [37]. As a result, the production of high-definition video by YouTube™, multi-channel network (MCN), and individual creators also increased significantly, and 4K real-time broadcasting was impossible for 4G mobile network. It is predicted that introduction of 5G and spread of 5G will lead to activation of 4K live video broadcasting. 5G can also be used for real-time broadcasting of ultra-high-definition video footage from broadcasting companies as well.

It is predicted that introduction of 5G will have huge impact in all aspects of consumption and production of ultra-high-definition video contents. As 5G network directly affects mobile and mobile-related industries and ecosystems rather than traditional fixed-line telecommunication, different areas of importance and ripple effects can be seen. For example, the main terminal for viewing ultra-high-definition images is currently a 4K TV, and 5G influence is bound to be limited because these devices are connected to wired internet and still have sufficient bandwidth for 4K image transmission. Smartphones, which are mobile terminals, have a problem that it is difficult for users to feel the benefits of 4K due to limitations that they are equipped with small screens. 4K images are only available in a completely unrestricted data system with no traffic and speed limits; however, there are concerns over how much 5G environment the mobile carrier will be able to handle. It is predicted that 5G will have greater ripple effect in terms of production than in terms of contents consumption at the beginning. In particular, while YouTube™ has been supporting real-time 4K broadcasting functions since 2016, the 4G network has been underutilized since most YouTube™ videos will be filmed on smartphones. With the introduction of 5G, ultra-high-definition and real-time broadcasting of individual creators is expected to be highlighted in the B2C area, with ultra-high-definition broadcasting transmission service of broadcasters/content developers in the B2B area.

4.6.2 Virtual Reality and Augmented Reality

Virtual reality (VR) typically requires a set of contents and devices such as headsets, goggles, 3D mouse, and a wired glove. VR devices and tools should provide high-quality contents, including high-resolution videos and realistic interactions. VR cameras are used to create VR photography using 360° video with various virtual elements through special effects. Although many VR applications such as 360° videos and pictures are already available, there exist some challenges such as low bandwidth in communications networks because of too much data traffic. Therefore, 5G is considered as a promising solution to provide full-blown VR services.

Augmented reality (AR) is the integration of various interactive digital elements such as visual overlays, sensor information projection, and haptic actuations into our real-world environment. There exist various successful AR applications and services such as Pokemon Go AR game and Google SkyMap. Various hardware and software are used to realize AR. For example, contact lenses, eyeglasses, and head mounted display can be worn on human body as a means for displaying AR.

In order to create efficient AR devices, a lot of sophisticated electronics need to be integrated in an AR device. As people do not want to wear a heavy AR device, the size of such electronics becomes small, and the computing power of such devices is not enough to process the integration of virtual information into real-world environment. Therefore, in order to satisfy users' demand on VR, 5G technologies are used to provide basic functionality to such devices via remote servers with the stability of the internet connection.

Realistic contents such as virtual reality, aided reality, and mixed reality generally have very large capacity requirement because they contain data such as images, stereoscopic sounds, and motion recognition that can be seen from 360° front. According to data analyzed by telecommunication equipment maker Ericsson, the monthly traffic generated by the 1080p VR is more than 10 GB or more than three times that of the normal 1080p video when actual contents are streamed for five minutes a day. AR content, which requires bandwidth of 25 Mbps, goes beyond 8K resolution images, generating nearly 30 GB of traffic. Due to this nature of high-capacity data, 5G is suitable for transmitting VR and AR contents on mobile platforms.

Another factor that greatly affects the quality of VR/AR content is latency. Users of VR content can see all views up to 360° depending on the direction of turning their heads. Considering that the visual response to movement of the field of vision is less than 7 ms and the time required for VR screen processing is taken into account, a very low delay of 1 ms is required for smooth screen movement. Because screen conversion does not take place quickly when one of the side effects of VR and AR contents is pointed out as one of the side effects when using VR and AR contents, it is expected that the delay time that is implemented in 4G network to be greatly reduced from 5G to 1–4 ms will greatly help implementation of high-quality real-life contents.

VR and AR markets are expected to see high growth in the future. Market research firm Statista [38] predicts the global market for VR and AR will grow 66.8% annually from $27 billion in 2018 to reach $209.2 billion in 2022. IDC [39], a global ICT market research firm, predicts that VR terminal shipments will reach 31.49 million units in 2022 and AR terminal 21.59 million units. Gartner [40], a global information and communication market research firm, has recognized its promise as one of the "Top 10 Strategic Technology in 2019" and cited its immersive experience, which is based on VR/AR/MR technology. The introduction of 5G is expected to support the high growth of VR and AR contents with the strength of ultra-fast and ultra-low-delay networks and contribute to the revitalization of realistic contents.

It is predicted that 5G will highlight realistic cloud games as a new business area through super-fast transmission speed and ultra-low-latency characteristics. Currently, high-quality VR contents, especially VR games that require large data capacities and low latency, are only available head-mounted displays such as Oculus Rift and HTC Vive [41]. They require massive data transfers with high-performance computing power (CPU/GPU), so they are delivered using the HMD terminal wired to high-end PCs. If VR cloud gaming that transmits game screens over a network is commercialized following the introduction of 5G, it can have huge ripple effect on markets. By transferring high-resolution images that are computed to cloud center PCs through wirelessly connected HMD, simple and efficient VR games will be possible. The advantage of this is that it can reduce the cost of purchasing high-end computers and significantly increase user convenience through wireless connectivity away from cumbersome and complex wired connections. Market research firm Ovum [42] predicted that the global VR/AR cloud gaming market will grow by 2400% over the next 10 years to reach $47.7 billion by 2028. The cumulative revenue contribution of 5G over the cited period is expected to reach $142 billion, which will play a key role in the commercialization of realistic cloud games.

Real-life advertisement that is provided through VR and AR images and interactive contents is also expected to emerge as a result of 5G. New media ads are mainly provided in a streaming format, which currently lacks the capacity to handle VR and AR contents and the advertiser's needs for realistic advertising. But, in order for high bandwidth to be secured through 5G, ad-based platform holders can generate new revenue streams, along with discriminatory advertising effects through VR and AR advertising. For example, HTC has introduced a targeting advertising solution that can detect if a user has seen or turned his or her head using sensor technology on its VR platform. As such, it is highly likely that new VR and AR advertising techniques or formats will be introduced in 5G environments.

4.7   Interworking with Internet-of-Things (IoT) Application Servers

Already in 3GPP Release 10, the 3GPP system architecture group (3GPP SA2) started to work on machine-type communications (MTC) for Universal Mobile Telecommunications System (UMTS) and long-term evolution (LTE) core networks [43]. The main focus at this time was basically the prevention of system overload and signaling congestion due to the expected huge number of devices (user equipment, UE). The feature evolved over the years with continuous improvements in every release up to Release 16. In this section, various 5G features supporting MTC and IoT are explained followed by how these features are used via third-party IoT application servers such as IoT service provider using their own IoT platforms. In particular, as one of the global standard initiatives for IoT, oneM2M, develops 3GPP interworking standards based on 3GPP Release 15 and 16, there is a section (see Sec. 4.7.2) explaining how 3GPP IoT features can be used by oneM2M standard-based IoT service layer platform.

4.7.1 MTC and IoT Features in 5G

There is a basic assumption that the end-to-end IoT communications, between the MTC application in the UE and the IoT application including MTC application in the external network, use services provided by the 3GPP system and optional services provided by a Services Capability Server (SCS). Typically, there is an application server (AS) that hosts various IoT applications in the external network. In order to support such IoT services, the 3GPP system provides various functions in terms of data transport, subscriber management, network management, and group data delivery.

The latest version of the MTC specification 3GPP TS 23.268 [44] has an extensive set of features. Some of MTC features are listed below (a list of full MTC features is described in Chap. 6):

- Device Triggering: It is the means by which a 3GPP Service Capability Server (SCS) sends information to the UE via the 3GPP network to trigger the UE to perform application-specific actions.

- Group Message Delivery: It allows an SCS/AS to deliver a message to a group of MTC UEs. This can be done via either MBMS, which is a broadcasting service that can be offered by cellular networks or unicast, which is a one-to-one communication.

- Monitoring Events: It allows an SCS/AS to monitor specific events in 3GPP system. Configuration and reporting of the monitoring events include UE reachability, location of the UE, loss of connectivity, and communication failure.

- High-Latency Communication: It is a function to handle unreachable UEs while using power-saving functions.

- Support of Informing about Potential Network Issues: It allows an SCS/AS to request the 3GPP core network for being notified about the network status in a geographical area.

4.7.2 IoT Service Layer Interworking with CIoT

In this section, an overview of interworking function and architecture between 3GPP and IoT service layer platform is introduced. In particular, this section is written from IoT service layer perspective which explains how IoT service layer sees 3GPP 5G CIoT and interconnects to the CIoT core for IoT services.

4.7.2.1 Background of 3GPP Interworking

Considering IoT devices having ubiquity of standards-based mobile connectivity, IoT service layer standards bodies are looking at an interworking service to be connected with 3GPP CIoT. oneM2M[2] is a partnership project for standardizing specifications for IoT common service layer platforms [45]. Similar to 3GPP, oneM2M is a project established based on an agreement between eight standards development organizations (SDOs): Telecommunications Technology Association (TTA), Association of Radio Industries and Businesses (ARIB), Alliance for Telecommunications Industry Solutions (ATIS), China Communications Standards Association (CCSA), Electronics

---

[2]http://www.onem2m.org

Telecommunications Standards Institute (ETSI), Telecommunications Industry Association (TIA), and the Telecommunications Technology Committee (TTC).

The work standardized in oneM2M to collaborate with 3GPP core network is to include the service capability exposure function (SCEF) API. This API exposes underlying network services to oneM2M service layer platforms and vice versa. Thanks to the SCEF API, IoT service layer and 3GPP core network can provide more intelligent services than before. For example, IoT applications running on top of oneM2M service layer platform can share application-specific information with the 3GPP Core Network to trigger sleeping IoT devices awake.

4.7.2.2 Overview of oneM2M Architecture

oneM2M system architecture design work is focused on providing both basic functionalities (e.g., registration and message handling) and various advanced functionalities (e.g., interworking with other systems). To achieve this, oneM2M defines a common service layer providing M2M services, which is independent of the underlying networks.

The oneM2M reference architecture [46] is illustrated in Fig. 4.9. The oneM2M system is formed by functional entities called nodes. These are known as the application dedicated node (ADN), application service node (ASN), middle node (MN), and infrastructure node (IN). Nodes consist of at least one oneM2M common services entity (CSE) or one oneM2M application entity (AE). A CSE is a logical entity that is instantiated in an M2M node and comprises a set of service functions called common service functions (CSFs). CSFs can be used by applications and other CSEs. An AE is a logical

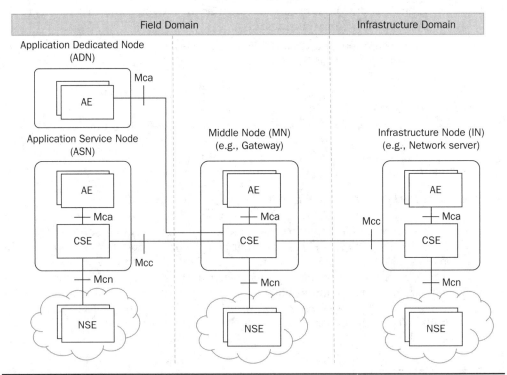

FIGURE 4.9 oneM2M reference architecture

entity that provides application logic, such as remote blood sugar monitoring, for end-to-end M2M solutions. oneM2M defines three reference points (i.e., Mca, Mcc, and Mcn), as indicated in Fig. 4.9. The Mca reference point enables AEs to use the services provided by the CSE. The Mcc reference point enables inter-CSE communications. The Mcc' reference is similar to Mcc but provides an interface to another oneM2M system. The Mcn reference point is between a CSE and the service entities in the underlying networks, such as device triggering service provided by Third Generation Partnership Project (3GPP) networks.

As mentioned earlier, oneM2M specifies a set of core common service functions (CSFs). Some CSFs provide administrative functions for the service layer and other CSFs. For example, the registration (REG) CSF provides a means for an AE or a CSE to register to a CSE and be able to use the services provided by that CSE. The security (SEC) CSF enables secure establishment of service connections and data privacy. An AE and a service layer management (ASM) CSF provide functions to configure, trouble-shoot, and upgrade CSEs and AEs. A device management (DMG) CSF manages device capabilities such as firmware updates. The communication management and delivery handling (CMDH) CSF is responsible for service layer message delivery. The network service exposure (NSE) CSF serves as the anchor point between the service layer and services provided by different underlying networks. Some CSFs provide value-added services to registered AEs and CSEs. The data management and repository (DMR) CSF is responsible for user data storage and processing. Users can subscribe and get notifications of changes in the data. The discovery (DIS) CSF provides a means to make the services and resources discoverable by other CSEs and AEs. A subscription and notification (SUB) CSF manages subscriptions to changes on the oneM2M platform. The service session management (SSM) CSF supports end-to-end service layer sessions. The service charging and accounting (SCA) CSF provides mechanisms to support service-layer-based charging. A group management (GMG) CSF supports bulk operations and manages group membership. The location (LOC) CSF allows M2M AEs to obtain geographic location information of an entity and receive location-based services. In addition to CSFs, a CSE includes a service enabler to ensure the extensibility of services.

4.7.2.3   Overview of 3GPP-oneM2M Interworking

In oneM2M, the common service entity (CSE) provides common service IoT control functions in IoT/M2M devices (e.g., cloud server, gateway, and end devices) while an application entity (AE) represents an IoT application. IoT servers, which are typically located on a cloud, are represented as an infrastructure node (IN). The service capability server (SCS) in the 3GPP reference architecture can be mapped to IN-CSE in oneM2M as shown in Fig. 4.10. IN-CSE connects to the 3GPP trust domain to communicate with UEs which is equipped with oneM2M applications via three interfaces, i.e., SGi, T8, and Tsp. These interfaces are designed to provide various protocols (IP for SGi, RESTful API for T8, and Diameter for Tsp) used by IoT applications.

In order to define interworking with 3GPP network, oneM2M [47] introduces how oneM2M entities can utilize IoT/M2M service functions provided by 3GPP Core Network. For example, as SCEF exposes "Monitoring," the specification describes how IN-CSE communicates with the 3GPP SCEF entity via T8 interface (which is Mcn interface in oneM2M). Similarly, a 3GPP UE hosted oneM2M AE and CSE can use a Group Messaging service (via MBMS) by the 3GPP Communications Unit.

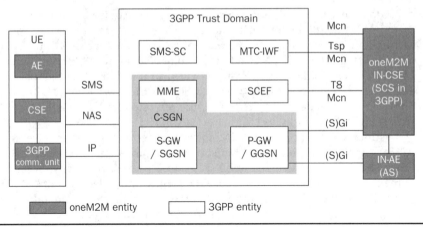

FIGURE 4.10   3GPP-oneM2M interworking architecture

As oneM2M service platform is designed based on RESTful, a set of APIs defining the related procedures and resources for the interaction between the SCEF and the IN-CSE (SCS) are standardized on T8 interface. Two specifications, 3GPP TS 23.682 [48] and 3GPP TS 29.122 [49], are developed for the T8 APIs to describe the architectural level description and the protocol level description of the T8 APIs, respectively. From the oneM2M side, a specification is developed defining how an IN-CSE interworks with a SCEF via the T8 APIs [48]. For the services provided by SCEF, detailed requests and responses exchanged between the IN-CSE and the IN-CSE are specified. In addition, detailed mechanisms how an IN-CSE generates and processes T8 requests and responses are described. In the next section, Group Management Interworking which is one of the interworking services defined in both 3GPP and oneM2M is explained to show how interworking is performed.

4.7.2.4   *Group Management Interworking*

The group management (GMG) CSF is defined in oneM2M to handle group-related IoT/M2M services. Various features such as creation of a group, management of group members, and performing fan-out operations to group member resources are defined in this group management service. In the group management interworking function, oneM2M application can get a benefit when the same content is sent to the members of a group that are located in a particular geographical area as 3GPP provides MBMS capabilities.

Let us consider a situation where a lot of UEs (with oneM2M AE/CSE) are located in a particular area, for example, a sport stadium with a big football match. If an IoT application should send an urgent message to the UEs at the stadium, the application can send the message using 3GPP MBMS multicast capability provided by SCEF and BM-SC as specified in 3GPP TS 29.122 [49]. Thanks to the MBMS capability, the group message from the IoT application does not need to be duplicated at the underlying network.

Figure 4.11 describes how a group message delivery is performed using MBMS interworking. In this procedure, it is assumed that an MBMS group has already been established so that 3GPP core network and oneM2M group hosting CSE knows required

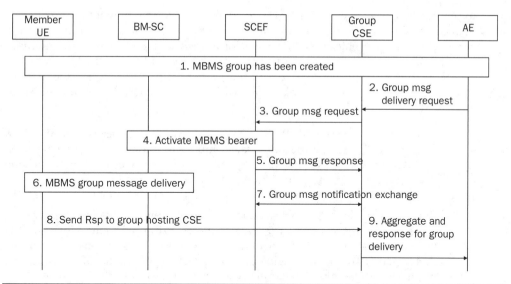

FIGURE 4.11   Group message delivery using MBMS

information to deliver a group message (Step 1). The IN-AE/CSE sends a group message delivery request with the group resource identifier and the payload to the group hosting CSE (Step 2). If this delivery request message contains proper parameters such as the multicastType (the indication of multicast type), which is configured with 3GPP_MBMS_group, the group hosting CSE sends the group message delivery request to the SCEF to activate the MBMS bearer (Step 3). The request shall contain the specified information in 3GPP TS 29.122 [49] including tmgi (specifying the MBMS network bearer), groupMessagePayload (the actual value to be delivered), and messageDeliveryStartTime (specifying the next start time of the time intersection). The MBMS bearer activation procedure is then processed by the 3GPP core network [50] followed by replying back the group message response to the group hosting CSE (Steps 4 and 5). Now, all the member UEs of the given group receive the group message based on the 3GPP MBMS delivery procedures (Step 6 and 7). After a successful delivery, the member hosting CSE replies response message within a certain time to the group CSE where the received response messages are aggregated and returned to the IN-AE/CSE (Steps 8 and 9).

4.7.3   End-to-End CIoT Use Case in Vehicle Domain

In the following section, an overview on the significance of CIoT to support V2X communications for the vehicular domain is presented. Then, how a traffic accident collection information system can be leveraged in a vehicular domain use case is presented. The use case highlights how we can integrate M2M platform with 5G cellular IoT to efficiently support V2X communications.

4.7.3.1   CIoT in Vehicular Domain

Lately, car manufacturers also referred as original equipment manufacturer (OEM) are equipping the modern vehicles with several sensors such as vehicle speed sensor (VSS), gas leakage sensor (GLS), vehicle noise sensor (VNS), and alcohol sensor (ALS) to provide smart vehicle services [51]. However, the information within the collected data is not fully exploited. Therefore, connecting the future vehicles to the internet and other peer devices (vehicles, smartphones, or infrastructures) to form IoT is a promising paradigm. Nevertheless, the vehicles come with additional challenges due to several factors including the speed and high mobility. In addition, the OEMs have to integrate the services and applications which currently face the following issues among others:

- Communication: The 3GPP TS 23.285 [52] within the Release 14 specifies V2X features that do not provide adequate solutions for the communication between the V2X application server (V2X AS) and the UE. Moreover, the communication between the V2X AS and V2X Control Function requires an efficient IoT platform to provide secure and efficient M2M communications.

- Interworking: In the conventional IoT, different standards bodies are providing guidelines and specifications to create tailored IoT applications. However, this approach isolates the interworking of the IoT applications due to the heterogeneity of the underlying platforms. Consequently, the OEMs for the car industry will face the same drawbacks for the internet of vehicles (IoV).

4.7.3.2   Traffic Accident Information Collection Use Case

The ultimate goal of the intelligent transportation system (ITS) is to guarantee a better environment for the road users including the vehicles users and pedestrians. For instance, the road accidents cause traffic jam due to the towing of the vehicles involved in the accident (for severe accidents) but also due to the waiting for the appropriate services (police and insurance companies) to collect useful information related to the accident. However, in the future IoV paradigm, such duties could be performed by the surrounding vehicles as illustrated in Fig. 4.12. Additionally, rescue teams can timely intervene to the accident scene if some preliminary data are forwarded to them beforehand. Likewise, the police would efficiently handle the affected area if a general idea of accident location is provided to them. All these data can be provided by other entities within the accident site including other vehicles, pedestrian's devices, or fixed infrastructures.

The vehicle domain consists of the following entities:

- ITS Authority: It is in charge of managing ITS applications and services using oneM2M platform. All the participants in the ITS services subscribe to oneM2M for accreditation.

- Police Station: It is registered for ITS services using M2M platform and is in charge of traffic control.

- Rescue Center: It registers to oneM2M platform for ITS services and is responsible for rescue operations.

- ITS-Station (ITS-S): It is an M2M device mounted in the vehicles with computing capabilities. It has a digital camera for capturing images or videos based on the driver commands. It can communicate with oneM2M platform using wireless network. It can use DSRC to communicate RSUs.

- Roadside Unit (RSU): It is a fixed infrastructure installed on roads. The RSU can communicate with the inbuilt vehicle's ITS-S using Dedicated Short-Range Communication (DSRC), or to oneM2M platform using wired or wireless communications.

When an accident occurs, the ITS-S mounted in the vehicle performs several steps. First, an accident report is generated by the ITS-S. This step can be automatically launched for instance upon the opening of airbags. Also, it can be manually activated by the driver. The traffic accident report is then signed by the ITS-S and sent to oneM2M platform. Note that the connection between the ITS-S and oneM2M platform can be done through a wireless network or via a nearby RSU using DSRC. The oneM2M platform verifies the signature on the accident report. The report will be securely sent to designated ITS service providers such as the rescue center or the police. Lastly, the ITS service providers (police or rescue center) might continue to communicate to ITS-S through oneM2M platform if additional data are required. Upon receiving all the essential data, the service providers agree on an intervention plan and head to the accident scene.

The oneM2M platform will facilitate the following in the described scenario as shown in Fig. 4.12:

- The platform will provide an interface that enables the communication between the V2X application server and ITS-S devices mounted in the vehicles.

- The oneM2M platform shall facilitate the communication between the V2X Control Function and the V2X Application Server.

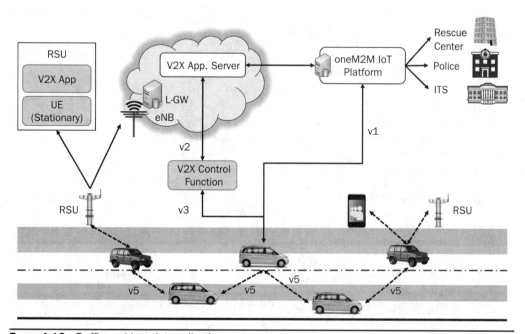

Figure 4.12 Traffic accident data collection

References

1. "Ericsson Mobility Report 2019," Ericsson, 2019.
2. "Verizon's Fixed 5G Improvements Will Boost the Appeal of Its Broadband Service," Business Insider, Sep 2019. [Online]. Available: https://www.businessinsider.com/verizons-fixed-5g-rollout-will-shake-up-home-internet-2019-9.
3. "Cable Giant Charter Tests Fixed 5G in 6 Cities, Plans Rural Broadband Push," VentureBeat, Jan 2018. [Online]. Available: https://venturebeat.com/2018/01/24/cable-giant-charter-tests-fixed-5g-in-6-cities-plans-rural-broadband-push/.
4. "Dish Confirms That It Will Become a Major US Mobile Carrier," The Verge, Jul 2019. [Online]. Available: https://www.theverge.com/2019/7/26/8931827/dish-carrier-plans-5g-network-tmobile-sprint-us-confirms.
5. "World Impressed by KT's 5G Technologies in PyeongChang," The Korea Herald, Feb 2018. [Online]. Available: http://www.koreaherald.com/view.php?ud=20180213000731.
6. "SKT, KT, LGU+ in Cut-Throat Competition to Install 5G Base Stations," The Korea Times, Apr 2019. [Online]. Available: https://www.koreatimes.co.kr/www/tech/2019/03/133_266159.html.
7. "KT Launches Commercial 5G Network in South Korea," Telecompaper, Apr 2019. [Online]. Available: https://www.telecompaper.com/news/kt-launches-commercial-5g-network-in-south-korea--1288704.
8. G. Yiming, "LG Uplus Leads 5G Service with AR, VR Features," Telecoms.com, Sep 2019. [Online]. Available: https://telecoms.com/intelligence/lg-uplus-leads-5g-service-with-ar-vr-features/.
9. M. Kapko, "Operators Dish on 5G Strategies for Enterprise," Sdx central, Sep 2019. [Online]. Available: https://www.sdxcentral.com/articles/news/operators-dish-on-5g-strategies-for-enterprise/2019/09/.
10. T. Lacoma and J. Kaplan, "What Is Fixed Wireless 5G? Here's Everything You Need to Know," Jan 2020. [Online]. Available: https://www.digitaltrends.com/computing/fixed-wireless-5g/.
11. NTT DOCOMO, "NTT Annual Report 2019," 2019.
12. IEEE ComSoc, "China Unicom to Launch 5G Consumer Services May 2019 in 7 Chinese Cities," 2019. [Online]. Available: https://techblog.comsoc.org/2019/04/24/china-unicom-to-launch-5g-services-may-2019-in-china/.
13. "China Rolls Out 5G Telecom Services," The Economic Times, Nov 2019. [Online]. Available: https://economictimes.indiatimes.com/news/international/business/china-rolls-out-5g-telecom-services/articleshow/71851743.cms?from=mdr.
14. F. Arena and G. Pau, "An Overview of Vehicular Communications," Future Internet, vol. 11, no. 2, 2019.
15. 3GPP, "TS 22.186 Service Requirements for Enhanced V2X Scenarios v16.2.0," 2019.
16. 3GPP, "TR 22.886, Study on Enhancement of 3GPP Support for 5G V2X Services," 16.2.0, 2018.
17. 3GPP, "TR 22.885, Study on LTE Support for Vehicle-to-Everything (V2X) Services," 14.0.0, 2015.
18. 3GPP, "TS 22.179, Mission Critical Push to Talk (MCPTT); Stage 1," 17.0.0, 2019.
19. Reuter, "Germany to Allocate Local 5G Frequencies Later This Year," 2019. [Online]. Available: https://www.reuters.com/article/us-germany-telecoms/germany-to-allocate-local-5g-frequencies-later-this-year-idUSKBN1QS1GR.

20. Ofcom, "Enabling Wireless Innovation Through Local Licensing," 2019.
21. A. David, S. Janette, and N. Chris, "Global Race to 5G," Analysis Mason, 2019.
22. 5GACIA, "Integration of Industrial Ethernet Networks with 5G Networks," 2019.
23. S. Singhal and S. Carlton, "The Era of Exponential Improvement in Healthcare?," McKinsey & Company, 2019.
24. Ericsson Consumer Lab, "From Healthcare to Homecare," 2017.
25. M. Mikulic, "Global Telemedicine Market Size from 2015 to 2021," Statista, 2019.
26. M. Shirer and C. MacGillivray, "The Growth in Connected IoT Devices Is Expected to Generate 79.4ZB of Data in 2025," IDC, 2019.
27. V. Këpuska and G. Bohouta, "Next-Generation of Virtual Personal Assistants (Microsoft Cortana, Apple Siri, Amazon Alexa and Google Home)," in IEEE 8th Annual Computing and Communication Workshop and Conference (CCWC), 2018.
28. S. Fernandes, J. Abreu, P. Almeida, and R. Santos, "A Review of Voice User Interfaces for Interactive TV," in Communications in Computer and Information Science, 2019.
29. M. Ignatius, N. H. Wong, M. Martin, and S. Chen, "Virtual Singapore Integration with Energy Simulation and Canopy Modelling for Climate Assessment," in IOP Conference Series: Earth and Environmental Science, 2018.
30. S. JaeSeung, "Six Cities in the 'CityHub' Project will Benefit from oneM2M Standards to Manage Semantic Data across Different Vertical Applications," oneM2M, 2019.
31. ZDNet, "SK Telecom to Develop 5G Smart Power Plant Solutions," 2019. [Online]. Available: https://www.zdnet.com/article/sk-telecom-to-develop-5g-smart-power-plant-solutions/.
32. Smart Energy International, "Uzbekistan Signs 30 Billion Smart Meter Deal," 2018. [Online]. Available: https://www.smart-energy.com/industry-sectors/business-finance-regulation/uzbekistan-signs-30-billion-smart-meter-deal-kt-corporation/.
33. Juniper Research, "The Value of 5G for Cities and Communities," O2, 2019.
34. ZDNet, "Arm and KEPCO to Co-develop Secure One-Chip for IoT Meter Project," 2018. [Online]. Available: https://www.zdnet.com/article/arm-and-kepco-to-co-develop-secure-one-chip-for-iot-meter-project/.
35. Huawei, "China's Super Grid: Huawei 5G Enables a World-leading Power Grid," 2019. [Online]. Available: https://carrier.huawei.com/en/success-stories/Industries-5G/Electric-power.
36. Ericsson, "Ericsson Mobility Report," 2017.
37. Ofcom, "New Service Developments in the Broadcast Sector and Their Implications for Network Infrastructure," 2014.
38. S. Liu, "Projected Size of the Augmented and Virtual Reality Market 2016–2023," Statista, 2019.
39. IDC, "AR Headset Prevalence Is Still a Few Years Out as Commercial Applications Slowly Build Momentum," 2018.
40. Gartner, "Gartner Identifies the Top 10 Strategic Technology Trends for 2019," 2019. [Online]. Available: https://www.gartner.com/en/newsroom/press-releases/2018-10-15-gartner-identifies-the-top-10-strategic-technology-trends-for-2019.
41. S. Stein and J. Goldman, "The Best VR Headsets for 2020," 2020. [Online]. Available: https://www.cnet.com/news/the-best-vr-headsets-in-2020/.
42. Ovum, "How 5G Will Transform the Business of Media & Entertainment," 2018.
43. 3GPP, "TS 22.386, Service Requirements for Machine-Type Communications (MTC); Stage 1," 15.0.0, 2019.

44. 3GPP, "TS 23.268, Architecture Enhancements to Facilitate Communications with Packet Data Networks and Applications," 16.5.0, 2019.

45. J. Swetina, G. Lu, P. Jacobs, F. Ennesser, and J. Song, "Toward a Standardized Common M2M Service Layer Platform: Introduction to oneM2M," IEEE Wireless Communications Magazine, vol. 21, no. 3, pp. 20–26, 2014.

46. oneM2M, "TS-0001, oneM2M Functional Architecture," v4.4.0, 2020.

47. oneM2M, "TS-0026, 3GPP Interworking," v4.3.0, 2019.

48. 3GPP, "TS 23.682, Architecture Enhancements to Facilitate Communications with Packet Data Networks and Applications," v15.4.0, 2018.

49. 3GPP, "TS 29.122, T8 Reference Point for Northbound APIs," v16.4.0, 2019.

50. 3GPP, "TS 23.401, General Packet Radio Service (GPRS) Enhancements for Evolved Universal Terrestrial Radio Access Network (E-UTRAN) Access," v16.5.0, 2019.

51. S. Olariu, I. Khalil, and M. Abuelela, "Taking VANET to the Clouds," International Journal of Pervasive Computing and Communications, vol. 7, no. 1, pp. 7–21, 2011.

52. 3GPP, "TS 23.285, Architecture Enhancements for V2X services," v16.2.0, 2019.

53. SAE, "J3016 Taxonomy and Definitions for Terms Related to On-Road Motor Vehicle Automated Driving Systems," 2014.

Problems/Exercise Questions

1. Compare 4G and 5G technologies in terms of five areas (i.e., connections, latency, data traffic, peak data rate and available spectrum).

2. Explain how 5G is different from 4G in terms of data speeds?

3. How can 5G be used in wired networks?

4. Provide at least three vertical industries where network slicing can be used.

5. Explain levels of automation defined by SAE J3016.

6. Provide at least three safety-related services applicable to vehicle-to-everything (V2X) and explain them.

7. What is vehicle-to-everything (V2X) communications and list four V2X services defined in 3GPP?

8. Explain what are virtual reality (VR) and augmented reality (AR) and why 5G is suitable for VR/AR services?

9. Provide at least three features for machine type communications (MTC).

10. What is oneM2M IoT standards and how oneM2M can be connected to 5G?

5 5G Radio Access Network Architecture

5.0   Introduction

The two main components of the overall 5G system (5GS), as defined and standardized by 3GPP, are the 5G radio access network (5G RAN) and the 5G core network (5G-CN or 5GC). The new radio technology developed for 5GS is called New Radio (NR), which is supported on the 5G RAN and developed in two phases—Phase 1 completed in 3GPP Release 15 (late 2018) and Phase 2 completed in 3GPP Release 16 (mid-2019). In this chapter, the overall 3GPP 5G RAN details mainly from an architecture perspective is provided. The 5GC details are captured in Chap. 6.

Firstly, the key requirements of 5G from the perspective of 5G RAN are discussed in detail. The rest of the chapter provides an overview of the 5G RAN architecture along with detailed discussion related to the overall 5G RAN design considerations, user- and control-plane designs, enhancements for enabling vehicular communication, and device-to-device communication. In addition, the key challenges faced during the 5G RAN design are described. Finally, this chapter provides a brief overview of some of the key 5G use cases where artificial intelligence (AI) technology is leveraged to add significant performance improvements in data throughput for RANs that are currently underutilized.

5.1   5G Radio Access Network Overview

5.1.1   5G Requirements

The next generation of the mobile and wireless communications, namely, the fifth-generation (5G) envisions new use cases, services, and applications, such as enhanced mobile broadband (eMBB), ultra-reliable low-latency communication (URLLC), and massive machine-type communication/massive internet-of-things (mMTC/mIoT). Any combinations of these use cases can also be possible, such as ultra-reliable communications, low-latency communications, or low-latency eMBB communications. A high-level overview of some of the key 5G RAN requirements, as specified by 3GPP, is shown in Fig. 5.1.

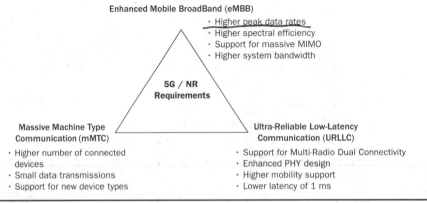

Figure 5.1   Overview of 5G RAN requirements

In 5G RAN, use cases originating from vertical industries (e.g., automotive, e-health, smart grid, etc.) are considered as drivers for 5G requirements due to the enhanced support for new device types and services. The 5G network has the capability to adapt based on the requirements of various services in terms of latency, reliability, security, QoS, etc. In particular, the following new requirements have been introduced in 3GPP for four main service types [2]:

- Enhanced Mobile Broadband (eMBB): It includes a number of different use case families related to higher data rates, higher density, deployment and coverage, higher user mobility, devices with highly variable user data rates, fixed-mobile convergence, and small-cell deployments. From RAN perspective, this will necessitate more dense access deployments, with densification of antennas and processing units, as well as the use of higher spectrum bands (e.g., mm-Wave radio) which can meet the high bandwidth requirements of such services.

- Massive Internet of Things (IoT): It focuses on use cases with a massive number of devices (e.g., sensors and wearables). This group of use cases is particularly relevant to the new vertical services, such as smart home and city, smart utilities, e-Health, and smart wearables. The requirement which is imposed in RAN by the deployment of millions of sensors may strongly affect the control-plane design to ensure meeting the ultra-high connection density key performance indicator (KPI).

- Ultra-Reliable and Low-Latency Communication (URLLC): It includes improvements within the RAN and CN to assist in delivering highly reliable data transmission not achievable in 3G and 4G. The main areas of improvements are latency, reliability, and availability to enable, for example, industrial control applications and the tactile internet, which are part of critical communications. These requirements can be met with an improved radio interface, optimized architecture, and dedicated core and radio resources. From RAN perspective, the simplification of protocol processing as well as the improvements in physical layer design is essential to meet ultra-low requirements in terms of latency (1 ms) and reliability (delivering 99.999% of the packets to the devices without losses) KPIs.

- Vehicular to Everything (V2X): It can be seen as special 5G service type, which can include both safety and non-safety applications [3]. One of the key requirements of V2X services is the critical latency (3–10 ms) and reliability (99.999% and higher), which may need to be adapted on demand, due to new application requests (e.g., level of automation change, dynamic group-UE formations) or network changes (network congestion to core network and/or access network entities, mode of transmission/operation change). One key challenge under these requirements is to ensure service continuity for V2X communications without any temporary loss of service or loss of data packets. Thus, the RAN of a 5GS will need to support V2X/URLLC services for user plane (UP), control plane (CP), or both to ensure meeting the reliability and coverage requirements.

As a general requirement for the latter two service types, 5G RAN shall be able to support the highly reliable (i.e., with low ratio of erroneous packets) and low-latency

services. The support for high-reliability services is subject to the following requirements and assumptions:

- High Reliability: It is about providing high likelihood of delivering error-free
  packets through the 3GPP system within a bounded latency. A performance
  metric for high reliability is the ratio of successfully delivered error-free packets
  within a delay bound over the total number of packets. The required ratio and
  latency bound may be different for different URLLC use cases.

- High Availability: It is related to a communication path through the 3GPP
  system providing reliable services. This communication path between the
  communication end points is made up of radio links as well as transport links
  and includes different hardware (HW) and software (SW) functions. The NR
  should provide high availability, in addition to deploy redundant components
  and links for these radio, transport, and HW/SW.

- Low Latency: 5G RAN should support latencies down to 0.5 ms uplink (UL)/
  downlink (DL) for URLLC.

Based on the above requirements, an important consideration in 5G RAN design is the
necessary evolution of traditional RANs, i.e., pre-5G RAN, toward highly densified and
heterogeneous deployments to meet the foreseen capacity and coverage demands [4].

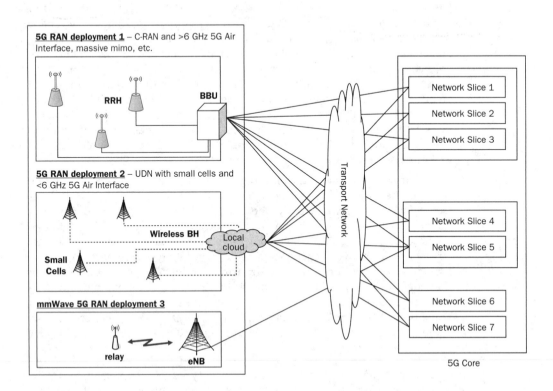

FIGURE 5.2   Slice-aware RAN deployments

An exemplary use case, as defined in [5], is the dense urban scenario, which assumes macro-cells and mixture of planned and unplanned small cells under the macro-cell umbrella, where both ideal and nonideal backhaul can be present. Here, ideal backhaul refers to very high-capacity and low-latency transport links usually provided by optical fiber or mm-Wave communications. On the other hand, nonideal backhaul is the case when wireless links (e.g., <6 GHz) are used for the communication between macro-cells and small cells with relaxed capacity and latency capability.

Small cells, typically densely deployed underlying the conventional macro-cellular RAN, are considered as a promising candidate of mm-Wave RAN architecture to cope with the adverse propagation conditions. The challenge with deploying a large number of small cells is that it may prove to be too expensive or impractical for equipping every cell with fiber connectivity. As an alternative and cost-efficient solution, wireless back-hauling can meet the desired coverage and capacity requirements. Such a paradigm may adopt the same spectrum for both backhaul (BH) and access links, reducing costs further but raising at the same time significant challenges in radio resource management (RRM) between BH and access links.

However, multiple limitations for backhaul/access might require certain handling of resource management, e.g., nonideal wireless backhaul between RAN nodes can be a limiting factor and will require extra RRM for the backhaul part. To this end, joint backhaul/access optimization can be used to meet high-throughput requirements for throughput demanding services. Another important factor is the excessive signaling which will be required in dense urban heterogeneous scenarios (with macro, numerous overlapping small cells) for wireless backhaul/access measurements. This is going to be more crucial by new RRM functional interactions which will be added by performing fast scheduling decisions in small cells, which could be considered as an essential feature for enabling URLLC.

A further consideration related to the introduction of network slices, which are logical end-to-end subnetworks corresponding to different verticals, is envisioned as a key 5G feature [4]. Network slices might impact the RAN design, and RRM is one of the key aspects which will be affected. Different slices aim at different goals, e.g., throughput, latency, or reliability. This affects how RRM functions work and also where these functions can be placed.

Figure 5.2 illustrates an exemplary system, where multiple access deployments can be mapped to different slices (of different or same vertical industries), based on operator's premises and per-slice service-level agreements (SLAs). In particular, the first deployment shows a centralized RAN (C-RAN) deployment, where a baseband unit (BBU) is a unit that processes baseband which is physically separated from the RF processing unit (remote radio head—RRH) via optical fiber. Another deployment is an ultra-dense network (UDN) with small cells and nonideal wireless backhaul, whereas the last deployment shows the scenario when mm-Wave access points (APs) are located as relays in hotspot areas. Different slices considering the vertical requirements may utilize different RAN deployments, which may have overlapping coverage. This requires that slice-aware mechanisms exist at RAN to ensure that traffic from different slices are logically isolated.

5.1.2 5G Spectrum

An overview of the various 5G spectrum bands and access techniques based on the discussions in [6], along with the possible use cases where the spectrum bands could be applied, is shown in Fig. 5.3. Here, it is important to note that the available spectrum bands would depend on the regulations within different geographies, and a contiguous spectrum assumption within all the possible frequency bands would not be realistic. In 4G, due to the limitations of the design of the physical layer numerology, deployments especially for the initial release of the standards were mainly limited to <6 GHz spectrum bands. This was also partly due to the wide-area coverage assumptions for the traditional mobile networks, which was already challenged in 4G with the development of heterogeneous networks with wide-area macro-cells and small cells coexisting with the same deployment area. Due to the higher data rate requirements of 5G, frequency bands >6 GHz have been considered as an essential enabler, with the utilization of flexible numerology [7] that provides improved spectral efficiency even with a wide range of bands defined for the system.

From the use case perspective, it is envisioned that majority of the services that would require massive connectivity to a large number of devices could be provisioned using the <6 GHz frequency band due to the wider coverage requirements. From a reliability perspective, using lower frequency bands would be advantageous due to the higher availability of the network and from a latency perspective having higher system bandwidth would enable high amount of redundancy to be incorporated within the data transmissions. Due to these dual considerations for URLLC, we assume that such services would be utilizing both below and above 6 GHz frequency bands, though it would be challenging to deploy such services using mm-Wave frequencies. For eMBB,

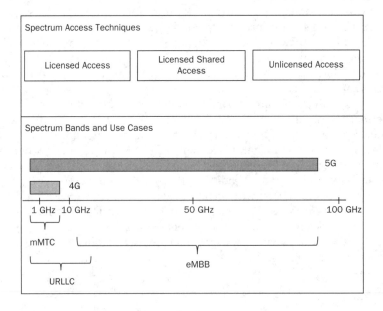

FIGURE 5.3   Overview of 5G spectrum aspects [6]

due to the requirements for higher data rates and resultant system bandwidth, it is expected that such services would be deployed mainly in the higher frequency bands. This would be essential for delivering services such as high-performance gaming, virtual/augmented reality, etc., which would require the availability of a significant amount of spectrum.

5G also provides a wide range of spectrum access techniques, which enables the network operators to deploy their networks, depending on the service/use case requirements. Here, it is important to note that the spectrum access techniques are uncorrelated with the spectrum bands being considered for 5G, mainly due to the difference in regulations between different countries in terms of how a particular spectrum band can be accessed by the network operators. The traditional licensed spectrum access technique where the operators obtain exclusive access to the spectrum is expected to continue similar to the previous generations of mobile networks. Licensed shared access is considered to be an enabler for higher spectrum use, where a licensee of a particular spectrum band can utilize the spectrum as long as the incumbent user is not using the spectrum. The shared user would need to stop using the spectrum when the incumbent becomes active, based on various conditions that the spectrum authority could determine. Unlicensed spectrum access was already incorporated into 4G standards [8,9], which is further enhanced in 5G particularly to gain access to the bands within the mm-Wave frequencies which are currently allocated for unlicensed use. An overview of the key 5G spectrum positions from GSM Association (GSMA), mainly from a network operator perspective, is presented in [10], including various services being considered utilizing the spectrum, spectrum bands, and spectrum access techniques.

5.1.3   5G RAN Interfaces and Functionalities

There are mainly two interfaces which are of key relevance to 5G RAN: (1) The NG interface which provides control- and user-plane connectivity between the RAN and core network, and (2) the Xn interface which provides connectivity between the NG-RAN nodes within the RAN. The control- and user-plane protocol stack of NG and Xn interfaces are shown in Fig. 5.4 [11].

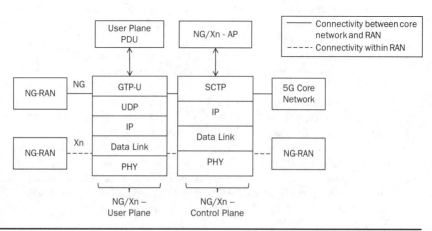

Figure 5.4   NG and Xn interface protocol stack [11]

The NG user-plane (NG-U) interface provides connectivity between NG-RAN and the user-plane function (UPF) to transport user-plane PDUs using GTP-U layer over UDP/IP protocol layer. The NG control-plane (NG-C) interface provides reliable connectivity—using Stream Control Transmission Protocol (SCTP) layer over IP, between the RAN and AMF in the 5G CN. The control-plane interface is enabled for the NG interface management, transportation of NAS messages between AMF and NG-RAN, UE context and mobility management, etc. The Xn user-plane (Xn-U) interface provides user-plane connectivity between NG-RAN nodes, with a similar protocol stack structure as the NG-U interface. Both user-plane interfaces support only non-guaranteed delivery of user-plane PDUs, with Xn-U also providing data forwarding and flow control. The Xn control-plane (Xn-C) interface provides control-plane connectivity between NG-RAN nodes while utilizing a similar protocol stack structure of the NG-C. The Xn-C enables the overall Xn interface management, dual connectivity feature, and mobility management, context transfer, and RAN paging for the UEs [11]. The application layer signaling for both the interfaces is based on the NG- and Xn-application protocol (NG-AP and Xn-AP). Application layer in this context means the management of the UE specific and nonspecific events mentioned above for NG-C between gNB and AMF or Xn-C between gNBs. The detailed functionalities and related signaling for the NG interface are available in [12] and those for the Xn interface can be found in [13].

5.1.4   3GPP 5G RAN Architecture Overview

In 5G Phase 1 the first features of RAN, which is commonly referred in 3GPP as Next Generation-RAN (NG-RAN) or New Radio (NR), have been specified mainly for the eMBB scenarios, focusing on (non-standalone, standalone) architecture and protocol aspects [14]. One of the key design considerations for RAN is the flexibility of RAN deployments, where the 5G RAN access nodes (aka gNB) can be split horizontally (CU–DU) or vertically (CP–UP) to allow for flexible and service-tailored virtualization of RAN functionalities in different nodes.

The network functions (NFs) of a wireless network are typically categorized into three groups. The UP is responsible for forwarding data from the source to the destination, including the corresponding processing. The CP controls the UP, for example, in terms of setting the routing path of a packet or of RRM. The CP also provides a set of other functionalities such as connection/mobility management and broadcasting of system information. Besides CP and UP, a wireless network also requires management and operation functions.

Based on the concept of virtualization, individual NFs within the RAN can be moved to the telecommunication cloud environments where computational resources are dynamically shared in order to increase energy efficiency and flexibility. The 5G RAN architecture will thereby most likely utilize both distributed cloud environment consisting of centralized clouds and edge clouds. The latter are also known as mobile edge clouds in the context of mobile radio networks.

The RAN protocol stack might be deployed virtualized in a telco cloud environment (cloud RAN). Even different NFs, such as the mobility management, resource scheduling, for special service types, such as V2X, or a different application might run and share with each other the same hardware resources. The gNB within the 5G RAN can consist of a CU and one or more DUs, as specified in [15].

In particular, the CU–DU split option specified by 3GPP introduces a split between Packet Data Convergence Protocol (PDCP) and radio link control (RLC) layer, as described in [15]. This implies that in terms of practical implementation of physical layer (PHY), medium access control (MAC) and RLC will be located in the DU while PDCP and Service Data Adaptation Protocol (SDAP) plus radio resource control (RRC) (see Sec. 5.2.2.1) will be located in the CU, SDAP for the UP stack, and RRC for the CP protocol stack.

Figures 5.5 and 5.6 show the protocol stacks for user plane and control plane, respectively, as depicted in [11]. The UPF and the access and mobility management function (AMF) as well as non-access stratum (NAS) protocol are explained in Chap. 6.

NG-RAN may either be deployed as a standalone network or may interoperate with E-UTRAN as non-standalone deployment. The overall architecture as provided in [11] is illustrated in Fig. 5.7, and shows the 5GC to NG-RAN interactions via NG interface, as well as the interfaces between RAN nodes. gNB is the term used in standardization for the 5G base station, whereas the term ng-eNB refers to the node providing E-UTRA user-plane and control-plane protocol terminations toward the UE, and connected via the NG interface to the 5GC.

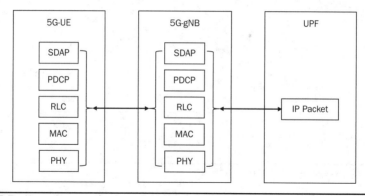

FIGURE 5.5   User-plane protocol stack [11]

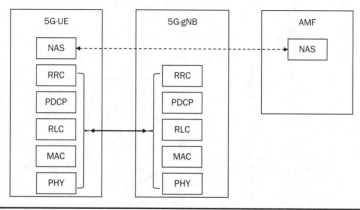

FIGURE 5.6   Control-plane protocol stack [11]

As can be seen in Fig. 5.7, NG-RAN consists of a set of gNBs connected to the 5G core (aka 5GC) through the NG interface. The 5G-gNB can support FDD mode, TDD mode, or dual mode operation, and gNB could comprise of a gNB-central unit (CU) and one or more gNB-distributed units (DUs). Also, for NG-RAN, the NG and Xn-C interfaces for a gNB consisting of a gNB-CU and gNB-DUs terminate in the gNB-CU.

In order to enable the balancing of computational resources between CUs and DUs within the gNBs in a telco environment comprising centralized and edge clouds, the F1 [45] interface [interface between CU and DU (see Fig. 5.8)] as well as the Xn-C interface need to carry information about computational resource usage, such as CPU, memory, and network interface utilization. Figure 5.8 illustrates the 5G RAN architecture and the CU–DU split of functionalities.

Furthermore, in [17], different architecture options have been introduced for providing NR access to UEs with different capabilities, taking into account different deployments of RAN-CN interface, and the interface between E-UTRA and NR radio access technologies. More specifically, following are the options for standalone and non-standalone deployments:

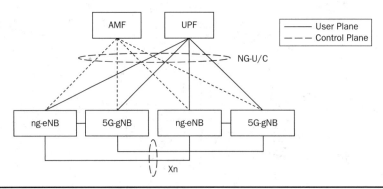

FIGURE 5.7 5G overall RAN architecture [11]

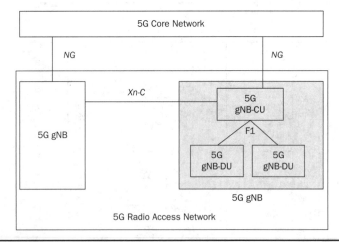

FIGURE 5.8 Illustration of 5G RAN architecture connecting to 5G CN, aka 5GC, over NG interface

- In deployment Option 2, the gNB is connected to the 5GC in a standalone deployment.

- In deployment Option 3/3A, the LTE eNB is connected to the EPC with non-standalone NR. The NR user-plane connection to the EPC goes via the LTE eNB (Option 3) or directly (Option 3A).

- In deployment Option 4/4A, the gNB is connected to the 5GC with non-standalone E-UTRA. The E-UTRA user-plane connection to the NGC goes via the gNB (Option 4) or directly (Option 4A).

- In deployment Option 5, the eLTE eNB is connected to the NGC.

- In deployment Option 7/7A, the eLTE eNB is connected to the 5GC with non-standalone NR. The NR user-plane connection to the 5GC goes via the eLTE eNB (Option 7) or directly (Option 7A).

The deployment options 4 and 7 are illustrated in Fig. 5.9 as key examples to show in detail the scenarios where the gNB or the eNB (denoted as ng-eNB if integrated with 5GC) are the anchor NG-RAN nodes, respectively.

NG-RAN also supports multi-radio dual connectivity (MR-DC) operation whereby a UE is configured to utilize radio resources provided by two distinct schedulers, located in two different NG-RAN nodes connected via a nonideal backhaul, one providing NR access and the other one providing either E-UTRA or NR access (more details of MR-DC operation can be found in [18]). One node acts as the master node (MN) and the other as the secondary node (SN). The MN and SN are connected via a network interface and at least the MN is connected to the core network. MR-DC is a generalization of the intra-E-UTRA dual connectivity (DC) described in [19], where a multiple Rx/Tx-capable UE may be configured to utilize resources provided by two different nodes

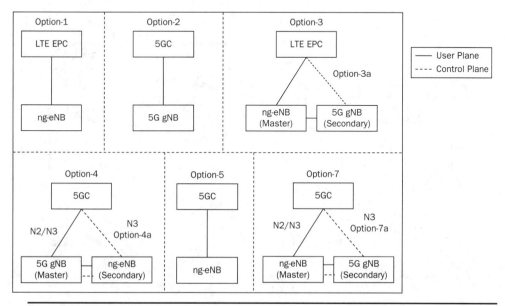

Figure 5.9 5G deployment architecture options

connected via nonideal backhaul, one providing NR access and the other one providing either E-UTRA or NR access.

There are multiple options defined in [18], based on whether an eNB or a gNB has the role of MN or SN. In case of EN-DC (dual connectivity, where the MN is an eNB), the S1-U and X2-C interfaces for a gNB consisting of a gNB-CU and gNB-DUs terminate in the gNB-CU. The gNB-CU and connected gNB-DUs are only visible to other gNBs and the 5GC as a gNB. The node hosting user-plane part of NR PDCP (e.g., gNB-CU, gNB-CU-UP, and for EN-DC, MeNB or SgNB depending on the bearer split) shall perform user inactivity monitoring and further informs its inactivity or (re)activation to the node having C-plane connection toward the core network (e.g., over E1 [43], X2). The node hosting NR RLC (e.g., gNB-DU) may perform user inactivity monitoring and further inform its inactivity or (re)activation to the node hosting control plane, e.g., gNB-CU or gNB-CU-CP. A detailed overview of the 5G network architecture is presented in [44].

To better capture some key architecture differences on the protocol design, we provide in Fig. 5.10 an overview illustration of the comparison between 4G and 5G RAN. Some key elements which are new in this picture are the RAN-CN interfaces, as well as the introduction of a new protocol and new functionalities. More details are presented in the following sections.

5.1.5   5G Centralized/Cloud-RAN Deployments

One of the most important design enhancements for 5G RAN from an architecture perspective has been flexibility in terms of both the availability of deployment options and various implementation possibilities for network infrastructure vendors. An illustration of this aspect is shown in Fig. 5.11, where various implementation options that are currently available for 5G base stations are highlighted.

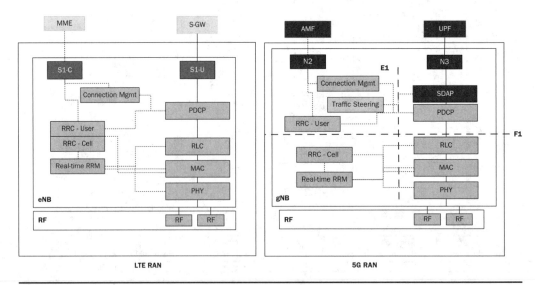

FIGURE 5.10   LTE RAN versus 5G RAN

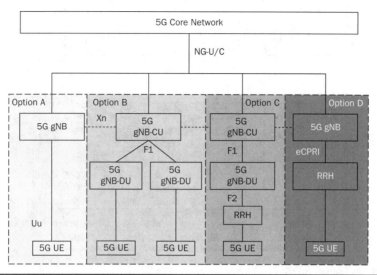

FIGURE 5.11 An overview of the various 5G RAN implementation and deployment options

Option A: This option corresponds to the basic distributed base station implementation and deployment option, where all the functionalities are implemented within a single gNB.

Option B: Centralized/cloud-RAN implementation with functionalities that are not latency critical would be located within the gNB-CU and functionalities that are latency critical such as scheduling/RRM would be located within the gNB-DU. The interface between the gNB-CU and DU is called the F1 interface [43] which consists of both the control- and user-plane RAN functions.

Option C: This option could be considered as a variant of option B, with the implementation of the gNB-DU node additionally simplified with the utilization of the RRH. Here, it is important to note that the 3GPP-based F2 interface provides connectivity between the gNB-DU and RRH.

Option D: This option could be considered as a variant of options A–C, where the RRH is connected directly to the distributed gNB using a common public radio interface (CPRI) [46], which has been enhanced for 5G called evolved (eCPRI).

The control- and user-plane functionalities and related interfaces within the gNB are shown in Fig. 5.12, which is essentially a functional decomposition of option C as shown in Fig. 5.11. From the figure, we can observe that the gNB-CU functionality is split into the user-plane and control-plane functions, gNB-CU-U and gNB-CU-C, which was linked by a logical interface called E1. The F1 and F2 interfaces also could be split into control- and user-plane logical interfaces, for transporting associated data between the CU and DU for F1, and the DU and RRH for F2. Such functional splits enable a significant amount of flexibility for implementing the functions within a cloud data server, especially in terms of where the functionalities are located.

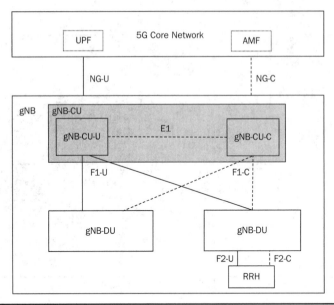

FIGURE 5.12   Control- and user-plane functions and interfaces within gNB-CU and DU

The detailed functional decomposition, especially in terms of the protocol architecture, is shown in Fig. 5.13. From the figure, we can observe that the SDAP and PDCP functionalities are located within the gNB-CU-U for the user plane, whereas the RRC and PDCP are located within the gNB-CU-C for the control plane. As mentioned before, the protocol functions within the gNB-CU do not have any strict latency constraints, which enables their implementation within the cloud or centralized data servers. The gNB-DU (based on option B) consists of the lower layers of the protocol stack such as the RLC, MAC, and PHY, all of which has strict time constraints in terms of various SLAs to be satisfied for over-the-air delivery of data traffic. The PHY layer within the gNB-DU could also be further decomposed into the higher (PHY-H) and lower (PHY-L) layers (option C), with the lower layer located within an RRH. A similar implementation could also be done for the distributed gNB, with all the layers above PHY-L located within the gNB (as shown in option D). A similar functionality split between the central and distributed unit has also been investigated in [16].

5.1.6   5G RAN Mobility

Mobility is the feature where the UE can move around either in RRC_CONNECTED mode, i.e., the UE has an active communication ongoing, or in RRC_IDLE/INACTIVE, mode i.e., the UE is currently not in an active communication and does not need resources.

The different RRC states, which are available for a UE in 5G, are shown in Fig. 5.14, based on the specification [11], with the addition of a new RRC_INACTIVE state as compared to LTE which only had the idle and connected modes. The functionality of RRC_IDLE mode is similar to LTE with the UE managing the mobility procedures based on the configurations made by the network. RRC_INACTIVE state enables the UE and RAN to maintain the UE context while allowing the UE to enter partially idle state with mobility updates not required to be done within the RAN notification area (RNA) which is configured by the RAN. The connection between the UE and core network

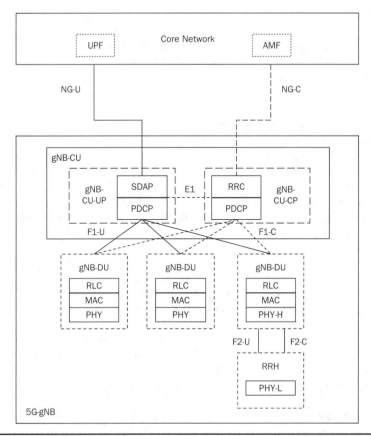

FIGURE 5.13   Detailed functional decomposition and protocol architecture

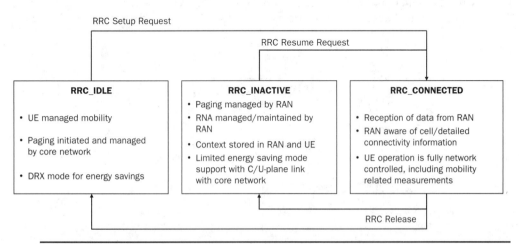

FIGURE 5.14   UE RRC states [11]

(AMF and UPF) is still maintained, thereby enabling the UE to enter connected mode with minimal delay. Since the RRC_INACTIVE state is managed by the RAN, the UE paging is managed by the RAN as compared to core network for the IDLE state. Within the INACTIVE state, the UE can enter into the power-saving mode efficiently as compared to the CONNECTED state.

RRC signaling initiated by either the UE or RAN is required for the UE to transition between different states. For transitioning from IDLE to CONNECTED mode, the UE needs to send RRC Setup Request to the NG-RAN, whereas the UE sends the RRC Resume Request for transitioning from INACTIVE to CONNECTED mode. The NG-RAN sends RRC Release message to the UE for transitioning from CONNECTED to IDLE or INACTIVE modes. Here, we will mainly focus on the connected mode mobility procedure within the 5G RAN, an overview of which is shown in Fig. 5.15. Here, the source gNB, which is the serving gNB for the UE, initiates a handover (HO) request to the target gNB (Xn handover) based on the UE measurements as well as the source gNB load and various other factors. The target gNB evaluates the HO request based on the admission control procedures defined within the gNB, e.g., in terms of having sufficient additional capacity for supporting the QoS flows of the UE. If the target gNB is able to accept the HO request, an acknowledge/ACK message is sent to the source gNB. The source gNB sends RRC reconfiguration message to the UE, which then synchronizes and connects to the target gNB. Once this process is completed, the UE would send an RRC reconfiguration complete message to the target gNB.

A detailed end-to-end handover procedure from the 5G RAN perspective, particularly in terms of handling the data flow, is shown in Fig. 5.16 [11]. Here, the UE is initially receiving the data from the UPF through the source gNB, with the AMF configuring the mobility information. As part of the normal operation of the UE, it will be sending measurement reports to the source gNB based on the configurations made by the network. Once the source gNB detects events that necessitate a handover, it sends HO request to the target gNB (similar to the procedure described earlier). In terms of data flow handling, the source gNB after receiving the HO ACK message from the target would forward the buffered data within the source gNB along with the new data received from the UPF to the target gNB. The target gNB based on this received data would schedule

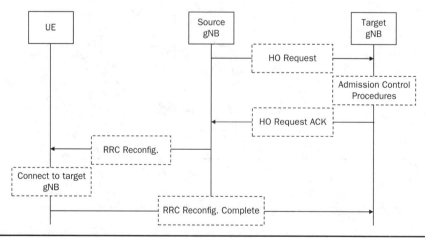

Figure 5.15   Overview of handover procedure in RRC_CONNECTED mode [11]

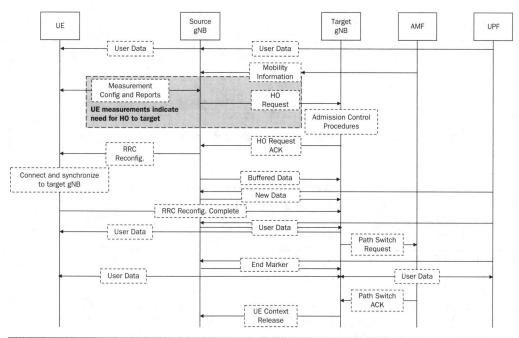

Figure 5.16 Detailed handover procedure from 5G RAN perspective [11]

over-the-air transmissions for the UE. The target gNB then initiates a path switch request to the AMF which prompts the UPF to end the session with the source gNB and release any remaining user-plane link with the source gNB. Once the target gNB receives the acknowledgment for the path switch request from the AMF, it also sends a UE context release message to the source gNB which completes the handover procedure.

5.1.7 5G RAN Quality-of-Service Handling

The quality-of-service (QoS) flow architecture from the perspective of 5G RAN is shown in Fig. 5.17, based on the architecture presented in [11]. In the figure, we assume three different application flows originating from three different application servers, which are then mapped into three QoS flows 1 to 3 by the UPF. The UPF tunnels the QoS flow to the gNB (similar rules are applicable to the NG eNB as well) using an NG-U tunnel. The 5G gNB maps the NG-U tunnel to an appropriate data radio bearer. In the figure, it is assumed that QoS Flows 1 and 2 are mapped to a single Radio Bearer-a, whereas the QoS Flow 3 is mapped to a different Radio Bearer-b, in order to illustrate the flexibility available within the RAN in terms of mapping the QoS flows to the radio bearer. Here, the radio bearer is assumed to have different characteristics in terms of data rates, latency, and reliability, which enables an efficient mapping of the applications to the related requirements within the RAN in terms of QoS.

The mapping of the QoS flows to the layer-2 protocol stack within the 5G gNB is shown in Fig. 5.18, based on the downlink layer-2 structure presented in [11]. Here, consider the same scenario as the one shown in Fig. 5.17, with three QoS flows received at the NG RAN through the NG-U tunnels. Here, the SDAP layer handles the QoS flows

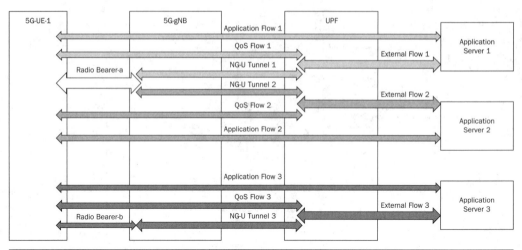

FIGURE 5.17   5G RAN QoS flow architecture [11]

and maps them into appropriate radio bearers and is delivered to the PDCP layer. The PDCP layer is responsible for robust header compression (ROHC) for the IP header of the data packet and ensuring the security of the data using ciphering and integrity protection, if needed. The radio bearers are mapped to appropriate RLC channels, which are then segmented within the RLC layer. The RLC layer also is responsible for retransmitting data that was received incorrectly using automatic repeat request (ARQ). ARQ and segmentation are applied only when the data is transmitted using acknowledge mode, where the network is expecting feedback from the device related to the successful reception of the transmissions. The RLC layer maps the data into appropriate logical channels which are then further processed by the MAC layer. The MAC layer is responsible for scheduling and priority handling, depending on the characteristics of the logical channel to which the data has been mapped. The MAC layer also maps the logical channels into physical channels and has the hybrid ARQ (HARQ) functionality, which ensures the reliable delivery of data between peer entities (within the base station and UE) within the PHY layer. In the considered example, since we assume only one UE, the MAC layer multiplexes the various logical channels into transport blocks which are then delivered to the PHY layer, which could be extended to multiple UEs as well.

An exemplary implementation of the QoS flows within a centralized 5G RAN is shown in Fig. 5.19, mainly from the user-plane perspective. Here, we assume that the QoS flows 1 (from Application Server 1) and 2 (from Application Server 2) have similar characteristics and hence are mapped to Radio Bearer-a which is delivered to the end through gNB-DU-1. The QoS flow 3 is delivered using Radio Bearer-b through gNB-DU-2, with the DUs potentially using different air interface variants (AIVs) in terms of frequency bands, numerology, etc. This could be considered similar to the assumptions made in [14], where traffic flows for different services such as high-priority URLLC flows, e.g., delivered using DU utilizing <6 GHz frequency band, and relatively lower priority eMBB flows delivered using mm-Wave frequencies, are delivered using different base stations/DUs. The implementation of the QoS flows described here indicates

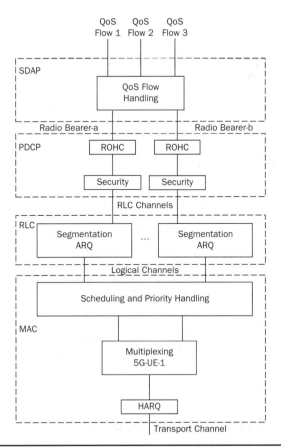

FIGURE 5.18   Mapping of QoS flows to Layer-2 protocol stack [11]

the level of flexibility that is available within 5G RAN in terms of enabling some of the key use cases and services.

5.1.8   Energy Saving Enhancements

Energy saving enhancements defined for 5G RAN enable the energy-efficient operation of the RAN nodes, while ensuring coverage and reliability for the UEs. The mechanism enables the network operators to minimize the energy consumed by the network particularly during low-load conditions. The network operators can use the O&M to configure the RAN with the configuration information required for autonomously deactivating the capacity booster cells, which are controlled by a non-capacity booster or coverage cell using the Xn interface. The capacity booster cells provide extra radio resources/capacity in case of high demand during peak hours and are fully covered by coverage cells, with higher radius/coverage area but with lower capacity. The configurations or policies for deactivation could include load conditions and time configurations, where the expected future load could be determined based on analyzing the

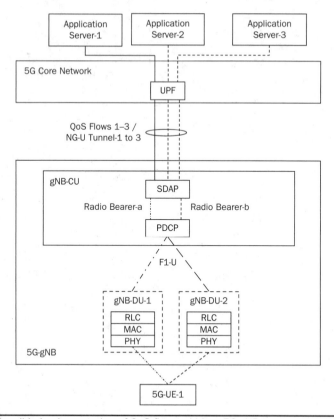

FIGURE 5.19 Possible implementation of QoS flows within a 5G-gNB

historical data consumption patterns, etc. Similarly, the O&M could also configure the configurations or policies for reactivating the capacity booster cells, once they are deactivated, which could also include the load thresholds for the coverage cell, and other information which indicates that the coverage cell cannot handle the expected load for the network without activating the booster cells.

The coverage cell signals the information related to activation and deactivation to the capacity booster cells using the Xn interface, based on the configurations and policies from the O&M. The coverage cell also needs to initiate handover of the UEs connected to the capacity booster cells that are going to be deactivated. For deactivating cells, the NG-RAN node configuration update procedure [22] is utilized by the gNB controlling the capacity booster cell. For reactivating the cell that has been deactivated, the cell activation procedure is used [22]. The deactivated cell would maintain the cell configuration data even during the inactive state. The performance gains from implementing such enhancements were investigated in [23] and [24], and it was shown that significant operational expenditure reductions could be achieved using optimal configurations or policies for activating and deactivating the capacity booster cells. An overview of the scenarios for 5G energy saving enhancements is as shown in Fig. 5.20.

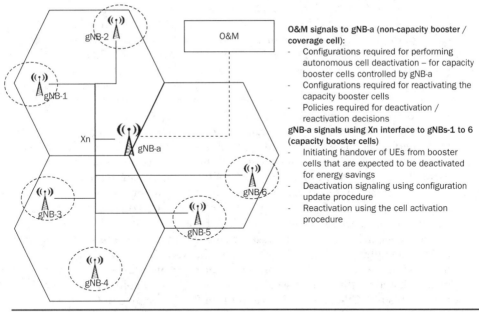

O&M signals to gNB-a (non-capacity booster / coverage cell):
- Configurations required for performing autonomous cell deactivation – for capacity booster cells controlled by gNB-a
- Configurations required for reactivating the capacity booster cells
- Policies required for deactivation / reactivation decisions

gNB-a signals using Xn interface to gNBs-1 to 6 (capacity booster cells)
- Initiating handover of UEs from booster cells that are expected to be deactivated for energy savings
- Deactivation signaling using configuration update procedure
- Reactivation using the cell activation procedure

FIGURE 5.20   5G scenario for energy saving enhancements [11,23,24]

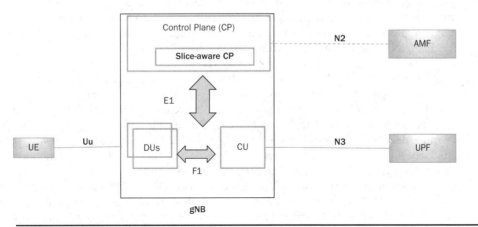

FIGURE 5.21   Overview of 5G architecture impact on UP design

5.2   5G/NR Design Considerations

The 5G architecture may provide some key considerations as well as limitations for the realization of the service-oriented NR design. Some key considerations will be due to the deployment of nodes and functionalities as well as the RAN-CN interface options. In particular, as can be seen in Fig. 5.21, in this section the effect of CP/UP split options, the slicing-awareness, and the RAN/CN protocol options are investigated as key factors that can affect the UP design considerations.

The separation of CP and UP according to the software-defined network (SDN) concept is a recent trend in the definition of the 5G architecture [20,21,25]. It requires categorizing all NFs as being either part of CP or UP, based on functional decomposition [26]. Any kind of interaction between CP and UP is supposed to happen through standardized interfaces. For the core network a separation of CP and UP functions is already partially realized, e.g., in LTE with the mobility management entity as the main CP element and the serving and packet gateways as UP elements. However, this separation is partial, e.g., the packet gateway contains CP functions.

The motivation for a full separation of CP and UP is threefold: Firstly, a standardized interface enables a consistent control over UP network elements and NFs from different vendors/manufacturers, e.g., in terms of interference management for UDNs. Secondly, a replacement or upgrade of a CP function often requires also the replacement of coupled UP functions and vice versa. Avoiding this might offer significant cost savings. Finally, the independent evolution of CP and UP could make the rollout of new NFs faster, thus enableing a more flexible network.

However, there are also significant challenges that make a full separation complex, mainly the tight coupling of CP and UP in the RAN [27]. In addition, extensive standardization is required. Integrating new in the future interfaces in a proprietary manner in combination with standardized ones is not a suitable solution, as it would destroy the benefits of a CP–UP split. For example, a flexible change of CP NFs in logical network elements would not be possible anymore if only selected UP NFs support certain proprietary interfaces. Additional effort in terms of testing is required to guarantee the interoperability of CP and UP functions from different sources (shifting the effort to system integrators supporting the operators instead of doing this work at a single vendor).

In RAN deployments, due to the strong coupling of UP with control logics, the complete C/U plane separation might not be feasible for all different deployments. In particular, we may have real-time scheduling (or fast RRM) functionalities which will require per transmission time interval (TTI) scheduling. Here, TTI refers to the duration of transmission on the radio link and typically of 1 ms (in 5G this can be even smaller). In the case of distributed-RAN (D-RAN), the decoupling of fast RRM will require ideal BH/FH between the control part of BS and the data processing part of BS. Hence, we show some cases where we partially split some C-plane functionalities. Since there might be multiple interactions between independent functionalities, we group them in three types of C-plane functionalities: (1) fast RRM or real-time scheduling, (2) slow RRM or non-real-time scheduling [cell re-selection, connection mobility control (CMC), load balancing (LB), and inter-cell interference coordination (ICIC)], and (3) RRC functions.

In the first option, the complete separation of CP–UP plane is provided. The main advantages of this option are the high gains due to centralized scheduling and the ability to support tight resource cooperation between multiple cells (e.g., virtual cell concept). On the other hand, one of the key drawbacks of this option is that it requires ideal BH, which supports extremely low latency over the backhaul/transport link to meet the tight latency requirements (for per TTI scheduling). In the second option, the CP–UP separation involved the separation of RRC and slow RRM; hence real-time/fast RRM remains at the DU. This is a good compromise for D-RAN deployments with nonideal BH, which supports transport of data with higher latency at a lower cost, since the real-time RRM is performed at the DUs. On the other hand, there might be dependencies between RRM functions that will require additional signaling. One example of such signaling is the requirement for exchanging ICIC information, such as high inter-ference indication (HII) from CU to DU, since this may affect the real-time scheduling at the DU.

In Fig. 5.22, the two different options are shown and how they can be mapped to dif-ferent RAN deployments. In particular, for a D-RAN deployment, the separation of RRC and slow RRM (option 2) seems more reasonable due to the wireless backhaul which can be nonideal. On the other hand, for an exemplary 5G scenario, where mm-Wave APs can provide hotspot coverage to enhance capacity, the separation of C-plane part at a BS operating in low frequencies (e.g., LTE coverage) will allow for efficient 5G opera-tion by centralizing/pooling resource gains and centralized resource management. This option, however, requires ideal BH/FH between the APs, as mentioned above.

5.2.1   User-Plane Design

The user-plane design is a key part of the 5G architecture and needs to be revisited in order to meet 5G service requirements with diverse and tight KPIs. Prior art on user-plane design for 5G networks is limited and has mainly focused on the virtualization of

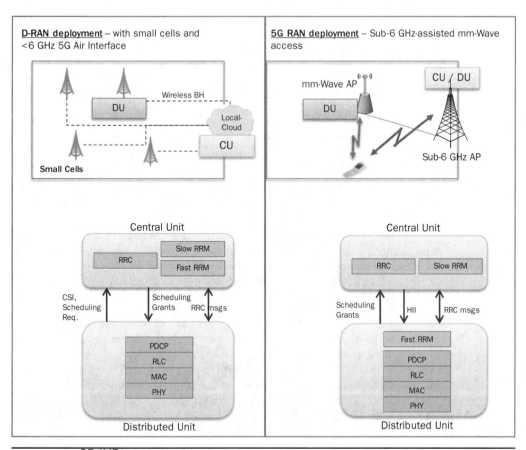

FIGURE 5.22 CP/UP options and mapping to exemplary RAN deployments

control- and user-plane functions at the RAN domain. In particular, the authors in [28] proposed a service-oriented virtual network auto-creation to select and deploy basic user and control functions at the cloud for different service types. The virtualization of radio NFs per service can be seen as a good candidate solution for C-RAN deployments where high centralization and pooling gains can be achieved. However, these functions are yet to be defined and the limitations at the wireless domain, given complex and different RAN deployments, will require designing a new user-plane framework which incorporates the AIV characteristics and the network capabilities. In this direction, the proposed a new framework explicitly describes the functional requirements and architecture for different 5G services, and we further analyze the impact of 5G architecture considerations to user-plane design realization.

3GPP has started a new study item "Study on NR Access Technology" [29] in the Radio Access Network Working Group. The objective is to develop NR access technology to meet a broad range of use cases including eMBB, mIoT, URLLC, and additional requirements including support for frequency ranges up to 100 GHz. The resulting normative specification would occur in two phases: Phase I (planned completion in June 2018) and Phase II (planned completion in December 2019). As part of this work, the following NR user-plane-related functions were defined:

- New Access Stratum sub-layer (handling QoS): New user-plane protocol layer above PDCP applicable for connections to the 5G core; single protocol entity is configured for each individual PDU session; mapping between a QoS flow and a data radio bearer; marking QoS flow ID in both DL and UL packets.

- PDCP: Sequence numbering; header compression and decompression, transfer of user data; reordering and duplicate detection; PDCP PDU routing; retransmission of PDCP SDUs; ciphering and deciphering; PDCP SDU discard; PDCP re-establishment and data recovery for RLC AM; duplication of PDCP PDU in case of multi-connectivity and CA.

- RLC: Transfer of upper layer PDUs, according to transmission modes AM, UM, and TM; sequence numbering independent of the one in PDCP; error correction through ARQ; segmentation and re-segmentation; reassembly of SDU; RLC SDU discard; RLC re-establishment.

- MAC: Mapping between logical channels and transport channels; multiplexing/de-multiplexing of MAC SDUs belonging to one or different logical channels into/from transport blocks (TB) delivered to/from the physical layer on transport channels; scheduling information reporting; error correction through HARQ; priority handling between UEs by means of dynamic scheduling; priority handling between logical channels of one UE; padding.

From PHY perspective, 3GPP [57] has specified the key PHY features for NR in Release 15 (NR phase 1) and Release 16 (NR phase 2). In NR Phase 1, there are common elements between LTE and NR, such as both using orthogonal frequency division multiplexing (OFDM). In particular, NR physical layer is based on OFDM with a cyclic prefix (CP). For uplink, Discrete Fourier Transform-spread-OFDM (DFT-s-OFDM) with a CP is also supported. To support transmission in paired and unpaired spectrum, both Frequency Division Duplex (FDD) and Time Division Duplex (TDD) are enabled. PHY is defined in a bandwidth agnostic way based on resource blocks, allowing the PHY to adapt to various spectrum allocations. Comparing to LTE numerology (subcarrier spacing and symbol length), the most outstanding difference is that NR supports multiple different types of subcarrier spacing (in LTE there is only one type of subcarrier spacing, 15 KHz). More details on the layer 1 aspects can be found in 3GPP [57]. For Release 16 the key focus is the NR support of vertical industries such as non-terrestrial networks (NTN), vehicle to everything (V2X), public safety, and industrial IoT. Subsequently, other trends and open study items include unlicensed access (NR-U), enhanced MIMO studies (in particular, >6 GHz), integrated access and backhaul (IAB), and non-orthogonal multiple access (NOMA) waveforms.

In NR, each AIV can be characterized by different sets of physical layer features (waveform, multiple access scheme, frame structure, numerology, etc.). Assuming multiple AIVs as part of NR, the AIV-to-resource mapping is a key control functionality closely coupled with UP design (which is exemplified in Fig. 5.23).

Semi-static AIV to Resource mapping: In this case, the mapping between AIV and spectrum is fixed for a given time period and is performed in PDCP. RAN allocates AIVs to time/frequency resources according to RRC or Operations, Administration & Maintenance (OAM) settings. However, the mapping can be updated based on the load, traffic volume, and radio condition changes. In this case, backhaul is not required to be ideal between BSs supporting different AIVs.

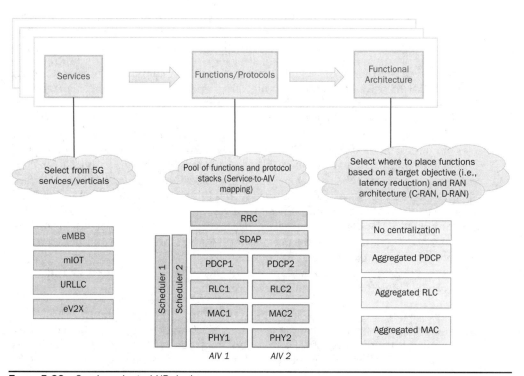

FIGURE 5.23   Service-oriented UP design

Dynamic AIV to Resource mapping: In each TTI, time/frequency resources can be allocated dynamically per AIV in MAC. This is ideal for high dynamic scenarios such that radio resource requirements for each AIV are very dynamic accordingly. MAC layer assigns radio resources to each AIV, and further determines the time/frequency resource allocation to services using certain AIV at per TTI basis. For this case, service multiplexing level can be high with proper RRM algorithms; nevertheless, this requires ideal backhaul between nodes (or colocation of nodes) supporting different AIVs.

Furthermore, as aforementioned, mm-Wave radio is envisioned as a key enabler for providing high capacity in 5G RAN. However, due to the path loss limitations in high frequencies, directional antennas with adaptive beam-forming and overlapping coverage are going to be used to exploit the benefits of operation in mm-Wave radio and ensure high coverage. In the mm-Wave radio, the main challenges regarding UP design are the following:

- High-penetration loss of mm-Wave frequencies can severely deteriorate the performance and hence, maintaining reliable connectivity is a challenge especially for delay critical services.

- Wireless channel conditions and link quality can change significantly during the movement of users, calling for fast RRM decisions and multi-connectivity support. User mobility also causes significant and rapid load changes and handovers due to small coverage areas of access nodes. Therefore, connection management and load balancing in conventional RRM functionalities need to be revisited to cope with the aforementioned challenges [30].

- Due to highly directional transmissions, crosslink interference characteristics become much different from <6 GHz systems. For example, there can be flashlight effects (an interfering beam hits a user).

From UP design point of view, it is recommended to adapt MAC design for high directivity for the case of operation in high frequencies. In 3GPP, beam management is discussed as a key functionality to meet the capacity and coverage requirements by performing resource-to-beam allocation and beam sweeping/steering at MAC/lower layer. The aim of the beam management is to maintain connectivity between UE and a serving access node during mobility and radio environment change. From a set of candidate beams, it would be dynamically decided in which beams each user can connect to and what should be the beam-to-resource allocation. To this end, one important aspect is the RAN/CN interface options which will provide requirements to the RAN design (e.g., the RAN design might need to be adapted in order to be able to meet different interface options, given the physical deployment). On the other hand, the service requirements might necessitate the use of certain RAN/CN interface options to be able to meet certain KPIs (e.g., low end-to-end latency).

Below, we present some key considerations for the configuration of functions, which can be tailored for four main 5G service types.

- In eMBB, in order to accomplish the ultra-high-throughput KPI, a key consideration is to enable the operation in higher frequencies, which can offer much higher bandwidth. Furthermore, high centralization of UP functions as well as coordinated multipoint transmission are envisioned as key technologies to meet the target KPIs.

- On the other hand, URLLC is a service type which requires ultra-high reliability and low latency for time critical use cases (e.g., vehicular safety, eHealth). For achieving ultra-high reliability, multi-connectivity/carrier aggregation (CA) can be seen as key technology enabler. In particular with the proposed layering, the upper layer will be able to connect to multiple lower layer entities, thereby potentially lowering the processing delays (taking also into account the retransmission processes). To this end, due to the small and fixed size of the packets, function like segmentation is not required; and fixed size logical channel prioritization (LCP) can be used at lower layer.

- In mIoT service type, one of the main KPIs is to maximize the connection density. Since the traffic requirement is low, the devices are expected to be static (e.g., sensors) and the density is expected to be high, group-based functionalities both in lower and upper layers are envisioned. Further, some functions related to mobility can be disabled.

- Enhanced V2X (eV2X) services (as introduced in [3]) can be seen as a mixture of previous service types, comprising both safety- and non-safety-related applications. One key aspect to be considered in this scenario is the fact that most V2X services are by nature group based and fully localized and they rely on D2D/V2V communications. So, the functions to be considered should take into account enhancements on PC5 (interface used for D2D communications), as well as group management mechanisms for resource control and transmission of resource in multicast groups. Furthermore, the same mechanisms should be provisioned as in URLLC; however, one key difference is the assumptions on the mobility of the users which may require fast mobility control mechanisms. One further difference is also the high transmission rate required for some high automation use cases, besides the high-reliability and low-latency requirement, which may stretch the radio resources and may require dynamic resource adaptations to meet the critical requirements.

5.2.2 Control-Plane Design

In this section, the control-plane design considerations in 5G RAN are discussed, which include mainly the RRC and management functionality framework. The functionality framework comprises the definition of new or enhanced control mechanisms, as well as the functional architecture which may be configured in a way that the optimization of resources is succeeded and the 5G KPIs are met.

In this section, the impact of some key 5G-related enabling technologies on control-plane design is investigated, e.g., the utilization of high frequencies, the slice awareness, backhaul/access integration, and multi-connectivity scenarios.

5.2.2.1 RRC Overview

The major role of RRC is to control and configure all the radio resources (PHY, MAC, RLC, and PDCP) to make it possible to communicate between UE and the base station (e.g., gNB, eNB, NB, BTS, etc.). Comparing to the LTE protocol stack, RRC has similar functionalities in NR. More details on the specification of RRC in LTE can be found in [31] and in NR in [32].

The main services and functions of the RRC sub-layer include the following:

- Broadcast of system information related to AS and NAS.
- Paging initiated by 5GC or NG-RAN.
- Establishment, maintenance, and release of an RRC connection between the UE and NG-RAN including the addition, modification, and release of carrier aggregation and dual connectivity in NR or between E-UTRA and NR.
- Security functions including key management.
- Establishment, configuration, maintenance, and release of signaling radio bearers (SRBs) and data radio bearers (DRBs).
- Mobility functions (including handover and context transfer, UE cell selection and reselection and control of cell selection and reselection, and inter-RAT mobility).
- QoS management functions.
- UE measurement reporting and control of the reporting.
- Detection of and recovery from radio link failure.
- NAS message transfer to/from NAS from/to UE.

The major difference from LTE RRC is the introduction of a new state, apart from RRC IDLE and CONNECTED state, namely the RRC INACTIVE which is beneficial for IoT service type. The benefits of the inactive state are that it significantly reduces latency and minimizes the battery consumption of both smartphones and IoT devices. The most important features of inactive state are the support for encrypted response messages, smart RAN paging, RAN architecture support, and fallback to legacy procedure. More information on the key procedures can be found in [32].

5.2.2.2   RRM Overview

Based on 3GPP LTE RRM functionalities and structure [19], given the RAN limitations, RRM can be grouped in three main groups given their output, their in-between interactions, and the time scale they operate: (1) slow RRM, which can trigger cell selection/re-selection, (2) fast RRM, which can change the resource utilization/restrictions, and (3) basic RRM for bearer admission and control. For the D-RAN scenario [33], we can also have a fourth type which is about the wireless backhaul resource management and wireless topology handling. Moreover, given the level of centralization of RRM, we can observe three different RRM types as described below:

Centralized RRM: RRM functions operate together in an entity for multiple access nodes in a group. This will provide fast and simple interaction between RRM functions, but on the other hand in heterogeneous networks (HetNets), we need ideal backhaul for some fast RRM functions (e.g., coordinated multi-point—CoMP; dynamic resource allocation—DRA). Moreover, signaling overhead can be very high in ultradense environments. Furthermore, for 5G systems, some solutions have been proposed that require controller for clusters of HetNets using cloud-based resource pooling and management (also called cloud-RAN/C-RAN) [34]. There, resource pooling and centralized management of resources can provide high gain in terms of capacity. Nevertheless, this requires ideal backhaul/fronthaul and can be seen as challenging task for DRA in fast changing environments assuming challenges related to interference from other C-RAN clusters.

Distributed RRM: In 3GPP LTE/LTE-A, we observe that RRM functions reside at the eNodeB. The main RRM functions are about DRA, ICIC and CMC, radio bearer admission and control (RAC, RBC), energy efficiency, and load balancing (LB). In LTE RRM structure, there can be interactions between RRM functions. An example is the cell on/ off function, which might require input from the resource restrictions due to interference management, and its output will require handovers which will affect mainly CMC and LB. Since all these functions reside at the eNodeB, there is no additional signaling specified in 3GPP for the RRM interactions. Also, in [11] similar functionalities are envisioned for NR as depicted in Fig. 5.24.

Semi-Centralized RRM: Other studies including the ones done in [35] and [37] discuss two levels of RRM, denoted as global and local scheduling. These studies mainly focus on centralized interference management and load balancing and distributed fast RRM functionalities. One of the main challenges in this case is the new interactions that will require additional signaling and complexity in various RAN nodes. Figure 5.25 illustrates these interactions in case of heterogeneous RAN with small cells under a macro-cell umbrella. Here, we observe that if RRM functionalities will be semi-distributed between the macro-cells and the small cells, extra signaling will be required since the duplication of some functions in different nodes will require an exchange of information between the involved nodes.

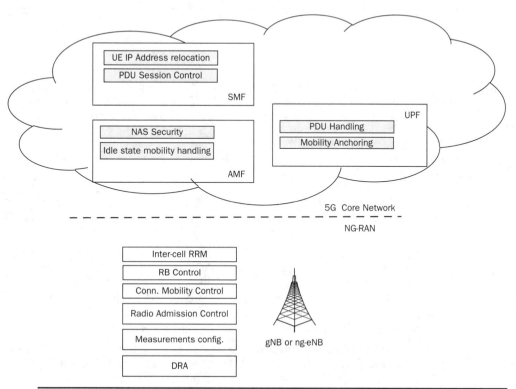

FIGURE 5.24 Functional split between NG-RAN and 5GC

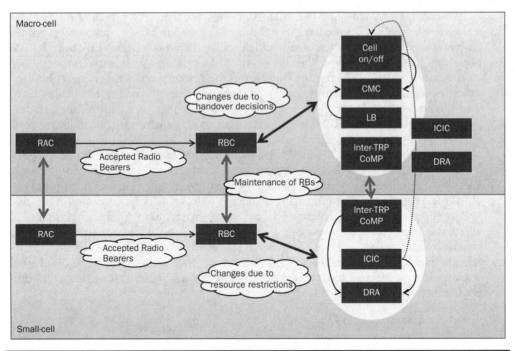

Figure 5.25   RRM possible interactions in 3GPP LTE-A with dual connectivity

Due to the tight interactions of RRM functions, a practical way to split the RRM is between slow and fast RRM given their time scale they operate. Hence, in this section, we discuss the split of RRM based on the categorization above.

5.2.2.3 Considerations for mm-Wave RAN

In mm-Wave radio, the main challenges regarding resource management are the following:

- High-penetration loss of mm-Wave frequencies can severely deteriorate the performance, and hence maintaining reliable connectivity is a challenge especially for delay critical services.

- Wireless channel conditions/link quality can change significantly during the movement of users, calling for fast RRM decisions and multi-connectivity support. User mobility also causes significant and rapid load changes and handovers due to small coverage areas of access nodes. Therefore, connection management and load balancing in conventional RRM functionalities need to be revisited to cope with the aforementioned challenges.

- Due to highly directional transmissions, the crosslink interference characteristic becomes much different from <6 GHz systems; there can be flashlight effect (interfering beam hits user). Advanced interference management is thus required.

For the particular categories of fast, slow, and topology resource management, we will provide some considerations on the functionalities which are required in mm-Wave RANs.

Fast RRM: Set of functions which require channel state information (CSI) measurements as input and have tight timing constraints (per TTI). The functions for the particular case of mm-Wave could be:

- Dynamic resource allocation: Similar functionality as in LTE, however, the TTI size in mm-Wave radio will be much smaller (around 100) and adaptation of DRA operation will be possible.

- Beam management: Dynamic beam alignment and corresponding resource allocation, and maintain connectivity between UE and a serving access node during mobility/radio environment change.

- Inter-node CoMP: Due to the high density of access nodes, the coordination between access nodes should consider large and dynamically changing clusters of cooperative nodes.

Slow RRM: Set of functions which require RRM measurements as input and have less tight timing constraints (100 TTIs).

- Load balancing: An existing function which will be modified to cope with the fast load fluctuations due to short mm-Wave range.

- Connection mobility control: Another function related to handover management between access nodes, which could be strongly coupled with beam management.

- Interference management: In addition to employing inter-cell interference management in time/frequency and power domain, in mm-Wave this should be also handled in spatial domain.

One further functionality is the topology RRM, which is a set of functions that require BH CSI/RRM measurements as input and have variable timing constraints. For mm-Wave BH, this mainly relates to how to steer beams in the BH subject to cell and traffic requirements. In this category, depending on the BH technology and topology, BH link scheduling and path selection is a functional block highly required. In the case of multi-hop mm-Wave BH, the proper path selection and the activation of access nodes in a way that target KPIs are met are key RRM processes to avoid BH bottleneck.

In dense mm-Wave RAN, multiple limitations for BH/access might require certain handling of resource management. In particular, nonideal wireless BH among RAN nodes can be a limiting factor and will require extra RRM for the BH part. To this end, joint BH and access optimization can be used to meet high-throughput requirements for throughput demanding services. Another important factor is the extensive signaling which will be required in HetNets for wireless BH and access measurements. In Table 5.1, we present a brief analysis of pros and cons for different RM placements.

As shown in Table 5.1, different placements of RRM functions can be good candidates for different mm-Wave RAN scenarios, subject to different parameters:

- Backhaul is an important factor, since the strict timing requirements for certain dynamic RRM functions can be a strong limitation toward centralization for particular cases (e.g., nonideal BH).

	RRM split A (centralized)	RRM split B (semi-centralized)	RRM split C (semi-centralized)
Description	All RRM centralized	Slow RRM centralized	Slow and fast RRM distributed, topology RRM centralized
Advantages	Per-TTI scheduling provides resource pooling gain and allow multi-connectivity	– Relaxed BH requirement (nonideal) – Ideal for low mobility and low/medium load scenarios	– Good for no mobility scenarios – Support flexible multi-hop backhauling
Limitations	Require ideal BH/FH	– Require fully functional small cells – Require extra signaling for interaction between fast and slow RRM	– Require fully functional small cells – For joint BH/access need extra signaling among access nodes

TABLE 5.1 Pros and cons for different RRM placements for mm-Wave

- Deployment is another key factor, as the deployment of multiple air interfaces for the non-standalone scenario (<6 GHz and mm-Wave radio access) will require centralization of certain slow functions (e.g., mobility control) to allow for multi-connectivity between different air interfaces.

- Furthermore, user mobility and cell density will also impact on the required centralization, since no/low mobility requires more distributed RM placements and more dense deployment needs higher centralization to exploit the gain of multi-connectivity.

5.2.2.4 Considerations for Slice-Aware RAN

Network slicing allows the deployment of multiple logical, self-contained networks, offering third parties and verticals customized services on the top of a shared infrastructure. More details are provided in Chap. 4 (market trends) and Chap. 6 (architecture). From RAN perspective, the slicing can be about provisioning radio resources (spectrum bands, lower layer configurations) to vertical segments facilitating diverse service requirements via the means of a virtualized AIVs, which support diverse service characteristics offering customization.

For allowing slice-awareness at RAN level, some principles have been discussed in [36], where an example illustration of the architecture enhancements is provided in Fig. 5.26. The network slice-awareness in 5G RAN will strongly affect the RAN design and particularly the CP design, where multiple slices, with different optimization targets, will require tailored access functions and functional placements to meet their target KPIs. On the other hand, the RAN deployment may provide some limitations on the efficiency of RRM due to the wireless channel, traffic load, and resource availability constraints, which may affect the overall performance (assuming numerous slices reusing the same RAN deployment). In particular, in dense urban HetNet scenarios, the signaling and complexity of RRM will be higher due to more signaling exchanges needed for passing RRM information to different entities. Moreover, the distribution of RRM functions in different radio nodes will provide new dependencies between RRM functions, which should be taken care of in order to optimize performance. In addition, in case of HetNet RAN deployments, nonideal backhaul (with limited capacity and non-negligible latency) between access nodes (macro-cells and small cells) will put some limitations on the RRM decisions and placement options to meet certain KPIs.

To ensure meeting the E2E slice requirements, assuming limited RCMs, which may be mapped to numerous slices, we introduce here a CP functionality framework, which is highly required to allow for slice-tailored optimization in RAN. In particular, the following RRM mechanisms are relevant to be considered:

Intra-RCM RRM: Slice specific resource management and isolation among slices, utilizing the same RAN is an open topic which is currently investigated. In literature [37], the conventional management of dedicated resources can be seen as intra-slice RRM, which can be tailored and optimized based on slice specific KPIs.

Inter-RCM RRM/RRC: On top of Intra-RCM RRM, inter-RCM RRM/RRC (which includes also inter-slice RRM and slice-aware Topology RRM for wireless self-backhauling) is defined as the set of RRM policies that allow for flexible sharing, isolation, and prioritization of radio resources among slices or slice types in coarse time scales. Inter-RCM RRM is an "umbrella" functional block to optimize the resource efficiency and utilization. In this direction, an inter-RCM RRM mechanism is proposed in [38], where slice-aware RAN clustering, scheduler dimensioning, and adaptive placement of intra-slice RRM functions are discussed in order to optimize performance in a dense heterogeneous RAN. Given the requirements of new access functions which can be tailored for different network slices, the distribution of RRM functionalities in different nodes is a key RAN design driver which allows for multi-objective optimization in a multi-layer dense RAN. The adaptive allocation of such functions is also envisioned as key feature to cope with the dynamic changes in traffic load, slice requirements, and the availability of backhaul/access resources. To this end, one further inter-slice/RCM RRM functionality is proposed in [39] which performs traffic forecasting of different slices and allocates resources to slices in a proactive manner.

Topology RRM: This can be seen as another category of inter-RCM RRM, mainly for distributed-RAN (D-RAN), where the resource allocation of wireless self-backhauling is essential to allow for joint backhaul/access optimization [40]. Thus, Topology RRM can be tailored for different slices [41] in order to allocate backhaul resources among RCMs in a slice-tailored manner in order to avoid backhaul bottlenecks.

Unified Scheduler: This is an overarching MAC Scheduler, where different slice types share the same resources and dynamic resource allocation and slice multiplexing is required on top of RCM-specific MAC.

Based on this categorization, an interesting aspect which may define the CP functionality requirements and the interface/signaling requirements between the CP functions is the functional split which is dependent on the CU–DU split options. Figure 5.26 shows the possible placement of inter-RCM and intra-RCM RRM and RRC functionalities. Depending on the placement, the interface requirements might be different due to the time/resource granularity of the CP functionalities and their possible interconnections.

In a slice-aware RAN, different RRM functionalities and placements are required to ensure that service-tailored KPIs are met. 5G tight requirements for delay and reliability (especially for URLLC) depend on fast and sophisticated RRM mechanisms.

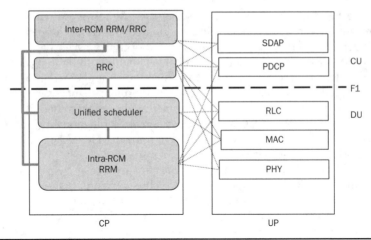

Figure 5.26   Exemplary functional deployment and interactions

Furthermore, different RRM procedures can be required for diverse slices and in different time scales. In RAN, the slice requirements are shaped by the target KPIs and key RAN characteristics (like user mobility and user/cell densities). Mobility may affect the backhaul resource allocation, handover, and interference management. Also, the user density can affect delays, signaling required, interference management, and resource availability. For each slice, these effects shall require different actions from access node point of view, to meet the end-to-end KPIs. In other words, different resource management and control placement of functions might be required from slice to slice.

An example can be given for the URLLC type of slice (e.g., for vehicular safety) where distributed RRM is preferred (due to low-latency requirement), so the motivation to centralize RRM will be low (e.g., for centralized mobility control in case of high mobility). In eMBB, on the other hand, we need high centralization to achieve higher throughput performance by centralized pooling of resources and exploit the benefits of multi-connectivity (e.g., no-cell concept). If nonideal BH exists, we choose the RRM split as centralized as possible, except if the mobility is very low and there is no overlapping between small cells.

Network slices will allow for flexible functional placements and tailored NFs to meet the per-slice SLAs. Hence, slice-specific resource management and isolation among slices, utilizing the same RAN, is an open topic which is currently investigated. In this context, inter-slice RRM can be defined as another functional block which dictates the RAN sharing and level of isolation/prioritization among Network slices. However, the impact of slicing on RRM functions which can trigger their adaptive placement in RAN nodes in a semi-distributed manner is not yet discussed.

Handling multiple and different resources in a dense urban 5G RAN, with different slice KPIs, will require a controller to orchestrate the resource management and control between slices. Centralized solutions will be required to meet required performance goals. In this context, an RRM controller can be defined as a logical entity which abstracts a set of access network functionalities and coordinates a group of access nodes to facilitate the resource management and control. The benefit of such controller is that

RRM can be optimized per KPI, e.g., for throughput (using sophisticated Interference Management), mobility, and reliability. This is desired since the different use cases have different KPIs as their prime goals. In this context, the RRM controller can be defined on a per-slice base and can be different from one slice to the other, taking into account the slice requirements and the RAN limitations toward achieving the target KPIs.

However, the physical placement and dimensioning of RRM controller plays an important role for the efficiency of RRM. The high number of controllers will provide more granular RRM; however, this might impose high delays for the communication between RRM controllers due to the functional dependencies in case these controllers reside in different entities. On the other hand, one flat RRM controller for RAN will not allow for efficient slice-tailored RRM, since the centralization impact may be different from slice to slice (e.g., for an IoT slice and eMBB slice, the impact of centralization of certain RRM functions may be different due to the diverse KPIs, the traffic situation, and coverage demands). Also, the backhaul capabilities will provide some limitations regarding the placement of such controllers.

So, the problem can be formulated as the way to initially create clusters from a pool of access nodes or BSs [denoted as transmit-receive-points (TRPs)], then select and configure one or more nodes as their controller(s) (denoted as RRM controller) and finally control the rest of the nodes of the cluster as the controlled entities (defined as slave-TRPs). Subsequently, how to decide which RRM functions will be decided at the RRM controller and which will be distributed at the other access nodes.

In Fig. 5.27, different splits can be illustrated. In our categorization, together with slow and fast intra-slice RRM which were discussed in the previous section, we can also have new RRM functions for the wireless BH (topology RRM) and RRM for inter-slice RRM.

Inter-slice RRM involves the functionalities which aim to decide on the efficient sharing of resources among different slices in case of shared spectrum utilization. There are scenarios within use cases such as V2X, which would require coordination and sharing of resources even between multiple operators, in order to ensure that the end user receives maximum coverage and network availability.

5.2.2.5   Considerations for BH/Access Integration Scenarios

One of the key 5G RAN deployments is the Integrated Access & Backhaul (IAB) deployment, which is investigated in 3GPP [42]. A key benefit of IAB is enabling flexible and very dense deployment of NR cells without densifying the transport network proportionately. This feature is considered essential for network operators who would like to

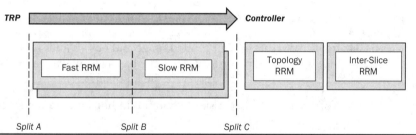

FIGURE 5.27   RRM split and grouping

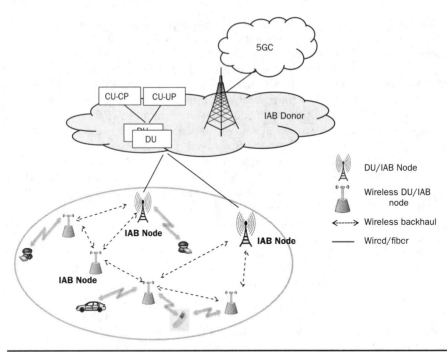

FIGURE 5.28   Reference diagram for IAB-architectures

minimize their investments related to transport networks, especially in dense urban areas. A range of deployment scenarios can be envisioned including support for outdoor small cell deployments, indoors, or even mobile relays (e.g., on buses or trains). The overview architecture is illustrated in Fig. 5.28, where the IAB donor may consist of the conventional gNB, which could be split in CU and DU parts; and the IAB nodes which can be seen as extension of gNB to enable connectivity to end users via multi-hop wireless backhaul means.

In this architecture, in-band and out-of-band backhauling with respect to the access link represent important use cases. In-band backhauling includes scenarios where access and backhaul link at least partially overlap in terms of spectral resources, thereby creating half-duplexing or interference constraints. Out-of-band backhauling on the other hand considers the use of orthogonal spectral resources for the access and backhaul link, thereby avoiding interference issues. The key challenge for out-of-band backhauling is the need for additional spectral resources. Furthermore, multi-hop backhauling is supported to provide more range extension than single hop. This is especially beneficial for >6 GHz frequencies due to their limited range and possibilities for frequency reuse.

To this end, in 3GPP study on IAB, the key objectives which may necessitate from optimization perspective RAN solutions are the following:

- Topology management for single-hop/multi-hop and redundant connectivity
- Route selection and optimization

- Dynamic resource allocation between the backhaul and access links
- High spectral efficiency while also supporting reliable transmission

Apart from the resource coordination required for interference management, load balancing, etc., additional resource coordination will be required to enable BH Resource Management or Joint BH/Access Resource Management (depending on the operation as in-band or out-of-band backhaul). BH RRM (or Topology RRM or BH/Access RRM) is proposed as a new functionality to enable the coordination of backhaul and access wireless resources.

As an example, some RRM functions (inter-gNB CoMP, LB, ICIC, and CMC) may need to be placed in IAB Donor or IAB Parent/Child Nodes either for service optimization purposes or due to BH/access channel conditions or due to the deployed architecture option. In this direction, two RRM split scenarios for IAB can be possible (as shown in Fig. 5.29), whereas the impact on interface/architecture requirements will be different:

- Scenario 1: Centralized Coordination of inter-node RRM functions to be placed in Donor IAB Node (CU/DU)
- Scenario 2: Distributed Coordination of inter-node RRM functions to be placed in IAB Node (Parent or Child)

Topology or BH RRM can be seen as an overarching backhaul and access RRM functionality block which includes functions that decide on the BH resource allocation, in terms of time, frequency, or spatial resources. Considering the backhaul technology

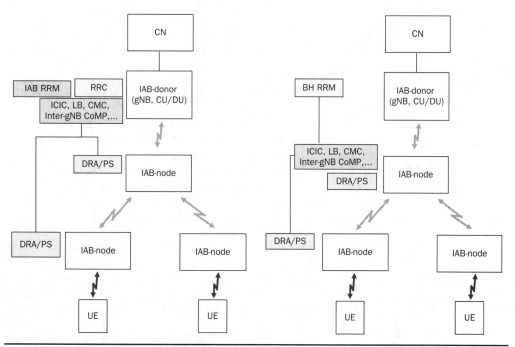

Figure 5.29   IAB coordination scenarios

and topology, as well as the spectrum consideration, BH RRM can optimize the IAB network performance, by selecting the best links and resources in a multi-hop deployment taking into account the traffic requirements for different service types, as well as the radio resource conditions.

To this end, the authors in [41] provide a solution focusing on two service types, eMBB and URLLC. URLLC slices require latency below 1 ms and 99.999% reliability; on the other hand, eMBB requires ultra-high user throughput. In a UDN with mm-Wave BH and access radio, the routing and scheduling policies should be adapted to meet the per-slice requirements. This involves the resource allocation of BH links, the number of hops, and the operation of transport nodes. Hence, the objective of this study was as follows: Firstly, to dynamically identify BH links and paths to be scheduled per a given time window, taking into account the target global objective for the particular service (in terms of maximizing BH capacity or minimizing latency or optimizing aggregate utility). Secondly, assuming different service types with conflicting KPIs, to identify whether and how the incoming flows are stored in the queues and forwarded to the next hops (or destinations), based on link selections in the previous step and the per-slice QoS requirements (delay, outage, or data rate).

5.3   5G/NR Enhancements for D2D/V2V Operation

As mentioned in Sec. 5.2.1, some 5G requirements which are imposed by the vertical industries may necessitate the enhancement of NR to support the direct communication of end devices/vehicles in close vicinity. This may include safety- and non-safety-related communications with diverse KPIs, spanning from URLLC to eMBB requirements.

Prior to 5G, device-to-device (D2D) access operation has been studied in [52]. Based on this, [53] specifies the Stage 2 of the Proximity Services (ProSe) features in EPS. ProSe Direct Communication enables establishment of communication paths between two or more ProSe-enabled UEs that are in direct communication range. The ProSe Direct Communication path could use E-UTRAN or WLAN for Public Safety-specific usage. ProSe-enabled Public Safety UEs can establish the communication path directly between two or more ProSe-enabled Public Safety UEs, regardless of whether the ProSe-enabled Public Safety UE is served by E-UTRAN.

The gap that NR D2D/V2V needs to fill comes from the stringent KPIs which are imposed by verticals (e.g., V2X, IIoT, Public Safety) for the direct, aka sidelink, communication among end devices. These new requirements are captured in [1,3].

In 3GPP Release 16, the enhancement of sidelink interface (aka PC5 interface) has been specified mainly for the V2X scenario, whereas for the other verticals the standardization progress is still ongoing. More specifically, in [54], a study how to support advanced V2X use cases identified in [3] is provided. One difference from 4G is that in NR V2X apart from the unlicensed ITS bands, also licensed bands in FR1 and FR2, up to 52.6 GHz will be used. Also, the QoS model is different, since it is flow based, rather than packet based of LTE-V. This will require more network control of resources and architecture enhancements to support such advancements. Also, it is eminent that NR V2X complements LTE V2X for advanced V2X services and support interworking with LTE V2X.

In particular, this specification addresses the following aspects:

- NR Sidelink (SL) design for V2X;
- Uu-based SL resource allocation/configuration by LTE and NR;

- RAT and interface selection;
- QoS management;
- Non co-channel coexistence between NR and LTE SLs.

5.3.1 Scenarios

The scenarios considered in the study are captured in [54]. The scenarios can be cat-egorized into standalone and MR-DC scenarios regarding the architecture. The study prioritized Scenarios 1, 2, and 3, and MN controlling/configuring both NR SL and LTE SL in Scenarios 4, 5, and 6 which is covered by Scenarios 1, 2, and 3, respectively.

As can be seen in Fig. 5.30, in Scenario 1 a gNB provides control/configuration for a UE's V2X communication in both LTE SL and NR SL, whereas in Scenario 2, an ng-eNB provides control/configuration for a UE's V2X communication in both LTE SL and NR SL.

5.3.2 Protocol Aspects

SL broadcast, groupcast, and unicast transmissions are supported for the in-coverage, out-of-coverage, and partial-coverage scenarios. The AS protocol stack for the control plane in the PC5 interface consists of at least RRC, PDCP, RLC, and MAC sub-layers, and the physical layer. The protocol stack of PC5-C is shown in Fig. 5.31.

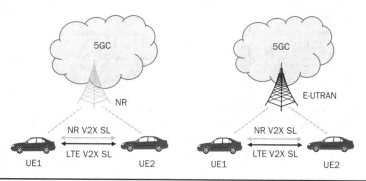

Figure 5.30   V2X Scenarios 1 and 2

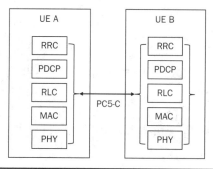

Figure 5.31   PC5 control-plane (PC5-C) protocol stack

The AS protocol stack for user plane in the PC5 interface consists of SDAP, PDCP, RLC, and MAC sub-layers, and the physical layer. The protocol stack of PC5-U is shown in Fig. 5.32.

More details on the protocol enhancements, based on [54], are captured in the subsections (from Sec. 5.3.2.1 to 5.3.2.6) below.

5.3.2.1 Physical Layer (PHY)

PHY Structures

The designs of a physical SL control channel (PSCCH), a physical SL shared channel (PSSCH), a physical SL feedback channel (PSFCH), and other matters related to physical layer structures were studied. The key aspects which are discussed are the following:

- Subcarrier spacing (SCS) and cyclic prefix: Both FR1 and FR2 frequency ranges are assumed for sidelink operation. In FR1, 15 kHz, 30 kHz, and 60 kHz SCS are supported with normal CP, and 60 kHz SCS with extended CP. In FR2, 60 kHz and 120 kHz SCS are supported with normal CP, and 60 kHz SCS with extended CP. In a given carrier, a UE is not required to receive simultaneously SL transmissions with more than one combination of SCS and CP, nor transmit simultaneously SL transmissions with more than one combination of SCS and CP.

- Channel coding: The channel coding defined for data and control in NR Uu are respectively the starting points for data and control on the NR SL.

- SL bandwidth parts (BWP) and resource pools: BWP is defined for SL, and the same SL BWP is used for transmission and reception. In specification terms, in a licensed carrier, SL BWP would be defined separately, and have separate configuration signaling, from Uu BWP. One SL BWP is (pre-)configured for RRC IDLE and out-of-coverage NR V2X UEs in a carrier. For UEs in RRC_CONNECTED mode, one SL BWP is active in a carrier. A resource pool is a set of time-frequency resources that can be used for SL transmission and/or reception. From the UE point of view, a resource pool is inside the UE's bandwidth, within an SL BWP and has a single numerology. Time domain resources in a resource pool can be non-contiguous. Multiple resource pools can be (pre-)configured to a UE in a carrier.

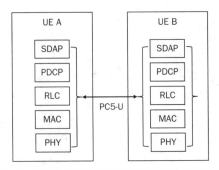

FIGURE 5.32 PC5 user-plane (PC5-U) protocol stack

PHY Procedures

- Multiplexing of physical channels: Three options for multiplexing of a PSCCH and associated PSSCH are studied. Option 1 assumes that PSCCH and the associated PSSCH are transmitted using non-overlapping time resources. On the other hand, Option 2 mentions that PSCCH and the associated PSSCH are transmitted using non-overlapping frequency resources in all the time resources used for transmission. Finally, in Option 3 part of PSCCH and the associated PSSCH are transmitted using overlapping time resources in non-overlapping frequency resources, but another part of the associated PSSCH and/or another part of the PSCCH are transmitted using non-overlapping time resources.

- HARQ procedures: For SL unicast and groupcast, HARQ feedback and HARQ combining in the physical layer are supported. HARQ-ACK feedback for a PSSCH is carried in SFCI format(s) via PSFCH in resource allocation Modes 1 and 2.

- Channel State Information (CSI) Acquisition: Examples of CSI information for V2X are CQI, PMI, RI, RSRP, RSRQ, pathgain/pathloss, SRI, CRI, interference condition, and vehicle motion. For unicast communication, CQI, RI, and PMI, or a subset among them, are supported with non-subband-based aperiodic CSI reports assuming no more than four antenna ports. The CSI procedure does not rely on a "standalone" RS. CSI reporting can be enabled and disabled by configuration.

- Power control: Open-loop power control (OLPC) procedures are supported for SL. When the transmitting UE is in-coverage, gNB can enable OLPC for a unicast, groupcast, or broadcast transmission based on the pathloss between the transmitting UE and its serving gNB.

5.3.2.2 Medium Access Control (MAC)
The MAC sub-layer provides the following functions for SL:

- Layer 2 packet filtering (at least for broadcast, if it is concluded that full identification is not used in L1 control information);
- SL carrier/resource (re-)selection at least for broadcast;
- SL HARQ transmissions without HARQ feedback and SL process at least for broadcast;
- SL-specific logical channel prioritization at least for broadcast;
- SL Scheduling Request for broadcast, groupcast, and unicast;
- SL Buffer Status Reporting for broadcast, groupcast, and unicast;
- UL/SL TX prioritization for broadcast, groupcast, and unicast.

The study also investigates whether and how to enhance SR procedure/configuration, MAC PDU format, HARQ/CSI feedback/procedure for groupcast and unicast, and configured SL grant transmission for MAC.

5.3.2.3   Radio Link Layer (RLC)

The RLC sub-layer provides the following functions for SL:

- Segmentation and reassembly of RLC SDUs for broadcast, groupcast, and unicast;
- RLC SDU discard function for broadcast, groupcast, and unicast.

Furthermore, a UM RLC entity is configured to submit/receive RLC PDUs for user packets of SL broadcast, groupcast, or unicast. If SBCCH is used for SL, a TM RLC entity is configured to submit/receive RLC PDUs for control information. RLC AM is supported for SL unicast and is not supported for SL broadcast and groupcast.

5.3.2.4   Packet Data Convergence Protocol (PDCP)

The PDCP sub-layer provides the following functions for SL broadcast, groupcast, and unicast:

- SL packet duplication and duplicated PDU discard at least for broadcast and groupcast;
- Timer-based SDU discard for broadcast, groupcast, and unicast.

5.3.2.5   Radio Resource Control (RRC)

RRC is used to exchange at least UE capabilities and AS layer configurations. For UE capability transfer, the information flow is triggered during or after PC5-S signaling for direct link setup, and can be done in a one-way manner. For AS layer configuration, the information flow is triggered during or after PC5-S signaling for direct link setup, and can be done in the two-way manner. There is no need for one-to-many PC5-RRC connection establishment among group members for groupcast.

5.3.2.6   Service Data Adaptation Protocol (SDAP)

The SDAP sub-layer provides the functions for SL unicast which apply to the per-flow QoS model described in [55] in the upper layers. In particular, it supports the QoS flow to SLRB mapping for SL unicast. The SDAP sub-layer does not apply to SL broadcast or groupcast.

5.3.3   Radio Resource Management Aspects

With respect to RRM, the study defines at least the following two SL resource allocation modes. The modes shown in Fig. 5.33 are as follows:

Mode 1: BS schedules SL resource(s) to be used by UE for SL transmission(s).

Mode 2: UE determines the SL resource allocation, i.e., BS does not schedule, SL transmission resource(s) within SL resources configured by BS/network or pre-configured SL resources. Resource allocation Mode 2 supports reservation of SL resources at least for blind retransmission. Also, sensing- and resource (re-)selection-related procedures are supported for resource allocation Mode 2.

The definition of SL resource allocation Mode 2 covers:

2(a). UE autonomously selects SL resource for transmission

2(b). UE assists SL resource selection for other UE(s), a functionality

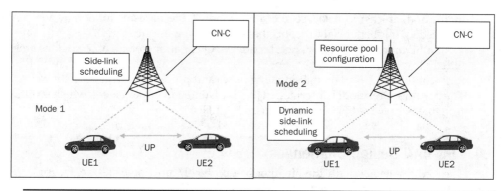

FIGURE 5.33   V2X Mode 1 versus Mode 2 operation

2(c). UE is configured with NR configured grant (Type-1 like) for SL transmission

2(d). UE schedules SL transmissions of other UEs

For Mode 2(a), the study considers SL sensing and resource selection procedures, in the context of a semi-persistent scheme where resource(s) are selected for multiple transmissions of different TBs and a dynamic scheme where resource(s) are selected for each TB transmission. The following techniques are studied to identify occupied SL resources:

- Decoding of SL control channel transmissions
- SL measurements
- Detection of SL transmissions

For Mode 2(c), for out-of-coverage operation, we assume a configuration of single or multiple SL transmission patterns, defined on each SL resource pool, whereas for in-coverage operation, Mode 2(c) assumes that gNB configuration indicates single or multiple SL transmission patterns, defined on each SL resource pool. If there is a single pattern configured to a transmitting UE, there is no sensing procedure executed by UE, while if multiple patterns are configured, there is a possibility of a sensing procedure.

For Mode 2(d), in the context of group-based SL communication, it is supported that for UE-A to inform its serving gNB about members UE-B, UE-C, and so on of a group, and for the gNB to provide individual resource pool configurations and/or individual resource configurations to each group member through UE-A. UE-A cannot modify the configurations, and there is no direct connection required between any member UE and the gNB. Higher-layer only signaling is used to provide the configurations.

5.3.4   QoS Aspects

QoS management is relevant to V2X in the context of its use in resource allocation, congestion control, in-device coexistence, power control, and SLRB configuration. Physical layer parameters related to QoS management are the priority, latency, reliability, and minimum required communication range (as defined by higher layers) of the traffic being delivered. Data rate requirements are also supported in the AS. An SL congestion metric and, at least in resource allocation mode 2, mechanisms for congestion control are needed. It is beneficial to report the SL congestion metric to gNB.

For SL unicast, groupcast, and broadcast, QoS parameters of V2X packets are provided by upper layers to the AS. For SL unicast, groupcast, and broadcast the SLRBs are (pre-)configured based on the signaling flows and procedures which are shown in [54]. The per-flow QoS model described in [56] is assumed in upper layers, which is different from the per packet QoS model discussed in LTE-V.

5.4   5G RAN Design Challenges

The key challenges with the air interface protocol/functional design focused on the entire protocol stack are highlighted as follows.

5.4.1   RAN Internal Challenges

In 5G, PHY and MAC configurations will need to be specific to service characteristics in many scenarios. If multiple services are active in a UE, this UE cannot schedule logical channels only using prioritized bit rate as it is currently done in LTE. For example, if the UL grant is for eMBB then URLLC packet should not be scheduled in this grant as URLLC packet transmission requirement cannot be met by the MAC layer configuration (e.g., HARQ) and physical layer configuration (e.g., numerology) for eMBB. So, some mechanism (other than prioritized bit rate as used in LTE) may be needed to map the UL grant to one or more logical channels. Moreover, the modeling of a MAC entity (as single or multiple MAC entities) is an open issue. In solutions where a single MAC scheduler cannot by design handle packets from multiple numerologies, multiple MAC entities may be required. However, a single MAC entity seems entirely appropriate if we assume a single MAC scheduler is used to handle packets from multiple numerologies. This solution has the benefit of forward compatibility, meaning that there is no need to add a new MAC entity whenever a new numerology is introduced. One other key outstanding issue is how to enable MAC to differentiate between different numerologies (the modeling of MAC entity will rely heavily on this). 3GPP has recently agreed that for multiple numerologies in PHY, at least the TTI length of the numerology(s) will be visible to MAC; however, which (if any) additional characteristics of the numerology are also visible to MAC is an open research area. The current version of LTE HARQ protocol can be fast; however, it is not reliable enough to meet 5G requirements. In order to compensate that, RLC is required to ensure reliability which renders additional latency. To be able to meet 5G requirements, the HARQ protocol has to be much faster, with lower overhead, be more reliable, and operate on a flexible timing base.

In 5G (unlike in LTE), HARQ parameters may need to be configured differently for different services. Similarly, not every access scheme is applicable for all services. Therefore, radio-bearer specific Layer 2 (L2) configuration in 5G may need to include the configuration of HARQ parameters and multiple access schemes.

Packet Data Convergence Protocol (PDCP)/RLC layering in LTE might not be flexible enough to meet diverse KPIs. Functionalities like redundant header processing, reordering, and duplication detection in both layers might provide additional overhead and might not be required by all services.

Finally, in case of multi-connectivity (which is a very key scenario in 5G, e.g., with below and above 6 GHz AIVs), different functional/bearer split options will introduce new challenges regarding the interaction of functions. For example, if retransmissions are handled in RLC and PDCP and PDCP has multiple RLC entities, there should be coordination for retransmissions in different protocol entities.

5.4.2   Architecture Challenges

To this end, some key challenges regarding the conventional architecture which necessitate the proposal of new service-oriented design are highlighted. One of the key challenges of legacy LTE architecture is that it cannot easily adapt to transport network performance, since due to the various access technologies that can be incorporated at 5G systems, we may require different provisioning of access network resources and functionalities based on the changing transport performance. A future RAN should be able to flexibly adapt to different characteristics of the underlying physical networks. This includes different options on where which parts of the processing should take place (at a central or a decentralized physical entity) and a flexible upgrade of selected NFs.

Another important challenge is the way that end-to-end QoS is handled, assuming multiple AIVs and service requirements. 5G QoS framework needs to be more flexible and dynamic with finer granularity to better support new requirements. Some examples include services with variable packet importance (i.e., different coding gain in multimedia services) or prioritizing specific AIV in multi-connectivity scenarios. Furthermore, to increase the system efficiency it may be beneficial to aggregate flows with the same QoS in a single RAN radio bearer (in LTE this happens at the CN). Finally, RAN-CN interface needs to be adapted to also consider the 5G service requirements. For example, for mIoT scenarios (e.g., sensors) new interface requirements need to be considered to cope with the high signaling cost of establishing dedicated tunnels in the 5G CN.

5.5   Toward AI-Driven RAN Deployments

With its inherent characteristics and with the availability of significant amount of data related to the operational parameters of the network, the RAN may greatly benefit from data analytics in the 5G era. As mentioned above, such inherent characteristics which motivate the use of data analytics are the scarce and expensive resources (e.g., spectrum and computational resources), the dynamic wireless channel conditions (e.g., fading and user mobility), the necessity for low complexity in RAN deployments, along with the wide-spread utilization of higher frequency bands that are even more susceptible to radio conditions. In addition, the RAN can be shared by a multitude of network slices, where the essential slicing objectives, such as slice isolation, SLA guarantee, and service continuity, shall be fulfilled. To this end, data analytics and its application through machine learning/AI can be utilized in the following use cases.

- Interference Handling: Data analytics such as diagnostic and prescriptive analytics, in particular, can be utilized to enhance the performance of interference mitigation techniques. This can include selection of uplink (UL) power control parameters (e.g., fractional versus full-compensation power control) and optimal number of blank sub-frames in case of time-domain interference coordination. Predictive analytics may support decisions in terms of the configuration of the initial parameter settings, e.g., considering a probable load increase due to group mobility. The utilization of data analytics can thus improve the resource utilization efficiency and reduce the need for frequent parameter adjustments.

- Prediction of Channel Quality: Wireless channels are inherently dynamic in nature, which can be impaired by short-term and long-term fading. Especially, in case of mm-Wave communications the need for directional transmission introduces additional challenges where transmission collisions, i.e., experiencing high interference due to concurrent mm-Wave transmissions, can easily result in radio link failures (RLFs) [47]. Therefore, predictive analytics can be employed to decide on the active mm-Wave links such that the transmission collisions are minimized. To this aim, context information, e.g., geographical locations of the mm-Wave APs, antenna configurations, and power control parameter settings, can be utilized.

- Control of Dynamic Radio Topology: One of the emerging 5G concepts is the use of unplanned dynamic small cells, where, for instance, a relaying functionality can be integrated into vehicles in the form of vehicular nomadic nodes (VNNs). VNN can be activated on-demand to adapt to the spatially and temporally changing traffic demands. Such deployment is dynamic in nature and requires tight network control, where the activation or deactivation of VNNs can depend on the channel conditions, vehicle battery, and the demand in a target service region. Descriptive analytics can provide the needed context information to a dynamic RAN control unit, which determines the active VNNs. Moreover, predictive analytics can process the parking statistics in a certain region, which can be in turn used to predict the parking duration that sets an upper limit for the VNN availability.

- Energy Efficiency Improvements: Turning off the under-loaded cells fully or partially used jointly with load balancing is an effective way for improving the network energy efficiency. Predictive analytics can be utilized to determine the expected load of a cell or cell group such that the optimal trade-off between resultant energy efficiency gains versus user performance reduction due to higher load of the active cells can be reached. The predictive analytics can, for instance, factor in the load history and activities in the neighbor cells.

- Multi-Slice Resource Management Enhancements: As mentioned above, it is envisioned that RAN will be shared by a multitude of network slices with possibly diverse performance requirements, business-driven additional requirements, and slice-specific protocol configurations. This requires a common multi-slice resource management coupled with a tight performance monitoring to fulfill SLAs of the network slices, while ensuring resource isolation between slices. The resource isolation can be in time, frequency, code, and computational domains. Besides, slices with mission-critical services impose further constraints on the SLA fulfillment. On this basis, all types of the aforementioned data analytics shall be utilized. Namely, descriptive analytics can be part of the SLA monitoring, where different performance thresholds can be defined to avoid an SLA violation. Diagnostic analytics on the SLA violations can be used as input to prescriptive analytics to adjust the performance thresholds for violation prevention. In addition, predictive analytics can provide insights into slice load variations over time and space exploiting, e.g., the history information, which may be used for semi-static resource allocation and cell range extension parameters.

The use of statistics for optimizing radio resources is not a new concept, since this was used in minimization of driving test (MDT) scenarios. Minimization of drive tests was a feature introduced in 3GPP Release 10 [48] that enables operators to utilize users' equipment to collect radio measurements and associated location information, in order to assess network performance while reducing the OPEX associated with traditional drive tests.

However, in 5G scenarios, the collection and more sophisticated processing of measurements in RAN for optimizing the resource is a necessary feature which will allow for meeting critical 5G KPIs.

To further analyze how data analytics can be applied to optimize the access network, the 3GPP has defined two novel study items in SA5 and RAN3 workgroups, respectively, namely "Study on SON for 5G" [49] and "RAN-Centric Data Collection and Utilization for long-term evolution (LTE) and NR" [50] for Release 16. The use cases (e.g., optimization of capacity/coverage, mobility and load sharing/balancing, energy saving, minimization of drive tests) were primarily inherited from legacy systems. The outcome of the study items should include the identification of requirements and new signaling and interfaces required for SON enhancements for different 5G use cases. In particular, the possible benefits and feasibility of introducing a logical RAN entity/function for data collection/utilization will be investigated in 3GPP RAN working group 3, which could be interpreted as the control or management functionality.

Furthermore, the authors in [51] provide an end-to-end analytics service-based architecture which defines an RAN Data Analytics Function (RAN-DAF). Real-time analytics are required for improving RAN NFs, like RRM. Since the RAN needs to provide fast decisions, the analytics based on the processing of real-time measurements may need to stay local for optimizing performance dynamically. Also, there are the business aspects, which may involve different stakeholders among RAN, CN, and Management. So, the storage and analysis of radio-related measurement may be restricted to be abstracted to CN or OAM.

References

1. S. C. Jha, K. Sivanesan, R. Vannithamby, and A. T. Koc, "Dual Connectivity in LTE small cell networks," IEEE Globecom Workshops (GC Wkshps), Austin, TX, 2014, pp. 1205–1210.
2. 3GPP TS 22.261, "Service requirements for next generation new services and markets," v16.5.0, September 2018.
3. 3GPP TS 22.186 "Enhancement of 3GPP support for V2X scenarios; Stage 1," v15.2.0, September 2017.
4. Next Generation Mobile Networks (NGMN) Alliance, "5G White Paper," February 2015.
5. 3GPP TR 38.913, "Study on Scenarios and Requirements for Next Generation Access Technologies," March 2016.
6. T. Wang, G. Li, J. Ding, Q. Miao, J. Li, and Y. Wang, "5G Spectrum: Is China Ready?," IEEE Communications Magazine, vol. 53, no. 7, pp. 58–65, July 2015.

7. S. Parkvall, E. Dahlman, A. Furuskar, and M. Frenne, "NR: The New 5G Radio Access Technology," IEEE Communications Standards Magazine, vol. 1, no. 4, pp. 24–30, 2017.

8. V. Valls, A. Garcia-Saavedra, X. Costa, and D. J. Leith, "Maximizing LTE Capacity in Unlicensed Bands (LTE-U/LAA) While Fairly Coexisting with 802.11 WLANs," IEEE Communications Letters, vol. 20, no. 6, pp. 1219–1222, June 2016.

9. M. Labib, V. Marojevic, J. H. Reed, and A. I. Zaghloul, "Extending LTE into the Unlicensed Spectrum: Technical Analysis of the Proposed Variants," IEEE Communications Standards Magazine, vol. 1, no. 4, pp. 31–39, December 2017.

10. "5G Spectrum: GSMA Public Policy Position," July 2019. Available online: https://www.gsma.com/spectrum/wp-content/uploads/2019/09/5G-Spectrum-Positions.pdf

11. 3GPP TS 38.300, "NR; Overall description," v15.6.0, June 2019.

12. 3GPP TS 38.410, "NG-RAN; NG general aspects and principles," v15.2.0, January 2019.

13. 3GPP TS 38.420, "NG-RAN; Xn general aspects and principles," v15.2.0, January 2019.

14. E. Pateromichelakis, et al., "Service-Tailored User-Plane Design Framework and Architecture Considerations in 5G Radio Access Networks," in IEEE Access, vol. 5, pp. 17089–17105, 2017.

15. 3GPP TS 38.401, "NG-RAN; Architecture description," v 15.1.0, April 2018.

16. A. Prasad, F. S. Moya, M. Ericson, R. Fantini, and O. Bulakci, "Enabling RAN Moderation and Dynamic Traffic Steering in 5G," IEEE 84th Vehicular Technology Conference (VTC-Fall), Montreal, QC, 2016, pp. 1–6.

17. 3GPP TR 38.801, "Study on new radio access technology: Radio access architecture and interfaces," v14.0.0, April 2017.

18. 3GPP TS 37.340, "NR; Multi-connectivity; Overall description; Stage-2," v15.6.0, June 2019.

19. 3GPP TS 36.300, "E-UTRA and E-UTRAN; Overall description," Release 13, v13.2.0, 2015.

20. S. R. Khosravirad, G. Berardinelli, K. I. Pedersen, and F. Frederiksen, "Enhanced HARQ Design for 5G Wide Area Technology," VTC-Spring 2016.

21. R. Trivisonno, R. Guerzoni, I. Vaishnavi, and D. Soldani, "SDN-based 5G mobile networks: architecture, functions, procedures and backward compatibility," Transactions on Emerging Telecommunications Technologies (ETT), November 2014.

22. 3GPP TS 38.423, "NG-RAN; Xn Application Protocol (XnAP)," Release 15, v15.6.0, 2020.

23. A. Prasad, A. Maeder, and C. Ng, "Energy Efficient Small Cell Activation Mechanism for Heterogeneous Networks," IEEE Global Communications Conference Wkshps (GCW), Atlanta, December 2013, pp. 754–759.

24. A. Prasad, and A. Maeder, "Energy Saving Enhancements for LTE-Advanced Heterogeneous Networks with Dual Connectivity," 80th IEEE Vehicular Technology Conference, Vancouver, Canada, Sept. 2014, pp. 1–6.

25. 5G PPP Architecture Working Group – View on 5G architecture, version 1.0, 2016-07; source: https://5g-ppp.eu/wp-content/uploads/2014/02/5G-PPP-5G-Architecture-WP-July-2016.pdf

26. P. Rost, A. Banchs, I. Berberana, M. Breitbach, M. Doll, H. Droste, et al., "Mobile Network Architecture Evolution toward 5G," in IEEE Communications Magazine, vol. 54, no. 5, pp. 84–91, May 2016.

27. 3GPP TS 23.214, Architecture enhancements for control and user plane separation of EPC nodes, Release 14, v14.2.0, March 2017.

28. H. Zhang, et al., "5G wireless network: MyNET and SONAC," in IEEE Network, vol. 29, no. 4, pp. 14–23, July–August 2015.

29. 3GPP TR 38.804. Study on New Radio Access Technology; Radio Interface Protocol Aspects V14.0.0, 2017.

30. Y. Li, E. Pateromichelakis, N. Vucic, J. Luo, W. Xu, and G. Caire, "Radio Resource Management Considerations for 5G Millimeter Wave Backhaul and Access Networks," Communications Magazine, IEEE, June 2017, Special Issue on Agile Radio Resource Management Techniques for 5G New Radio.

31. 3GPP TS 36.331, "Evolved Universal Terrestrial Radio Access (E-UTRA); Radio Resource Control (RRC); Protocol specification," v15.6.0, June 2019.

32. 3GPP TS 38.331, "NR; Radio Resource Control (RRC); Protocol specification," v15.6.0, June 2019.

33. T. O. Olwal, K. Djouani, and A. M. Kurien, "A Survey of Resource Management towards 5G Radio Access Networks," in IEEE Communications Surveys & Tutorials, vol. PP, no. 99, pp. 1656–1686.

34. D. Sabella, et al., "RAN as a service: Challenges of designing a flexible RAN architecture in a cloud-based heterogeneous mobile network," Future Network and Mobile Summit (FutureNetworkSummit), 2013, Lisboa, 2013, pp. 1–8.

35. D. H. Lee, K. W. Choi, W. S. Jeon, and D. G. Jeong, "Two-Stage Semi-Distributed Resource Management for Device-to-Device Communication in Cellular Networks," in IEEE Transactions on Wireless Communications, vol. 13, no. 4, pp. 1908–1920, April 2014.

36. 5G PPP WG Architecture, Architecture White Paper v2.0, "View on 5G Architecture," December 2017.

37. F. Zarrar Yousaf, and T. Taleb, "Fine-grained resource-aware virtual network function management for 5G carrier cloud," IEEE Network 30.2, pp. 110–115, 2016. TS 23.791.

38. E. Pateromichelakis, and C. Peng, "Selection and Dimensioning of slice-based RAN Controller for adaptive Radio Resource Management," Wireless Communications and Networking Conference (WCNC), 2017.

39. V. Sciancalepore, K. Samdanis, X. Costa-Perez, D. Bega, M. Gramaglia, and A. Banchs, "Mobile Traffic Forecasting for Maximizing 5G Network Slicing Resource Utilization," IEEE INFOCOM, 2017.

40. Y. Li, et al., "Resource Management Considerations for 5G millimeter-Wave Backhaul/Access Networks," IEEE Communications Magazine Special Issue on Agile Resource Management in 5G, July 2017.

41. E. Pateromichelakis, K. Samdanis, Q. Wei, and P. Spapis, "Slice-tailored Joint Path Selection & Scheduling in mm-Wave Small Cell Dense Networks," IEEE Globecom, 2017.

42. 3GPP TS 38.874, "NR; Study on integrated access and backhaul," V16.0.0, January 2019.

43. 3GPP TS 38.460, "NG-RAN; E1 general aspects and principles," v15.2.0, January 2019.

44. A. Sutton, "5G Network Architecture," J. Inst. Telecommun. Professionals, vol. 12, no. 1, pp. 9–15, 2018.

45. 3GPP TS 38.470, "NG-RAN; F1 general aspects and principles," v15.7.0, January 2020.

46. A. de la Oliva, J. A. Hernandez, D. Larrabeiti, and A. Azcorra, "An overview of the CPRI specification and its application to C-RAN-based LTE scenarios," IEEE Communications Magazine, vol. 54, no. 2, pp. 152–159, February 2016.

47. P. Marsch, O. Bulakci, O. Queseth, and M. Boldi, "5G System Design: Architectural and Functional Considerations and Long Term Research," Wiley, April 2018, ISBN:978111942514.

48. W. A. Hapsari, et al., "Minimization of Drive Tests Solution in 3GPP," IEEE Communications Magazine, vol. 50 (6), pp. 28–36, 2012.

49. 3GPP TR 28.861, "New Study on Self-Organizing Networks (SON) for 5G," v.0.2.0, December 2018.

50. 3GPP RP-182105, "Study on RAN-centric data collection and utilization for LTE and NR," September 2018.

51. E. Pateromichelakis, et al., "End-to-End Data Analytics Framework for 5G Architecture," in IEEE Access, vol. 7, pp. 40295–40312, 2019.

52. 3GPP TR 36.843, "Study on LTE Device to Device Proximity Services," V12.0.1, March 2014.

53. 3GPP TS 23.303, "Proximity-based services (ProSe); Stage 2," V15.1.0, June 2018.

54. 3GPP TS 38.885, "Study on NR Vehicle-to-Everything (V2X)," v16.0.0, March 2018.

55. 3GPP TS 37.324, "Evolved Universal Terrestrial Radio Access (E-UTRA) and NR; Service Data Adaptation Protocol (SDAP) specification," V15.1.0, September 2018.

56. 3GPP TS 23.287, "Architecture enhancements for 5G System (5GS) to support Vehicle-to-Everything (V2X) services," V2.0.0, August 2019.

57. 3GPP TS 38.201, "NR Physical layer; General description," V16.0.0, December 2019.

Problems/Exercise Questions

1. What are the key 5G requirements from a radio access network perspective?

2. Can 5G support a multitude of requirements simultaneously to the end users such as URLLC and eMBB?

3. What are the key differences between 4G and 5G in terms of spectrum bands that are supported? What was the underlying reason for this difference to occur?

4. What is the difference between eNB and ng-eNB?

5. Provide an overview of the 5G RAN architecture, including a detailed description of the radio protocol stack.

6. Describe in detail the various deployment options that are available for the network operators utilizing centralized 5G RAN architecture.

7. Describe the connected mode mobility procedure for 5G RAN.

8. Describe the QoS flow handling in 5G in detail. What are some of the key advantages of the 5G QoS architecture for delivering highly diverse set of services?

9. Describe briefly the key functions of the layer-2 protocol stack.

10. Explain the signaling aspects related to energy saving enhancements for 5G RAN.

11. Provide an overview of the 5G RAN user- and control-plane design.

12. What does service-oriented user-plane design imply? Provide an architectural overview of the service-oriented design considerations in 5G RAN.

13. What are some of the key advantages and disadvantages of different RRM placements for mm-Wave networks?

14. What is "slice-awareness"? Provide an overview of some of the key applications for slice-awareness.

15. What is slow and fast RRM in the context of mm-Wave networks?

16. Describe briefly the two scenarios considered for NR V2X.

17. Briefly explain the QoS aspects of the 5G V2X communication feature.

18. What are some key enhancements for D2D and V2V operation?

19. What are the key 5G RAN design challenges?

20. Briefly explain the minimization of drive test (MDT) feature.

21. What is the role of AI in future RAN deployments?

22. Describe the multi-slice resource management enhancements that are possible with the utilization of advanced machine learning techniques.

5G Core Network Architecture

6.0   Introduction

The two main elements of the operator's overall mobile network infrastructure discussed in this textbook are the radio access network (RAN) and the core network (CN). Then there is a third element, called the Transport Network, which is a transmission medium for interconnecting spatially separated entities of the overall operator's mobile network infrastructure—interconnecting remote radio cell towers to the radio network controller (RNC), the RNC to remote CN entities, or the CN entities to remote data networks, and so on. The details of the transport network are not covered in this textbook.

The goal of this chapter is to first provide high-level technical analysis of the early 3GPP CN architectures followed by more detailed technical analysis of the key 3GPP 5G Core Network (5G CN or 5GC) functions/capabilities. The evolution of CN technologies has continued from the first generation (1G) to the fifth generation (5G) in parallel with the evolution of radio technologies, as highlighted in Fig. 6.1 (see also Chap. 3 for additional details). The key driver for necessitating the gradual development of CN technologies over time is the continued introduction of new requirements, which in turn emerged the employment of new capabilities. The intent here is to give readers a high-level understanding of how mobile CN technologies have evolved from 1G to 5G, and the rationale behind this continued evolution. To this end, the overall 5GC functions/capabilities will be covered in high level; and subsequently some selected key 5GC capabilities will be explicitly described due to their increased significance.

Figure 6.1 provides a high-level view on how the key CN entities have evolved over time. The various 3GPP CN entities are shown in this figure, and their capabilities will be further elaborated in later sections of this chapter.

6.1   Mobile System Domains

To better capture the overall makeup of the mobile system, Fig. 6.2 [1] illustrates the domain constituents of the mobile systems and shows how its major components are partitioned into different domains. The mobile system domains have been defined since 2000 as part of the general 2G/3G Universal Mobile Telecommunications Service (UMTS) architecture, consisting of two main domains—the user equipment domain (basically the mobile phone including SIM or USIM card) and the infrastructure domain. The infrastructure domain is further divided into the access network domain and the core network domain. The access network domain relates to radio access network architecture discussed in Chap. 5. The focus of this chapter is to provide details of the core network domain.

The SIM or USIM cards are provided to the end users by the mobile operator at subscription time, so as to be inserted into their mobile devices (see Fig. 6.3 and Chap. 3 for details). These cards consist of a small microprocessor chip which is capable of storing network information (e.g., mobile network identifier) and subscriber information

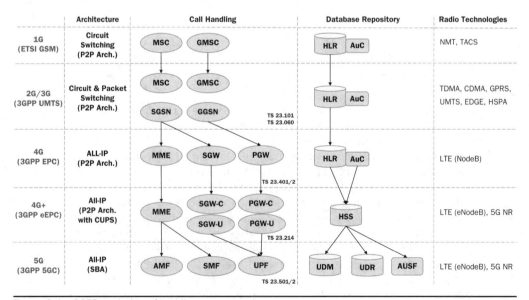

FIGURE 6.1   3GPP evolution of mobile core network

(e.g., IMSI). SIM stands for *Subscriber Identity Module* which was introduced to operate on GSM networks providing user connectivity, pre-smartphone days. USIM stands for *Universal Subscriber Identity Module* which was introduced to operate on 2G/3G UMTS networks providing user connectivity and able to support smartphones. The SIM cards were converted into USIM cards with the introduction of UMTS. USIM is advanced version of SIM with a much larger storage capability. Consumers know them as a SIM card or smart card.

The interface between SIM/USIM and mobile equipment is not covered in this chapter. For further details, one can refer to 3GPP TS 11.11 [2] and 3GPP TS 31.101 [3].

The core network domain consists of the serving network domain, the home network domain, and the transit network domain:

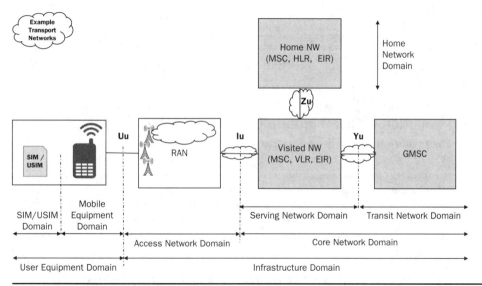

FIGURE 6.2   Domains of the mobile system

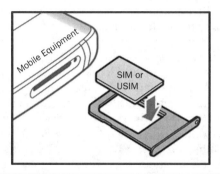

FIGURE 6.3   SIM or USIM usage

- The serving network domain connects to the access network domain that provides access for the users/subscribers. The serving network domain keeps track of the location of all active users and is responsible for routing calls and transporting user data/information from source to destination.

- The home network domain represents the part of core network functions that are resident at a permanent location in the area where the user resides (called the home network). The USIM card is provided to the user at subscription time, and the information contained in the USIM card is permanently stored in the home network domain for verification and safe keeping. The home network domain, therefore, is responsible for management of the user's subscription information.

- The transit network domain provides a communication path between the serving network domain and the destination party. The transit network domain does not come into play if the remote party is located inside the same network as the originating party.

The aforementioned mobile system domains interact with each other, as the user data flows through the different domains from source to destination. This will become evident in different sections of this chapter, specifically in the core network procedures (Sec. 6.7).

6.2 Evolution of 3GPP Mobile Core Network Architectures

The evolution of mobile core network has continued in 3GPP, starting from the GSM-based Release 1999 standard. With the completion of each 3GPP release, the mobile system undergoes some capability enhancements, which may lie within a technology or introduce a technology advancement. The following sections describe the major milestones in mobile core network evolution as defined and specified in the different 3GPP releases. Only the 3GPP releases that had major technology advancement are explicitly mentioned and complement the Chap. 3 background information from the architecture perspective.

6.2.1 3GPP Release 1999 (UMTS—to Support 2G/3G UTRAN Radio Technology)

The first 3GPP-defined mobile core network standard known as Release 1999 UMTS was completed in the first quarter of 2000, which is also referred to as 2G/3G. It was essentially a consolidation of the underlying GSM specifications (standardized by ETSI) along with the development of the new UMTS Terrestrial Radio Access Network (UTRAN) radio air interface. A new radio air interface technique was developed for UTRAN called WCDMA, instead of adopting the TDMA technique of the GSM air interface. The WCDMA air interface technique provided much faster speeds of up to 384 kbits/s in the downlink (network to user) direction and speeds of 64–128 kbits/s in the uplink direction (user to network). The Release 1999 standard contained a radio access part (UTRAN), which was a new radio interface, and the core network part (UMTS), which was inherited from ETSI-based GSM core network. The radio access part consists of (1) the base stations (aka NodeBs), which provide the radio signaling, and (2) the radio network controller (RNC), which serves as the centralized intelligence

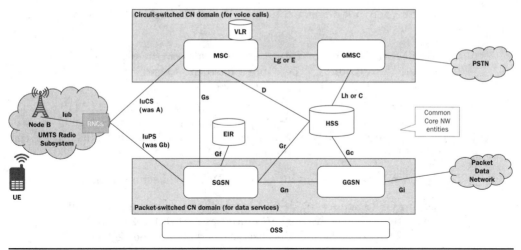

FIGURE 6.4   2G/3G UMTS architecture

to manage the NodeBs. RNC also manages the mobile signaling handover between NodeBs as the mobile device crosses NodeB cell boundaries.

The ETSI-based GSM circuit-switched (CS) core network technology supporting voice traffic, along with the General Packet Radio Service (GPRS) technology supporting data traffic, was used as a basis of Release 1999 UMTS. The core network consisted of two parts—a CS domain and a packet-switched (PS) domain. The CS domain provided circuit connection between two endpoints and was mostly used for real-time services (e.g., two-way voice calls, three-way voice conferencing). The PS domain consisted of routers for routing data packets between two endpoints and was mostly used for non-real-time services (e.g., messaging, web access). A further constituent of the core network was the home location register (HLR) which contained user subscription information, user profile, and the current location of the user and authentication center (AuC). In the UMTS architecture, the HLR and AuC were combined to form the home subscriber server (HSS). The main enhancements for UMTS were the interfaces required for interconnecting the mobile switching center (MSC) and the serving GPRS support node (SGSN) with the RNC.

The Release 1999 UMTS core network supported 2G/3G radio technologies. The UMTS architecture is shown in Fig. 6.4, identifying its key core network entities. The biggest difference in UMTS, compared to its ETSI GSM/GPRS counterpart, was the Gb interface from UE[1] to base station (not shown), renamed as IuPS, which uses ATM instead of frame relay on lower layers of the protocol stack. The other core network entities remained the same.

---

[1]There are two terms used in this chapter when referring to a mobile device. These are MS (mobile station) or UE (user equipment). Both can be used interchangeably but UE is most commonly used now. The term MS was used as a 2G terminology, mostly in GSM networks standardized by ETSI, prior to networks based on 3GPP standards. So, MS was an ETSI terminology for a mobile device, while the term UE was introduced by 3GPP in Release 1999 UMTS, to differentiate from terms used by ETSI.

The CS core network domain entities in the UMTS architecture are described as follows:

- Mobile Switching Center (MSC): This provides circuit switching and manages all voice calls requiring a permanent fixed end-to-end circuit connection from source to destination for the duration of the voice call.
- Gateway MSC (GMSC): This is a gateway MSC connecting the UMTS network to external CS networks in PSTN. For incoming calls from PSTN, this entity is used to query the HLR/HSS to find out which MSC a UE is attached to. The GMSC then routes the call towards that MSC.

The PS core network domain entities in the UMTS architecture are described below:

- Serving GPRS Support Node (SGSN): As the name implies, this entity was first developed when GPRS was introduced as part of the ETSI GSM standard to handle PS calls, and its use has been carried over to the 2G/3G UMTS mobile network architecture defined by 3GPP. The SGSN provides numerous functions within the UMTS mobile network architecture, as follows:
  - Mobility Management (MM): When an MS (or UE) attaches to the PS domain of the UMTS core network, the SGSN provides subscriber location information based on the mobile's current location, which is used by the MM functions to trace physical presence of the subscriber. This information is used to deliver an incoming voice call or an SMS text message to the subscriber using the paging function of the system.
  - Session Management (SM): It is used to manage the UE data sessions by applying the SGSN-provided quality of service required for the data sessions. It also manages what are termed as the PDP (Packet Data Protocol) contexts, i.e., the pipes over which the data is sent.
  - Charging: The SGSN also collects charging-related data (Charging Data Record—CDR) for the data sessions in progress for subscriber billing purposes. CDR is a formatted collection of information about the charges applicable to the session (or event) in progress.
- Gateway GPRS Support Node (GGSN): Like the SGSN, this entity was also first introduced into the GPRS network. It handles interworking between the UMTS PS mobile core network and external PS networks (e.g., internet), and is basically a data packet router. In operation, when the GGSN receives data addressed to a specific user, it checks if the user is active and then forwards the data to the SGSN serving the particular UE. It also handles data packet address conversion.

The shared common entities of the 2G/3G UMTS mobile core network architecture are as follows:

- Home Location Register (HLR): This is the database containing all the administrative information about each subscriber along with their last known location. The other entities of the mobile core network interact with HLR to access subscriber information in order to route calls to the relevant RNC/NodeB where the subscriber is currently located. The subscriber location information is populated in the HLR when subscribers switch on their UE to register.

Even when the UE is idle but switched on, it re-registers periodically to ensure that the HLR is populated with the latest location of the UE. The HLR is also equipped with an authentication function as a separate entity or integrated, which is handled by the AuC.

- Equipment Identity Register (EIR): The EIR is the entity that decides whether a given UE may be allowed onto the network based on a valid International Mobile Equipment Identity (IMEI). This equipment identifier is installed in the mobile equipment at the time of manufacture and can be checked by the network during registration or at other times which are configured by the operator.

- Authentication Center (AuC): The AuC is a secure protected database which is closely linked to HLR providing the generation of secret keys, based on subscriber's IMSI. These secret keys are used to encrypt the user data for transmitting over the radio interface, so it cannot be spoofed.

- Operational Support System (OSS): This provides the overall network management of the core network entities, fault management, service provisioning, etc. The details of OSS are not provided in this textbook.

6.2.2   3GPP Release 4 (UMTS All-IP CN to Support 3G Radio Access Technology UTRAN)

The 3GPP Release 4 specifications completed in second quarter of 2001 and consisted of enhancements done on the Release 1999 UMTS architecture [1]. One of the major enhancements was the bearer independent CS architecture (also called BICC—Bearer Independent Circuit-Switched Core Network) [4], as shown in Fig. 6.5. This enhancement allowed the transport and the control signaling to be disassociated for better transport resource efficiency and a convergence with the PS domain transport. Also, this enabled the use of one single set of GSM layer-3 protocols, e.g., Direct Transfer Application sub-Part (DTAP) outlined in 3GPP TS 24.008 [5], or Mobile Application Part (MAP) described in 3GPP TS 29.002 [6], on top of different transport resources, such as ATM, IP, STM (Synchronous Transfer Mode), or even new ones. DTAP is used to transfer messages between the MSC and the UE. MAP is used between two connecting MSCs or MSC and HSS to transfer subscriber profile information (IMSI, telephone number, services subscribed to, etc.) and authentication-related information.

This implies that both UMTS CS and PS domains offer the same bearer and teleservices (see Chap. 2 for details), and have the same external behavior for the handling of call control, related supplementary services, application services, and mobility support. Also, none of the protocols used on the radio interface were modified by this enhancement, requiring existing legacy terminals to be supported. It should be noted that the support of legacy terminals is required for several years with any mobile system enhancement.

The basic principle of Release 4 BICC architecture is that the MSC is split into an MSC server and a CS Media Gateway (CS-MGW), with external interfaces remaining the same as the monolithic MSC. Similarly, the GMSC is split into a GMSC server and a CS-MGW. The GMSC server provides the call control and mobility management functions, and the CS-MGW provides the bearer control and transmission resource functions.

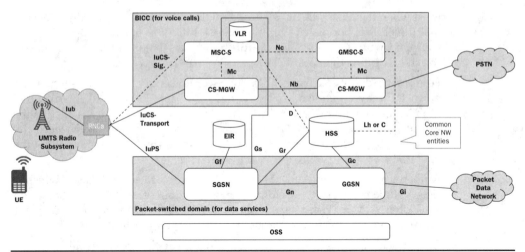

Figure 6.5   Release 4 UMTS BICC core network architecture

6.2.2.1   BICC Core Network Entities

The BICC core network entities are described below.

MSC Server: The MSC server comprises the call control and mobility control parts of an MSC, including the VLR which holds the mobile subscriber's supported services and CAMEL (Customized Applications for Mobile networks Enhanced Logic)[2]- related data. As such, it is responsible for the control of mobile-originated (MO) and mobile-terminated (MT) CS domain calls. It terminates the user-to-network signaling [5] and translates it into the relevant network to network signaling [6]. The MSC server controls the parts of the call state that pertain to connection control for media channels in a CS-MGW.

GMSC Server[3]: The GMSC server provides similar functionality as the MSC server. It provides control of MT CS domain calls as the Gateway MSC server is the entry point into the network from the PSTN.

CS-MGW: The CS-MGW is the interfacing entity for the transport control part of the UMTS radio subsystem with the core network, over the Iu transport interface. It interacts with the MSC and GMSC servers for resource control. It may also terminate bearer channels from a CS network and media streams from a PS network (e.g., RTP streams in an IP network). As the interfacing entity between the access and the core network, the CS-MGW operates the requested media conversion (it contains, e.g., the TRAU—Transcoder and Rate Adaptation Unit that performs transcoding function for speech channels and rate adaptation for data channels), the bearer control, and the payload processing (e.g., codec, echo canceller, and conference bridge). It supports the different Iu options for CS services [for both ATM Adaptation Layer 2 (AAL2)/ATM-based as well as RTP/UDP/IP-based protocols]. The CS-MGW bearer control and payload processing capabilities also need to support mobile-specific functions such as Serving

---

[2]CAMEL is a set of standards based on Intelligent Network and designed to work on the older ETSI-based GSM core network and on the 3GPP-based UMTS network.
[3]An MSC server may also be a GMSC if the mobile originated call being handled by the MSC server is calling another UE on the same network.

Radio Network Subsystem (SRNS) relocation/handover and anchoring applicable during mobility scenarios. This is enabled by applying current H.248.1 [7] standard mechanisms. Further tailoring (i.e., packages) of the H.248.1 may be required to support additional codecs and framing protocols, etc.

6.2.2.2   BICC Core Network Interfaces and Protocols

The BICC core network interfaces and protocols used are described below.

Mc: uses H.248.1/IETF Megaco protocol [7] jointly developed by ITU-T and IETF, with the parameters and options specified in 3GPP TS 29.232 [Media Gateway Controller (MGC)–Media Gateway (MGW) interface] [8].

Nc: supports IP and ATM transports in a bearer-independent manner for the ISDN service set, allowing the physical separation of the call control entities from the bearer control entities, hence the name "Bearer-Independent Call Control." Refer to 3GPP TS 29.205 [9].

Nb: supports bearer control and transport, as defined in 3GPP TS.29.414 [10]. It can be IP bearer control protocol, BICC tunneling protocol, or ATM Adaption Layer Type 2 (AAL2) signaling protocol as specified in ITU-T Recommendation Q.2630.1 [11].

6.2.3   3GPP Release 8 (EPC to Support 3G Radio Access Technology LTE)

As mentioned above, in ETSI GSM (1992), the architecture relied on CS network technology and provided voice services along with Short Message Service (SMS). With CS a dedicated voice path between the calling and called party is established. In 1998, PS network known as General Packet Radio Service (GPRS) was added on top of the GSM CS network to support mobile packet data services. With GPRS technology, data could be transported in packets without the establishment of dedicated circuits, providing flexibility and efficiency in data transmission. Therefore, the core network was composed of two domains: circuit (GSM) and packet (GPRS). In 3GPP UMTS (2G/3G—Release 1999), this dual-domain concept was still maintained on the core network side.

When designing the evolution of the 3G system, the 3GPP community decided to use IP (Internet Protocol) as the key transport protocol for all services. It was therefore agreed that the Evolved Packet Core (EPC—2008 Release 8) would not have a CS domain anymore and that the EPC should be an evolution of the PS architecture used in GPRS/UMTS. This was a major enhancement in core network technology.

6.2.3.1   Architecture of the EPC

EPC network standard was completed by 3GPP in Release 8 (fourth quarter of 2008) and is embodied in the 3GPP TS 23.401 specification [12] to support E-UTRAN CDMA-based 3G radio access technology (known as LTE or eNodeB). It was decided to have a "flat architecture" for EPC to allow efficient handling of payload (the data traffic) from performance and costs perspective. It was designed to support only PS services (including VoIP), and provided seamless IP connectivity between user equipment (UE) and the packet data network (PDN), without any disruption to the end users' applications during mobility. It was also decided to separate the user data traversing the user plane from the signaling data traversing the control plane to make the scaling independent. Thanks to this functional split, the operators could dimension and adapt their core networks easily. The mobile core networks based on EPC standard are deployed widely around the world and have been continuously enhanced until now with each new 3GPP release.

Figure 6.6 shows the evolved packet system (EPS = radio subsystem + EPC) architecture comprising the UMTS and E-UTRAN (called eNodeB) radio subsystems connected to the EPC. The Evolved NodeB (eNodeB or E-UTRAN) is the base station for supporting LTE radio technology. The key entities of the EPC consist of the Serving GateWay (SGW), the Packet data network GateWay (PGW), the mobility management entity (MME), the home subscriber system (HSS), and the policy and charging rules function (PCRF). Another function called traffic detection function (TDF) was also introduced in EPC later on which is used to enforce traffic policies based on preset rules or dynamically determined rules by the PCRF on data flows in real time. The EPC is connected to the external data networks, which can include the IP-multimedia subsystem (IMS), which is explained in Sec. 6.9.

6.2.3.2   EPC Entities

The EPC architecture consists of the following entities. Mobility Management Entity (MME): This provides handling of control-plane signaling related to mobility and security for E-UTRAN access. It is responsible for tracking and paging of the UE in idle mode. It is the termination point of the Non-Access Stratum (NAS) protocol for mobility and session management signaling.

Home Subscriber Server (HSS): This is the database residing in the home network which is based on the UMTS HLR/AuC database. It contains mobile subscriber-related information. It provides support for mobility management, call and session setup, user authentication, and access authorization.

Serving GateWay (SGW): This is the serving gateway residing in the user plane for interconnecting the radio subsystem with the EPC. It is the gateway for routing incoming and outgoing traffic to/from the UE and routes to P-GW. It is the anchor point for the intra-LTE mobility (i.e., in case of handover between eNodeBs) and between LTE and other 3GPP accesses.

Packet GateWay (PGW): This is the PDN gateway used as the point of interconnect between the SGW and the external IP networks. The PGW routes packets to/from the external PDNs or IMS, and performs necessary functions for routing traffic, i.e., IP address/IP prefix allocation. It also interacts with PCRF to apply policy control and charging on the routed traffic. Both S-GW and P-GW reside in the user plane and are responsible for routing IP traffic data between the UE and the external networks.

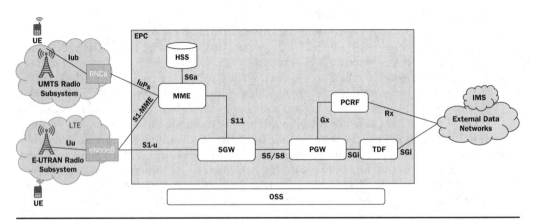

FIGURE 6.6   Release 8 EPC architecture

Traffic Detection Function (TDF): This entity provides the traffic detection function to enforce traffic policies based on preset rules or dynamically determined rules by the PCRF on data flows in real time.

Policy and Charging Rules Function (PCRF): This entity holds operator-controlled policy and charging information to be applied to various types of subscriber data traffic, based on type of services and other factors.

6.2.3.3 *Support of Multiple Access Technologies*

As shown in Fig. 6.6, the EPC supports E-UTRAN (LTE) radio access along with legacy UMTS radio access. With any core network evolution, the legacy radio accesses support is required to cater for support of legacy mobile devices. Along with the support of multiple access technologies, the handover between these accesses is also supported. 3GPP specifications define how the interworking is achieved between an E-UTRAN (LTE and LTE-Advanced), GERAN (radio access network of GSM/GPRS), and UTRAN (radio access network of UMTS-based technologies WCDMA and HSPA). As illustrated in Fig. 6.7, the EPS also allows non-3GPP access technologies to interconnect the UE with the EPC. Non-3GPP means that these accesses were not specified by 3GPP. These technologies include WiMAX®, cdma2000®, WLAN, or fixed networks.

Non-3GPP accesses can be split into two categories—trusted and untrusted. Trusted non-3GPP accesses can interact directly with the EPC. However, untrusted non-3GPP accesses interwork with the EPC via a network entity called the evolved Packet Data Gateway (ePDG). The main role of the ePDG is to provide security mechanisms such as IPsec tunneling for connections with the UE over an untrusted non-3GPP access. 3GPP does not specify which non-3GPP technologies should be considered trusted or untrusted. This decision is left to the operator.

6.2.4   3GPP Release 10 (Enhanced EPC to Support 4G Radio Access Technology LTE Advanced)

The 3GPP Release 10 specifications were completed in the first quarter of 2011 to support International Mobile Telecommunications (IMT) Advanced radio access requirements issued by ITU-R [13], enhancing the existing LTE radio access which was renamed to "LTE Advanced" and marketed as 4G. With LTE Advanced the existing services, such

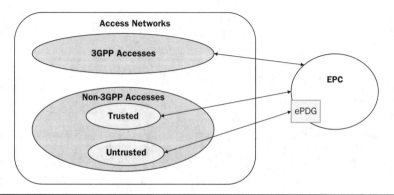

FIGURE 6.7 Support of 3GPP and non-3GPP access networks

as mobile broadband, MMS, and video chat, were enhanced to accommodate higher data rates along with support of new services like HDTV. To cater for enhanced services enabled by LTE Advanced radio technology required EPC to be upgraded as well with enhanced QoS. The basic architecture of Release 8 EPC remained the same for Release 10, but its network functions were enhanced resulting in an evolved EPC. The evolved EPC has been in operation for a long period and is currently deployed worldwide.

6.2.5   3GPP Release 14 (CUPS Feature for Enhanced EPC)

3GPP Release 14, which was completed in 2017, entailed control- and user-plane separation of EPC nodes (CUPS) [14]. The nodes impacted were SGW, PGW, and TDF, as shown in Fig. 6.8. The reasoning behind the separation of control plane (CP) from user plane (UP) was to allow independent scalability of CP and UP functions allowing ease of network resources management for the operator. This split did not impact the design of UE or RAN, and made interworking with other networks easier in roaming scenarios. The CP functional entities in the architecture handle control-plane signaling procedures (e.g., network attachment procedures) and manage creation of paths in the user plane required for delivery of user data traffic. The UP functional entities in the architecture handle packet forwarding of user data traffic based on rules specified by the CP functional elements (e.g., packet forwarding for IPv4, IPv6, or Ethernet).

The -C or -U suffix appended to S5 and S8 reference points only indicates the control-plane and user-plane components of those interfaces. This is a simplified architecture to convey the idea of control- and user-plane separation. The three new interfaces added with the CUPS feature are the following:

Sxa: Reference point between SGW-C and SGW-U.

Sxb: Reference point between PGW-C and PGW-U.

Sxc: Reference point between TDF-C and TDF-U.

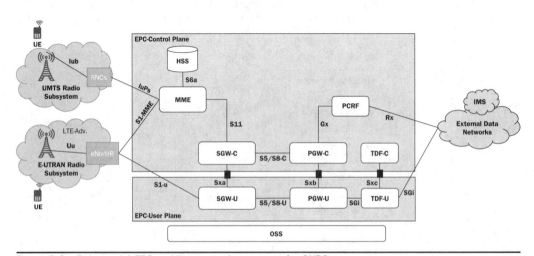

Figure 6.8   Release 14 EPC architecture enhancement for CUPS

Instead of using GTP-U-based protocol for the new Sxa, Sxb, and Sxc interfaces, 3GPP introduced a new protocol called PFCP (Packet Forwarding Control Plane) for these interfaces. PFCP is used between the control-plane and the user-plane functions, and this protocol is carried over for use in 5GC as well. See 3GPP TS 29.244 specification [15] for more details on the use of PFCP for Sxa, Sxb, and Sxc.

6.2.6   3GPP Release 15 (5GC to Support 5G New Radio Access Technology NR)

The 5G mobile core network (5GC) architecture is significantly different from its pre-decessor EPC, utilizing virtualization and cloud native software design. In EPC, the session, connection, and mobility management tasks were all done by MME, but in 5GC these tasks are segregated among AMF and SMF. In addition, 5GC user-plane function (UPF) is quite unique allowing decoupling of packet gateway control- and user-plane functions. The details of 5GC architecture are provided in Sec. 6.5.

Since the 5GS (new radio plus 5GC) is brand new, the evolution of 4G to 5G will be a slow migration based on operator needs. It is not possible to simply deploy 5G new radio (NR) technology without picking a network platform that can deliver end-to-end well-aligned capabilities across devices, RAN, CN, and management systems. Unlike previous generations which required that with each introduction of new radio technol-ogy the core network of the same generation be deployed (e.g., UTRAN with UMTS core or LTE with EPC), but with 5G it is possible to integrate elements of 4G and 5G in different configurations, namely:

- Standalone using only one radio access technology, and
- Non-standalone combining 4G and 5G radio access technologies.

The reason for the gradual migration to 5GC is that the current EPC deployed by operators around the world has been enhanced over several years and is performing well. Furthermore, EPC can partially support the new services (eMBB, eV2X, eCIOT), when connected to NR.

Therefore, it would be important to better capture the motivation which necessi-tates the migration to 5GC. By analyzing the two core network technologies, it can be concluded that 5GC is an evolution of EPC. 5GC will provide enhanced QoS and support URLLC more efficiently and massively, compared to EPC. In the longer run when 5GC will be enhanced and become feature rich, it will eventually replace EPC. But at the time of writing this textbook, EPC will remain and interwork with 5GC. The 3GPP 5G specifications allow provisions for incremental steps in the migration from 4G to 5G giving operators several deployment options [16]. These options are depicted in Fig. 6.9, allowing operators to deploy NR on existing LTE/EPC infrastructure.

6.2.6.1   EPC Evolution to 5GC

As mentioned earlier, the 5GC is essentially an evolution of EPC. EPC was enhanced by CUPS feature in 3GPP Release 14 which separated control plane and user plane. In the development of 5GC architecture, the CUPS architecture was adopted as its basis. Also, EPC nodes over time have been virtualized providing reduction in CAPEX for operators. The virtualized concept was taken a step further in 5GC by moving toward a service-based architecture using cloud *native* technology (described later). This allows 5GC to be a service-based application platform opened up to third-party applica-tion developers. Using this approach, operators can focus on their core services

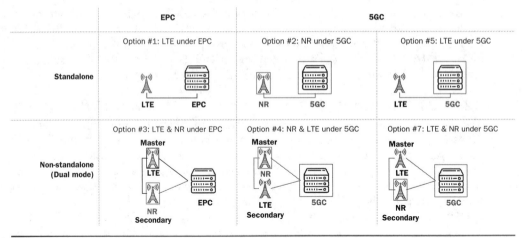

Figure 6.9 4G/5G architecture deployment options [16]

(i.e., transport, QoS, network usage optimization, network management, etc.) leaving the creation of advanced services/applications to third-party application developers allowing utilization of 5GC network capabilities in innovative ways.

The CUPS feature in 3GPP Release 14 enabled the separation of control plane and user plane. Three network entities were impacted in EPC as follows:

- SGW → SGW-C and SGW-U
- PGW → PGW-C and PGW-U
- TDF → TDF-C and TDF-U

Figure 6.10 provides a high-level representation of how the Release 14 EPC architecture has been used as a basis for the 5GC architecture.

As shown in Fig. 6.10, the 4G EPC components have been reorganized into service-oriented functions in 5GC. For details of 5GC architecture, see Sec. 6.4.

6.3 Network Function Virtualization (NFV) and Software-Defined Network (SDN)

Mobile operators embarked on migrating their IP networks from dedicated hardware/software boxes for each NF performing specialized tasks to virtualized, software-based platforms to create programmable networks using NFV and SDN technologies. This allows operators to lower CAPEX and OPEX and allows them to manage their networks with agility to meet market needs and opportunities as well as to meet competitive challenges. SDN and NFV are complementary but increasingly codependent in order for the benefits of software-defined networking to be fully realized.

Virtualization, also called server virtualization, allows physical hardware boxes in a mobile network doing specific tasks to run on virtual machines (VMs) on a cloud-based platform, thus eliminating the need for hardware boxes which make up the mobile network. The premise of 5G is to make use of cloud computing easier, by utilizing NFV

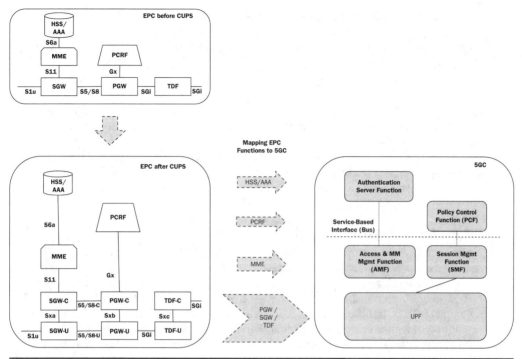

FIGURE 6.10   EPC-to-5GC architecture evolution [16]

and SDN standards. The concept of server virtualization is not new. It was pioneered by companies such as AT&T Bell Labs, General Electric (GE), and International Business Machines (IBM) back in the early 1960s. But now its use in mobile networks is becoming prevalent.

6.3.1   Network Function Virtualization (NFV)

NFV is a means of virtualizing mobile network nodes (or NFs) operating in a cloud as VMs. It abstracts a set of logical (virtual) entities for (1) compute, (2) storage, and (3) networking on top of physical compute, storage, and networking assets. The motivation behind NFV is to eliminate dedicated hardware for each NF to be virtualized to allow cost saving for dedicated hardware (CPU) for running software and allow sharing of NF easier across many software domains by their instantiation, and of course easier evolution. NFV is the technology comprising of hypervisors (e.g., Linux Kernel-based VMs), which provide the abstraction of multiple logical network entities that can be instantiated on top of a single physical network. Hypervisor is basically a building block for hosting the applications that reside in a VM. Hypervisor may or may not use guest OS to run each application. Another technology used for NFV is container technology, known as container-based virtualization, to store and process data onto the cloud servers as packages of applications (containers). Container can also use a guest OS residing on the host machine to run the containerized applications in a distributed fashion which can be accessed from several isolated systems.

6.3.2 Software-Defined Network (SDN)

On the other hand, SDN is all about how to manage packet routing in the network, which in our case is the operator's virtualized NFs (VNFs) infrastructure (see Fig. 6.11). SDN, via its control layer, manages the connectivity between the VNFs of the control plane and the user plane using open APIs. SDN technology provides a network management tool for Mobile Network Operators (MNOs) providing complete view of all the network nodes (or VNFs) and their topology. So, while the NFV virtualizes the NFs into software components allowing them to reside in a cloud architecture, the SDN is used to manage the routing of the data between these VNFs based on operator policies. There are many advantages that come with SDN, such as centralized provisioning, scalability, reliability, reduced CAPEX/OPEX, and adaptable; however, there are some disadvantages as well, such as latency and security. But the advantages of using SDN outweigh the disadvantages.

6.3.3 Coexistence of NFV and SDN

NFV and SDN are distinct technologies used in IT. The main components of NFV are softwarization, virtualization, management/orchestration, and automation. SDN can be useful for implementing virtual networks. VMs are the key building blocks for NFV. NFV and SDN go hand in hand and each does a completely different function. NFV leverages SDN to handle network reachability between the virtualized nodes, i.e., traffic routing. It is the centralized command and control of the entire network. It manages both control plane and data plane (or user plane) separately. Rather than each node taking its own decision on how to route the traffic to its next best hop (i.e., next node), the SDN has view of the entire network and can make decisions for the entire network on how to route traffic on the best possible path. This makes the routing decision much

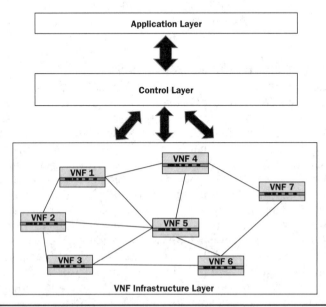

FIGURE 6.11 SDN architecture view

faster and error free. So, SDN keeps an eye on the entire network (holistic view) and has knowledge of where the bottlenecks are for routing data packets and how to avoid them. The challenge is to ensure SDN meets the demands of the 5G key requirement for ultra-reliable low-latency communication (URLLC) without introducing latency.

The ETSI NFV Industry Specification Group (ISG), founded in November 2012, develops the NFV standards.[4] The Open Networking Foundation (ONF) organization develops standards for SDN[5] and works in close collaboration with ETSI NFV ISG. Both NFV and SDN technologies have been used by MNOs to virtualize their NFs in 4G LTE networks, and now in 5G these technologies will play an even bigger role.

6.4   5G CN Architecture

The mobile CN previous to 3GPP Release 15 has always been a point-to-point (p2p) architecture consisting of functional entities (FEs) and well-defined standardized interfaces connecting the different FEs. When modifications or enhancements are required in the mobile CN architecture, the affected FEs and interfaces are modified accordingly. This is how the mobile CN has evolved over the years. In Release 15, it was decided to move away from a p2p architecture to a service-based architecture consisting of logical network functions (NFs) using the standard API framework. The reason for this decision was to provide a flexible, adaptable, and scalable architecture. The flexibility is a result of using programmable virtualized NFs based on cloud computing technology. These NFs can be easily tailored to adapt to specific customer needs and can be easily scaled (replicated) depending on market size. By using web-based standard APIs between NFs, the common off-the-shelf service components can be used.

6.4.1   Origins of Service-Based Architecture (SBA)

SBA's roots can be found in Service-Oriented Architecture (SOA) published in 2008 by Gartner Research [17]. SOA gained popularity due to advances in web services and due to the continued advances in software engineering. In the past, systems were based on distributed objects/components tightly coupled to each other and each object tasked to do a specific job. Then service-oriented systems became the norm moving away from tightly coupled platforms to more robust loosely coupled platform-independent systems made up of services (or applications). Therefore, SOA specifies and describes services and how these services interact in a distributed computing architecture. There are many literatures available on SOA if readers are interested in additional background (e.g., "A Comparison of Service Oriented Architecture with Other Advances in Software Architectures" [18]).

Another related architecture to SOA is Micro-Services Architecture (MSA) [19]. MSA is made of many independently developed microservices that can be executed independently to handle a specific dedicated task or can be combined to create a service or an application. MSA is more suitable for greenfield projects where legacy issues are nonexistent making it easier to control specific tasks via microservices. For example, a new power plant control system may consist of multitude of tasks, where each task can be controlled by a microservice, having its own database. However, for a 3G/4G mobile

---

[4]For NFV standards, refer to ETSI: https://www.etsi.org/technologies/nfv.
[5]For SDN standards, refer to ONF: https://www.opennetworking.org/software-defined-standards/.

core network which has a monolithic architecture made up of a handful of FEs and a centralized monolithic database, it is not easy to break down its FEs into microservices. This requires breaking apart the existing code, moving to a distributed database design, using DevOps best practices, having to deal with many disperse software development organizations, difficulties in managing the software integration of the microservices, etc. This poses some difficulty in delivering a reliable system. Therefore, for 5G, adopting SBA for the p2p mobile CN architecture doesn't require nearly as much change. The different 4G FEs are converted to NFs of SBA each made up of a set of services. These NFs interact with each other over a common web-like interface, allowing consumers to call the different NF services in building new unique services. So, rather than having a multitude of p2p interfaces between NFs, the system can use a common web-like interface where all network entities can interact with each other easily. This makes the mobile CN based on SBA much more flexible. The advantages of SBA are as follows (not listed in order of priority):

- The network is composed of modularized NFs dealing with specific network capabilities, making it easier to provide support to key 5G features such as network slicing.

- NFs become loosely coupled, allowing them to be easily upgraded without impacting other NFs.

- Use of an SBI message bus (like an API) allows each NF service to have a uniform and consistent interface exposed to any other consumer NF services, rather than many p2p interfaces which constantly require upgrade with any new addition of functionality.

- It enables NFs to be easily exposed to third-party application developers in a secure manner opening the door for the development of plethora of unique applications using existing application methods (i.e., leveraging standardized software APIs via API gateways).

- It allows operators to become platform providers, in addition to owning and maintaining their network, enabling third parties to use the platform to easily create and offer services to their customers using the services of SBA NFs. And also be able to deliver specific QoS.

- It reduces system maintenance and upgrade costs.

The cloud computing and virtualization technologies are already used in 4G EPC deployments. For 5GC deployments, SBA enables NFs to use *cloud native* rather than just using *cloud computing* and virtualization technologies.

So, what is the difference between *cloud computing* and *cloud native*? In simplest terms, cloud computing is coined with virtualization of services on data servers residing in the cloud which can be securely accessed via internet. The 4G NFs have been virtualized by operators enabling them to be accessed and replicated from anywhere, as needed. On the other hand, cloud native is the architecture to assemble (or develop) applications built with granular services (called microservices) residing in the cloud and packaged in containers or VMs that can be managed on software development platforms. These containers or VMs can then be reproduced and dynamically orchestrated for use in different applications. Therefore, cloud native computing is an open-source software stack to create services using finer granular microservices, allowing applications to use stateless building components that can be dynamically scaled and tailored. The goal in 5G is to use cloud native computing to breakdown the NFs into microservices that can be exposed for use by other NFs. This will allow fast creation of new features without impacting the old ones.

6.4.2   Details of SBA

SBA, as described in 3GPP TS 23.501 [20] and TS 23.502 [21], provides a modular frame-work from which common NF services can be deployed using components of vary-ing sources and suppliers. The SBA adopted for 3GPP Release 15 is a control-plane functionality with interconnected NFs each with authorization to access each other's services using common data repositories of a 5G network. The step to adopt SBA for the 5G core network was taken to enable 5G network functionality to become more granular and decoupled, where NFs are divided into multiple NF services. Figure 6.12 shows the 5G core network architecture in which the control-plane NFs are based on SBA and communicate with each other via the service-based interface (SBI), while the user-plane NFs use p2p architecture. All the logical interfaces are shown. Also, shown are the p2p interfaces between the control-plane NFs, to mesh with the p2p user plane for completeness. However, all 5GC control-plane deployments will be using 3GPP SBA normative specifications going forward.

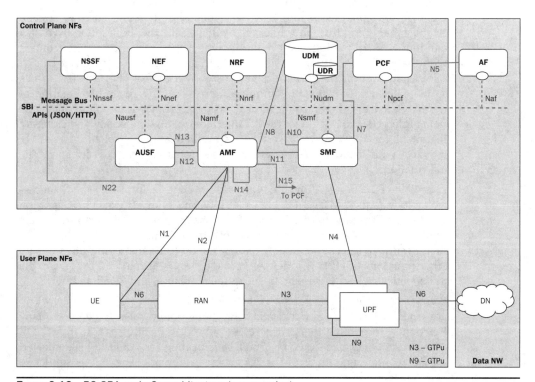

Figure 6.12   5G SBA and p2p architecture (non-roaming)

The 5GC SBA control-plane NFs are self-contained, independent, and reusable, assuming the role of either service consumer or service producer. Each NF exposes its functionality through a Service-Based Interface (SBI), which employs a well-defined Representational State Transfer (REST) interface using IETF Hyper Text Transfer Protocol 2 (HTTP/2) using Java Script Object Notation (JSON) bodies. To mitigate issues around TCP head-of-line blocking, the Quick UDP Internet Connections (QUIC) protocol may be used in the future. HTTP/2 and JSON were selected by 3GPP just because they are well understood in the software industry. For further details on these, check out these references: HTTP/2 [22], JSON [23], TCP [24], and QUIC [25].

The 5GC SBA NFs are described as follows:

- Access and Mobility Management Function (AMF): AMF terminates RAN control-plane interface N2 providing a direct entry point for radio access control, and terminates the NAS N1 interface from UE for provision of NAS ciphering and integrity protection. AMF supports registration management, connection management, reachability management, and mobility management functions for all 3GPP accesses as well as non-3GPP accesses such as WLAN. However, it forwards any UE session requests to the best-suited SMF over the N11 interface. Also, AMF interacts with PCF to gain access to mobility-related policies (e.g., mobility restrictions) and forwards them to the UE. In addition, AMF fully supports interoperability with 4G EPC (for handover purposes) by interfacing to 4G MME node (not shown in Fig. 6.12).

- Session Management Function (SMF): SMF is responsible for management of UE PDU sessions across the UPF on a per session basis, such as session establishment, modify, and release. AMF selects the best-suited SMF for handling session management requests, which then manages session context with the UPF, and authorizes, allocates, and manages IP addresses to UEs. It selects and controls the UPF for data transfer over the N4 interface. It also acts as the external point for all communication with UPF related to steering of traffic related to different data services to their proper destination, and in so doing interacts with PCF over the N7 interface on how the policy and charging treatment for different data services is applied and controlled. Also, SMF interacts with UDM over the N10 interface for accessing subscriber profile information.

- User-Plane Function (UPF): UPF is an anchor point for intra/inter radio access technology (e.g., UMTS, LTE, etc.) mobility and is an external PDU session point interconnecting to data network (DN) via N6 interface. It performs packet routing and forwarding—the support of Uplink Classifier (UL-CL) used to divert traffic to selected data networks based on traffic matching filters. Some other functions of UPF include packet inspection (was handled by TDF in 4G EPC), policy rule enforcement, traffic usage reporting, and QoS handling.

- Policy Control Function (PCF): PCF supports a unified policy framework to govern network behavior in 5GS. It provides policy rules for control-plane function(s) to enforce them. It accesses subscription information from UDR relevant for policy decisions. The 3GPP specification TS 23.503 [26] is dedicated for PCF containing details of the 5G policy framework. The PCF is similar to PCRF in 4G EPS.

- Network Exposure Function (NEF): NEF is required to securely expose the capabilities and events of different NFs within the control plane of the 5GC network to outside third parties. NEF receives information from other network functions (based on exposed capabilities of other NFs), and stores the received information as structured data using a standardized interface to the UDR. This stored information can be accessed and "re-exposed" by the NEF to other NFs and AFs, and used for other purposes such as analytics. The NEF is similar to SCEF in 4G EPS.

- Network Repository Function (NRF): NRF maintains a record of available NF instances and their supported services. It allows other NF instances to subscribe and be notified of registrations from NF instances of a given type. NRF supports service discovery by receipt of NF discovery request from NF instance. It maintains the NF profile of available NF instances and their supported services, and also notifies about newly registered/updated/deregistered NF instances along with its NF services to the subscribed NF service consumer.

- Unified Data Management (UDM): UDM stores subscriber subscription data. It provides generation of 3GPP AKA authentication credentials, and user identification handling for storage and management of Subscription Permanent Identifier (SUPI) for each subscriber in the 5GS. It also handles service/session continuity of ongoing sessions. UDM is analogous to HSS in 4G EPC.

- Unified Data Repository (UDR): UDR supports storage and retrieval of subscription data in the UDM, policy data in the PCF, NF Group ID corresponding to subscriber identifier (e.g., IMPI and IMPU), and structured data for exposure. It is the master database located in the same PLMN as the NF service consumers' home location. It allows storing in and retrieving data from it using intra-PLMN interface Nudr (shown in Fig. 6.12).

- Authentication Server Function (AUSF): AUSF supports authentication for both trusted 3GPP and untrusted non-3GPP accesses. It informs the UDM that a successful or unsuccessful authentication of a subscriber has occurred. Further details of AUSF are provided in Chap. 7.

- Application Function (AF): AF, as its name implies, deals with treatment of application traffic. It may send requests to influence SMF routing decisions for traffic of PDU session for a specific application. It may also influence UPF (re) selection and allow routing user traffic to a local access to a Data Network (identified by a DNAI). AF keeps data regarding active applications. An NF that needs to collect this data provided by an AF uses the NEF capability to subscribe/unsubscribe to notifications regarding data collected from an AF. AF also interacts with PCF if any policy decisions have to be taken on the application data. In addition, AF interacts with IMS. Based on operator deployment, AF considered to be trusted by the operator can be allowed to interact directly with relevant NFs; otherwise, it has to go through NEF.

- Network Slice Selection Function (NSSF): Network slicing is a unique new capability on 5G to provide operators flexibility in deployment of services with efficient resource utilization. NSSF selects the set of network slice instances serving the UE, determines the allowed NSSAI, and determines the AMF set to be used to serve the UE, or, based on configuration, a list of candidate AMF(s), possibly by querying the NRF. Through use of operator service-based platform, third parties like Netflix or Amazon can create specific network slices to serve their own customers via customized applications.

There are two capabilities in 5GC handled by AMF and SMF that need to be explained as these come into play often. These are connection management and session management. An example of using these capabilities would be when a voice call in 5GC is originated. The UE should be connected to NR and must have an established session. Therefore, both connection management and session management functions are needed prior to voice call origination.

Connection management: It comprises the functions of establishing and releasing an NAS signaling connection between a UE and the AMF, comprising AN signaling connection between the UE and the AN (RRC Connection over NR) and the N2 connection for this UE between the AN and the AMF. Two CM states are used to reflect the NAS signaling connection of the UE with the AMF: CM-IDLE and CM-CONNECTED.

A UE in CM-IDLE state has no NAS signaling connection established with the AMF over N1. A UE in CM-CONNECTED state has an NAS signaling connection with the AMF over N1. The AMF shall enter CM-CONNECTED state for the UE whenever an N2 connection is established for this UE between the AN and the AMF. The reception of initial N2 message (e.g., N2 INITIAL UE MESSAGE) initiates the transition of AMF from CM-IDLE to CM-CONNECTED state.

Session management: This relates to PDU connectivity service, i.e., a service that provides exchange of PDUs between a UE and a data network identified by a DNN. The PDU connectivity service is supported via PDU sessions that are established upon request from the UE. Session establishment entails involvement of SMF, which is responsible for managing the PDU session(s).

The above two functions will be required for voice call setup. This requires that all 5G cell phones will support 4G radios as well in order to make voice calls. The serving PLMN AMF shall send an indication toward the UE during the registration procedure over 3GPP access to indicate if an IMS voice over PS session is supported or not supported in 3GPP access and non-3GPP access. A UE with "IMS voice over PS" voice capability over 3GPP access should take this indication into account when performing voice domain selection.

6.4.3 NF Services

Each NF in the 5GC has a set of services which are self-contained, reusable, and use management schemes independently of other NF services. The NF services might have dependencies between NF services within the same NF due to sharing some common resources, e.g., context data, but are still managed independently of each other. Each NF service can be exposed to other NFs via the SBI of the 5G SBA for use in constructing procedures. Figure 6.13 [20] depicts how a procedure can be constructed. In 5GC, all procedures depict an NF service calling another NF service as opposed to an NF calling another NF.

Table 6.1 lists example 5G services supported by some 5GC NFs [20].

Table 6.1 also lists some example services and their service operations to give readers a glimpse of the level of granularity provided to allow consumer NF service to call another NF service in another NF in the development of system service procedures.

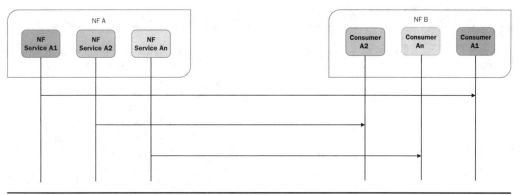

FIGURE 6.13   NF–NF interaction

6.4.4   URLLC

The support of URLLC is a key requirement in 5GS to make time-sensitive machine–machine communications (e.g., autonomous driving) possible with five-nine reliability. In order to support URLLC the 5GC has to do its part. It supports dual redundant paths for data transfer to ensure that there is zero loss in end-to-end communication. With dual paths data delivery end-to-end is able to better meet URLLC requirements. These URLLC capabilities can be selected by the UE depending on the type of service being provided.

6.4.4.1   *Dual Connectivity-Based End-to-End Redundant User-Plane Paths*

With this capability, a UE can request NG-RAN to set up user-plane paths of two redundant and disjoint end-to-end PDU sessions. This indication is provided to SMF from UDM. To enable duplicate paths for redundant traffic delivery end-to-end, the upper layer protocols [e.g., IEEE Time-Sensitive Networking (TSN) and Frame Replication and Elimination for Reliability (FRER)] must be relied upon that can ensure delivery of these duplicate paths spanning both the 5GS and if necessary the fixed network. To enable this capability the UE and NG-RAN must be able to support dual connectivity, and 5GC must be aligned with such NG-RAN deployments.

Figure 6.14 [20] illustrates the dual PDU sessions capability. One PDU session spans from the UE via Master NG-RAN to UPF1 acting as the PDU session anchor, and the other PDU session spans from the UE via secondary NG-RAN to UPF2 acting as the PDU session anchor. Based on these two PDU sessions, two independent user-plane paths are set up. UPF1 and UPF2 connect to the same data network (DN), even though the traffic via UPF1 and UPF2 may be routed via different user-plane nodes within the DN.

In order to establish two redundant PDU sessions and associate the duplicated traffic coming from the same application to these PDU sessions, URSP (policy used by the UE to determine how to route outgoing traffic) or UE local configuration is used as specified in TS 23.503, clause 6.6.2 [26].

The dual connectivity can also be achieved by using only a single RAN.

Service name	Service description	Service operation examples
AMF		
Namf_ Communication	Enables an NF consumer to communicate with the UE and/or the AN through the AMF.	- Namf_Communication_UEContextTransfer: Provides the UE context to the consumer NF. - Namf_Communication_RegistrationCompleteNotify: Used by the consumer NF to inform the AMF that a prior UE context transfer has resulted in the UE successfully registering with it.
Namf_ EventExposure	Enables other NF consumers to subscribe or get notified of the mobility-related events and statistics.	- Namf_EventExposure_Subscribe: The consumer NF uses this service operation to subscribe to or modify event reporting for one UE, a group of UE(s) or any UE. - Namf_EventExposure_UnSubscribe: The NF consumer uses this service operation to unsubscribe for a specific event for one UE, group of UE(s), any UE.
SMF		
Nsmf_ PDUSession	Manages the PDU sessions and uses the policy and charging rules received from the PCF. The service operations exposed by this NF service allows the consumer NFs to handle the PDU sessions.	- Nsmf_PDUSession_Update: Update the established PDU session. - Nsmf_PDUSession_Release: It causes the immediate and unconditional deletion of the resources associated with the PDU session.
Nsmf_ EventExposure	Exposes the events happening on the PDU sessions to the consumer NFs.	- Nsmf_EventExposure_Notify: Report UE PDU session-related event(s) to the NF which has subscribed to the event report service.
PCF		
Npcf_ AMPolicyControl	Provides access control, network selection and Mobility Management-related policies, UE Route Selection policies to the NF consumers.	- Npcf_AMPolicyControl_Create: NF service consumer can request the creation of an AM Policy Association and by providing relevant parameters about the UE context to the PCF.
UDM		
Nudm_UECM	1. Provides the NF consumer of the information related to UE's transaction information, e.g., UE's serving NF identifier, UE status, etc. 2. Allows the NF consumer to register and deregister its information for the serving UE in the UDM. 3. Allows the NF consumer to update some UE context information in the UDM.	- Nudm_UECM_Update: Consumer updates some UE-related information (e.g., UE capabilities, Intersystem continuity context, etc.). - Nudm_UECM_Deregistration: The NF consumer requests the UDM to delete the information related to the NF in the UE context.

TABLE 6.1   5G example services

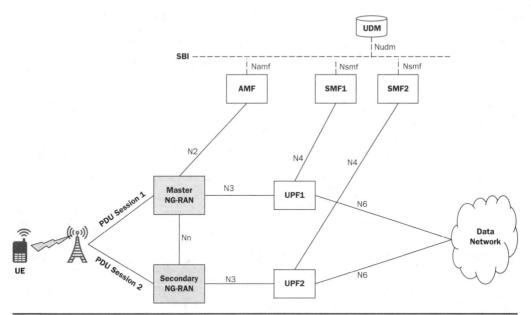

FIGURE 6.14   End-to-end redundant user-plane paths using dual connectivity [20]

6.4.4.2   Support of Redundant Transmission on N3/N9 Interfaces

When the reliability of NG-RAN and 5GC NFs is sufficient to meet URLLC requirements but the backhaul network connecting the UPFs is not considered sufficient, redundant transmission may be deployed between PDU Session Anchor (PSA) UPF and RAN via two independent N3 tunnels, which are associated with a single PDU session, over different transport layer paths to enhance the reliability. The SMF and PSA UPF should provide different IP addresses to establish two disjoint transport layer N3 tunnels, as shown in Fig. 6.15 [20]. N3 tunnel is between NG RAN and UPF.

Another option supported is two intermediate UPFs (I-UPFs) between the PSA UPF and the NG-RAN for redundant transmission based on two N3 and N9 tunnels between a single NG-RAN node and the PSA UPF. An N9 tunnel is between UPF and UPF2.

In Fig. 6.16 [20], there are two N3 and N9 tunnels between NG-RAN and PSA UPF for the same PDU session for redundant transmission. The PSA UPF duplicates the downlink packet of the QoS flow from the DN and assigns the same GTP-U sequence number to them. These duplicated packets are transmitted to I-UPF1 and I-UPF2 via N9 Tunnel 1 and N9 Tunnel 2 separately. Each I-UPF forwards the packet with the same GTP-U sequence number which receives from the PSA UPF to NG-RAN via N3 Tunnel 1 and N3 Tunnel 2, respectively. The NG-RAN eliminates the duplicated packet based on the GTP-U sequence number. In case of uplink traffic, the NG-RAN duplicates the packet of the QoS flow from the UE and assigns the same GTP-U sequence number to them. These duplicated packets are transmitted to I-UPF1 and I-UPF2 via N3 Tunnel 1 and N3 Tunnel 2 separately. Each I-UPF forwards the packet with the same GTP-U sequence number which receives from the NG-RAN to PSA UPF via N9 Tunnel 1 and N9 Tunnel 2, respectively. The PSA UPF eliminates the duplicated packets based on the GTP-U sequence number.

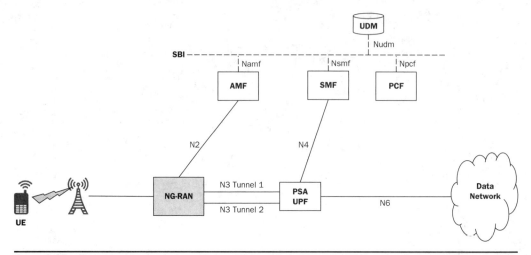

FIGURE 6.15 Redundant transmission with two N3 tunnels between the UPF and a single NG-RAN node [20]

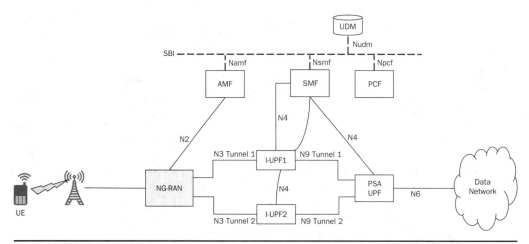

FIGURE 6.16 Two N3 and N9 tunnels between NG-RAN and UPF for redundant transmission [20]

6.4.5 5G User-Plane Management

In 4G EPC, the CUPS feature provided separation of control-plane signaling from user-plane signaling enabling efficient handling of data traffic with low latency. The CUPS concept has been carried over in the design of the 5GC architecture with the creation of UPF which is totally devoid of any control-plane signaling. The UPF is a key component of 5GC SBA deployed in the cloud using the cloud native technology providing a dynamic packet processing foundation for efficient handling of packet routing and forwarding, policy enforcement, data buffering, and interconnection to the DN.

3GPP specifications support a single UPF or multiple UPF deployments for a given PDU session. The selection of UPF is performed by SMF by considering UPF deployment scenarios such as centrally located UPF and distributed UPF located close to or at the Access Network site. The UPF handles the UE IP address management which includes allocation and release of the UE IP address along with renewal of the allocated IP address. The selection of the UE-requested PDU IP session type is dependent on different UE configurations, e.g., UE local configuration containing PDU session type of "IPv4," "IPv6," or "IPv4v6," or UE Route Selection Policy (URSP) rule containing which IP session type the UE can support.

The SMF to UPF interface N4 is a busy interface. It is used by SMF to send information to UPF related to UE IP address management, management of core network tunnel info (Tunnel ID and the endpoint IP address of 5G Core), traffic detection (how to detect user data traffic and what rules to apply), user-plane traffic forwarding based on policy rules, activation of traffic usage reporting in UPF, QoS flow policing based on PCC rules, sending of "end marker" (to assist the reordering function in the Target RAN), buffering at UPF, etc. The N4 interface is also used by UPF to send information to SMF related to traffic usage, etc.

For additional details on 5G user-plane management, refer to 3GPP TS 23.501, clause 5.8 [20].

6.4.6   5G Protocol Stack

The 5GAN/5GC UPF protocol stack is based on LTE/EPC protocol stack with a few additions. The protocol stacks for user plane and control plane are shown in Figs. 6.17 and 6.18, respectively.

- PDU layer: This layer corresponds to the PDU carried between the UE and the DN over the PDU session. When the PDU session type is IPv4 or IPv6 or IPv4v6, it corresponds to IPv4 packets or IPv6 packets or both of them; when the PDU session type is Ethernet, it corresponds to Ethernet frames; and so on.

- GPRS tunneling protocol for the user plane (GTPU): This protocol supports tunneling user data over N3 (i.e., between the 5G-AN node and the UPF) and N9 (i.e., between different UPFs of the 5GC) in the backbone network (for details, see TS 29.281 [27]). GTP shall encapsulate all end user PDUs. It provides encapsulation on a per PDU session level. This layer carries also the marking associated with a QoS flow.

    For the lower layers of the 5G user-plane protocol stack (SDAP, PDCP, and RLC), see Chap. 5, Sec. 5.3.2 for additional details.

- NAS-MM: The NAS protocol for MM functionality supports registration management functionality, connection management functionality, and user-plane connection activation and deactivation. It is also responsible of ciphering and integrity protection of NAS signaling. 5G NAS protocol is defined in 3GPP TS 24.501 [20].

- 5G-AN protocol layer: This set of protocols/layers depends on the 5G-AN. In the case of NG-RAN, the radio protocol between the UE and the NG-RAN node (eNodeB or gNodeB) is specified in 3GPP TS 36.300 [28] and TS 38.300 [29].

The user-plane and control-plane protocol stacks shown above provide an overview of the different layers used between various network entities. For full details, refer to 3GPP TS 23.501 clause 8 [20]. The SBI protocol stack is based on HTTP/2 which is a

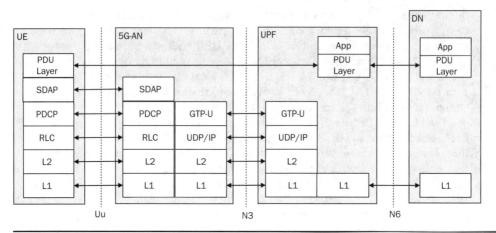

FIGURE 6.17   5G user-plane protocol stack

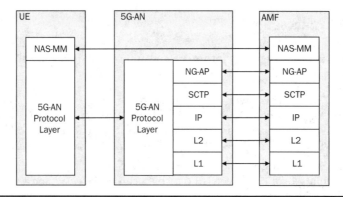

FIGURE 6.18   5G Control-plane protocol stack

standard protocol commonly employed in cloud native web applications, and in common use by third-party application developers.

6.5   Quality of Service in Mobile Networks

One of the primary functions of the mobile system, or for that matter any telecommunication or data network, is to ensure that the voice and data traffic traverses through it efficiently from source to destination with minimum delay. This is particularly important for real-time communications, for example, emergency calls, driver safety info, etc. There are many different types of traffic handling required for data that traverses the mobile network, dependent on the service data requirements. This differentiated treatment of traffic is handled using the various quality-of-service (QoS) mechanisms in the mobile system, as depicted in Fig. 6.19. The QoS mechanisms are also needed for efficiently managing the limited network resources of a mobile system that have to be shared among huge number of subscribers.

In 1994, ITU-T took on the task to define the terms related to QoS, used in the field of telephony, to help avoid confusion in the use of these terms in different standards development organizations. Refer to ITU-T Recommendation E.800 [30] for more details. There are several terms defined in this recommendation, for example, packet loss, jitter, bit error rate, latency, service-level agreement, etc. QoS is a technology that controls and manages the traffic flow through the network in order to reduce packet loss, latency, and jitter, specifically the traffic traversing the user plane. In controlling and managing the traffic flow, there are specific types of QoS priorities assigned to specific types of traffic on the network. These QoS priorities assist in different treatment of low versus medium versus high-priority traffic resulting in efficiently managing network resources and at the same time ensuring adequate service quality is maintained to satisfy customer needs. Therefore, all traffic traversing the mobile network is not treated equally; otherwise there will be huge bottlenecks.

Traditionally, four types of characteristics are attributed to traffic flow as follows:

- Reliability: This relates to how reliable end-to-end data transmission is for delivering data packets belonging to the same traffic flow. If there is packet loss in data transmission, it means retransmission of data packets will be required which adds to the delay. Reliability is application dependent. Some applications require highly reliable data transmission (e.g., electronic mail), while others can tolerate delay (e.g., audio conferencing).

- Delay: This relates to the amount of time it takes for a data packet to reach its destination.

- Jitter: This relates to delay variation between deliveries of data packets belonging to the same traffic flow. High jitter means that data packets belonging to same traffic flow arrive at their destination in variable times from when they departed (i.e., sporadically). Low jitter means the data packets arrive their destination with low variation. For example, if four packets depart at times 0, 1, 2, and 3, and arrive at 10, 11, 12, and 13, all have the same delay of 10 units of time (low or no jitter). On the other hand, if the above four packets arrive at variable times 11, 15, 13, and 18, they will have different delays of 11, 14, 11, and 15, respectively (high jitter).

- Bandwidth: Simply put, it means the acceptable level of bandwidth required for transmission of data packets for an application. Applications like video streaming require high bandwidth.

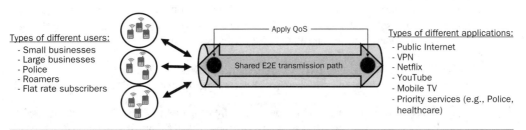

FIGURE 6.19 Differentiated QoS

6.5.1   QoS in UMTS

The 3GPP specification TS 23.107 [31] has specified technical QoS requirements for the 3G UMTS, also incorporating the QoS concept from GPRS, the so-called packet data protocol (PDP) context and its mapping between GPRS and UMTS systems. UMTS is specifically geared for treatment of data packets flowing end-to-end through the entire mobile system for each application session belonging to a network service. Here, end-to-end entails mobile device, RAN, CN, and up to the CN gateway node. The CN is responsible for providing adequate control mechanisms as the traffic for different application session flows through various entities of the mobile system. It provides a mapping between application requirements and UMTS network services, and is able to modify QoS levels even during active sessions.

The QoS control mechanisms provided in the mobile network are designed to be robust and capable of providing reasonable QoS based on user experience. The mobile system is unable to provide QoS similar to a fixed network due to inherent limitations of the radio air interface. But every effort is made to ensure that satisfactory level of user experience is maintained for a moving mobile device.

6.5.1.1   *Diffserv-Based QoS Mechanism for UMTS Core Network*

The Diffserv (DS) standard developed by IETF RFC 7657 [32] for classifying and managing network traffic on IP networks based on a Per-Hop Behavior (PHB) was adopted by 3GPP as one of the QoS mechanisms for UMTS PS domain. The PHB is the basic DS building block.

The UMTS PS core network domain comprising SGSN and GGSN supports IP bearer service with guaranteed QoS. As shown in 3GPP QoS architecture in Fig. 6.20, the end-to-end service goes from UE to terminating equipment (TE). But the 3GPP UMTS mobile network is responsible for delivering guaranteed QoS on the UMTS bearer service. It ensures that DS is applied at each node in the traffic path in managing per packet flow. The DS codepoints have been standardized by 3GPP in order for nodes to use standardized attributes for different QoS traffic classes in delivering the end-to-end QoS, per service-level agreements (SLAs).

The UMTS bearer consists of two parts: the radio access bearer service and the CN bearer service. In this chapter, the focus will be on CN bearer service, where DS framework is applied. The radio access bearer service is under RAN control, and the external bearer service involves multiple external networks, which are not described here. However, the CN bearer service does interwork with radio access bearer service and external bearer service to achieve the expected QoS for a specific bearer service based on the pre-negotiated SLA. In addition, the radio access bearer service provides confidential transport of signaling and user data between MT and SGSN with the QoS adequate to the negotiated UMTS bearer service or with the default QoS for signaling. This service is based on the characteristics of the radio interface and is maintained for a moving UE.

The UMTS packet core network supports four different QoS traffic classes depending on how delay sensitive is the traffic. These are as follows:

1. Conversational class: This scheme applies for use in voice telephony in GSM. However, as the internet and multimedia have grown there are a number of new applications requiring the use of this scheme such as VoIP and video

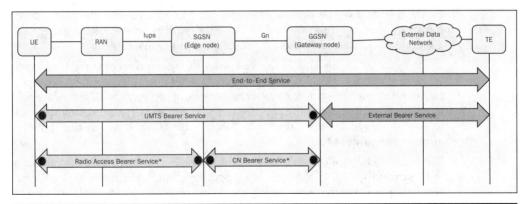

FIGURE 6.20   UMTS QoS architecture

conferencing which are like real-time applications. This is the only scheme where the required characteristics are strictly given by human perception.

2. Streaming class: This scheme applies when users are listening to the audio of a real-time video. It is a one-way transport aimed at a human destination. The delay variation of the end-to-end flow shall be limited to preserve the time relation (variation) between information entities of the stream. But as the stream normally is time aligned at the receiving end (in the user equipment), the highest acceptable delay variation over the transmission media is given by the capability of the time alignment function of the application.

3. Interactive class: This scheme applies when remote data is requested by either a human or a machine, for example, during web browsing or polling of data by a remote electric metering device. Round trip delay time is, therefore, one of the key attributes of this scheme.

4. Background class: This scheme applies when a computer sends and receives data-files in the background, for example, e-mails, SMS, or download of measurements records from databases requested by applications running in the background.

The above QoS traffic classes support different UMTS bearer service attributes in order to make assumptions about the traffic source and optimize the transport for that traffic type. The following is the list of bearer service attributes:

- Traffic class: It is used to optimize the transport for type of UMTS bearer service.

- Maximum bitrate (kbps): It is the maximum number (upper limit) of bits delivered to/from UMTS and within a period of time, divided by the duration of the period. All UMTS bearer service attributes may be fulfilled for traffic up to the maximum bitrate depending on the network conditions.

- Guaranteed bitrate (kbps): It describes the bitrate the UMTS bearer service shall guarantee to the user or application. It is the guaranteed number of bits delivered by UMTS within a period of time (provided that there is data to deliver), divided by the duration of the period. The traffic is conformant with the guaranteed bitrate as long as it follows a token bucket algorithm where token rate equals guaranteed bitrate and bucket size equals maximum SDU size.

- Delivery order (y/n): This is derived from the user protocol (PDP type) and specifies if out-of-sequence SDUs are acceptable or not. This information cannot be extracted from the traffic class. Whether out-of-sequence SDUs are dropped or reordered depends on the specified reliability. Delivery order should be set to "no" for PDP Type = "IPv4" or "IPv6." The SGSN shall ensure that the appropriate value is set.

- Maximum SDU size (octets): The maximum SDU size for which the network shall satisfy the negotiated QoS. It is used for admission control and policing and/or optimizing transport (optimized transport in, for example, the RAN may be dependent on the size of the packets).

- SDU format information (bits): List of possible exact sizes of SDUs for RAN to be able to operate in transparent RLC protocol mode, which is beneficial to spectral efficiency and delay when RLC retransmission is not used.

- SDU error ratio: It indicates the fraction of SDUs lost or detected as erroneous. It is used to configure the protocols, algorithms, and error detection schemes, primarily within RAN.

- Residual bit error ratio: It indicates the undetected bit error ratio in the delivered SDUs. It is used to configure radio interface protocols, algorithms, and error detection coding.

- Delivery of erroneous SDUs (y/n): It indicates whether SDUs detected as erroneous shall be delivered or discarded. It is used to decide whether error detection is needed and whether frames with detected errors shall be forwarded or not.

- Transfer delay (ms): It indicates maximum delay for 95th percentile of the distribution of delay, tolerated by the application, for all delivered SDUs during the lifetime of a bearer service, where delay for an SDU is defined as the time from a request to transfer an SDU at one SAP to its delivery at the other SAP.

- Traffic handling priority: This is needed within the interactive class to differentiate between bearer qualities.

- Allocation/retention priority: It specifies the relative importance compared to other UMTS bearers for allocation and retention of the UMTS bearer. The allocation/retention priority attribute is a subscription attribute which is not negotiated from the mobile terminal, but the value might be changed either by the SGSN or the GGSN network entities. It is used for differentiating between bearers when performing allocation and retention of a bearer.

- Source statistics descriptor ("speech"/"unknown"): It specifies characteristics of the source of submitted SDUs, which helps RAN, SGSN, and GGSN, and also UE to calculate a statistical multiplex gain for use in admission control on the relevant interfaces.

- Signaling indication (y/n): Signaling traffic can have different characteristics to other interactive traffic, such as higher priority, lower delay, and increased peaks. This attribute permits enhancing the RAN operation accordingly. This indication is sent by the UE in the QoS IE.

- Evolved allocation/retention priority: It enhances the allocation/retention priority attribute with an increased value range of the priority level and additional information about the pre-emption capability and the pre-emption vulnerability of the bearer. The pre-emption capability information defines whether a bearer with a lower priority level should be dropped to free up the required resources. The pre-emption vulnerability information whether a bearer is applicable for such dropping by a pre-emption capable bearer with a higher priority value.

Based on the above UMTS traffic classes and UMTS bearer service attributes, the DS mechanism uses these on PHB for traffic conditioning, traffic scheduling, and queue management of each traffic flow passing through the nodes.

6.5.2   QoS in LTE

Long-term evolution (LTE) or Evolved Universal Terrestrial Radio Access Network (EUTRAN) is the 4G mobile broadband technology standardized by 3GPP in Release 8 technical specifications. The Evolved Packet Core (EPC) supporting E-UTRAN is an all-IP-based network for PS voice and data, along with IP Multimedia System (IMS) for handling IP-based multimedia services. However, EPC lacks native support for CS services as provided in UMTS since there is no MSC, but Voice over LTE (VoLTE) service is supported due to the addition of IMS in order to fulfill this gap. When IMS is not deployed, another option to support voice is by using over-the-top (OTT) third-party providers such as Google Talk or Skype. But the use of OTT third-party providers does not guarantee QoS, and no service continuity is provided using this method when mobile moves outside the LTE coverage area.

With the introduction of the all-IP-based EPC, the support of QoS became a challenge due to the lack of dedicated connection channels. With the growth of multimedia data services/applications offered on smart phones, the need for new QoS solutions was required for maintaining satisfactory user experience. The EPS (EUTRAN and EPC) supports different QoS data flows for each service/application. The EPS QoS architecture is shown in Fig. 6.21.

The EPS QoS architecture supports real-time and non-real-time services. This architecture is based on data flows concept. The data flows are then mapped to three bearers—radio, S1-U, and S5/S8—which in combination provide the end-to-end QoS support in EPS. Each bearer is assigned a unique scalar value called the QoS Class Identifier (QCI). The QCI addresses the shortcomings of the Release 5 3GPP QoS profile with the addition of this new information element QCI between RAN and the CN. The QCI is used as a pointer to a pre-configured Traffic Forwarding Policy (TFP). Until Release 5, the expected QoS was initiated by the UE, but with QCI it does not need to be available at the UE.

Table 6.2 illustrates the standardized QCI values for each class of service as specified in 3GPP TS 23.203 [33].

The standardized characteristics are not signaled on any interface. They are provided as guidelines for the pre-configuration of node-specific parameters for each QCI. The goal of standardizing a QCI with corresponding characteristics is to ensure that services/applications mapped to that QCI receive the same minimum level of QoS in multi-vendor network deployments and in case of roaming. A standardized QCI and corresponding characteristics are independent of the UE's current 3GPP or non-3GPP access. For further details on usage of QCIs, refer to TS 23.203 clause 6.1.7.2 [33].

6.5.3   QoS in 5G

In 4G EPS, with the introduction of CUPS feature, the control plane and user plane were decoupled which improved efficiencies in user data traffic management and improved QoS. However, in EPS the user plane is centralized, so the user traffic always has to be routed through the PGW, which is a bottleneck as data traffic increases. Moreover, in EPS, the QoS is managed based on EPS bearer in the CN (EPC) and RAN. That is, all types of traffic mapped to the same EPS bearer receive the same level packet forwarding treatment.

To meet the demands of massive growth in data traffic generated by mobile devices and by demands of ultra-reliable machine-to-machine communications, 5G is poised to deliver E2E user data traffic with significantly improved QoS. It is expected that mobile networks built on the basis of 5G technologies will provide data transfer speed of more than 10 Gb/s. Previous 4G technologies (LTE/LTE Advanced) provide flexible QoS management schemes based on the division of data transfer characteristics into nine QCI classes. These classes cover both 4G quality principles for data transfer not requiring quality assurance (i.e., best effort or non-GBR) and data transfer requiring GBR.

The QoS management mechanisms in 5G will be controlled at the network level similar to 4G, but the QoS concept is flow based, where packets are classified and marked with QoS Flow Identifier (QFI). Both types of QoS flows are supported in 5G, the ones that require guaranteed flow bit rate (GBR QoS flows) and QoS flows that do not require guaranteed flow bit rate (non-GBR QoS flows). Each QoS flow is based on its QoS profile, which is sent to the (R)AN containing QoS Identifier (5QI) and Allocation and Retention (ARP) priority attributes.

The 5G QoS model also supports reflective QoS. Reflective QoS enables the UE to map uplink user-plane traffic to QoS flow rules derived from the received downlink traffic without involvement of SMF. Hence, its name reflective QoS. Reflective QoS

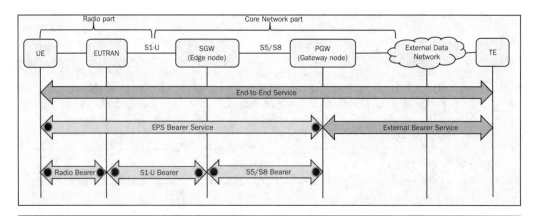

Figure 6.21   EPS QoS architecture

QCI	Resource type	Priority level	Packet delay budget (ms)	Packet error loss rate	Example services/applications
1	GBR	2	100	10^{-2}	Conversational Voice
2		4	150	10^{-3}	Conversational Video (Live Streaming)
3		3	50	10^{-3}	Real-Time Gaming
4		5	300	10^{-6}	Non-Conversational Video (Buffered Streaming)
5		1	100	10^{-6}	IMS Signaling
6		6	300	10^{-6}	Video (Buffered Streaming) TCP-based (e.g., www, e-mail, chat, ftp, p2p file sharing, progressive video, etc.)
7	Non-GBR	7	100	10^{-3}	Voice, Video (Live Streaming) Interactive Gaming
8		8	300	10^{-6}	Video (Buffered Streaming) TCP-based (e.g., www, e-mail, chat, ftp, p2p file sharing, progressive video, etc.)
9		9			

TABLE 6.2   3GPP standardized QCI characteristics

applies to IP PDU session and Ethernet PDU session. Reflective QoS and non-reflective QoS can be applied concurrently within the same PDU session. If the 3GPP UE supports reflective QoS functionality, the UE should indicate support of reflective QoS to the network (i.e., SMF) for every PDU session. Reflective QoS is controlled on per-packet basis by using the Reflective QoS Indication (RQI) in the encapsulation header on N3 and N9 reference points.

There are two types of QFIs: one with standardized QoS profiles and the other with operator-specific QoS profiles. For the first one, only the QFI value is used in the network. For the latter one, QoS attributes are also signaled between the network entities. The 5G QoS flows are mapped in the Access Network to Data Radio Bearers (DRBs), unlike 4G where the mapping is 1:1 between EPC and radio bearers.

The UE performs the classification and marking of UL user-plane traffic, i.e., the association of UL traffic to QoS flows, based on QoS rules. These QoS rules may be explicitly provided to the UE (i.e., explicitly signaled QoS rules using the PDU session establishment/modification procedure), pre-configured in the UE, or implicitly derived by the UE by applying reflective QoS (refer to 3GPP TS 23.501, clause 5.7.5 [20]). A QoS rule contains the QFI of the associated QoS flow, a Packet Filter Set (refer to 3GPP TS 23.501, clause 5.7.6 [20]) and a precedence value (refer to 3GPP TS 23.501, clause 5.7.1.9 [20]). An explicitly signaled QoS rule contains a QoS rule identifier which is unique within the PDU session and is generated by SMF.

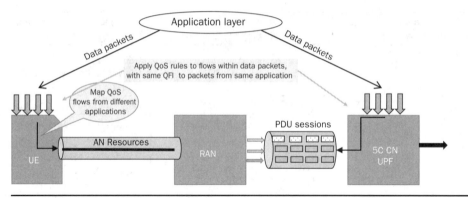

Figure 6.22   UPF- and UE-level classification and marking for QoS flows [20]

The following concepts are important in 5G QoS:

- **QoS flow**: This is the finest granularity for QoS management in the 5G core network. A QoS flow can either guarantee the bit rate or not, i.e., Guaranteed Bit Rate (GBR) QoS flow or non-GBR (NGBR) QoS flow. The radio bearer concept is maintained in 5G and by introducing the QoS flow concept, granularity for QoS treatment (per flow) becomes different from tunneling (per PDU session, basically), and it will enable more flexible QoS control.

- **QoS rule**: 5G defines QoS rule, which consists of QoS profile, packet filters, and precedence order. The QoS profile is composed of QoS parameters and QoS marking. The value of the QoS marking indicates the type of QoS profile, i.e., A-type or B-type. A-type QoS profile has standardized QoS parameters where B-type QoS profile has QoS parameters dynamically signaled over NG2. Packet filter is used for the purpose of binding a flow to a specific QoS marking. Precedence order represents the priority to adapt QoS rule to a flow.

The QoS in 5GC is applicable to the user plane, which is responsible for carrying user data traffic. In 5GC, the UPF is an evolution of EPC SGW-U and PGW-U. The UPF performs routing and forwarding of data packets, and data packet inspection (e.g., deep packet inspection) which enables UPF to control QoS. Figure 6.22 shows the classification of packets and end-to-end QoS handling.

Table 6.3 lists 3GPP Standardized 5QIs showing each one's capability/characteristics and example services supported by each.

For further details on 5G QoS, refer to [20]. The work on 5G QoS continues to be refined in 3GPP Release 17.

6.6   5GC Policy and Charging Control Framework

6.6.1   Background

As mentioned in the QoS section, all traffic traversing the mobile system does not receive the same treatment. Why? Because the mobile system (RAN+CN) resources are limited. In order to utilize the mobile system resources efficiently, the traffic traversing the mobile system has to be managed based on the type of traffic and the treatment it should receive. This is why Policy and Charging Control (PCC) Framework is needed for the mobile core network to assist in efficiently managing the traffic flow.

The main functions of PCC framework are:

- to provide appropriate mechanisms for charging and online credit control for service data flows and application traffic flowing through the network, and

- to control traffic flow in the network by applying QoS policies in real time.

5QI value	Resource type	Default priority level	Packet delay budget	Packet error rate	Default maximum data burst volume	Default averaging window	Example services
1	GBR	20	100 ms	10^{-2}	N/A	2000 ms	Conversational Voice
2		40	150 ms	10^{-3}	N/A	2000 ms	Conversational Video (Live Streaming)
3		30	50 ms	10^{-3}	N/A	2000 ms	Real-Time Gaming, V2X messages Electricity distribution — medium voltage, Process automation— monitoring
4		50	300 ms	10^{-6}	N/A	2000 ms	Non-Conversational Video (Buffered Streaming)
65		7	75 ms	10^{-2}	N/A	2000 ms	Mission Critical user plane push-to-talk voice (e.g., MCPTT)
66		20	100 ms	10^{-2}	N/A	2000 ms	Non-Mission-Critical user-plane push-to-talk voice
67		15	100 ms	10^{-3}	N/A	2000 ms	Mission Critical Video user plane
71		56	150 ms	10^{-6}	N/A	2000 ms	"Live" Uplink Streaming

(Continued)

5QI value	Resource type	Default priority level	Packet delay budget	Packet error rate	Default maximum data burst volume	Default averaging window	Example services
72		56	300 ms	10^{-4}	N/A	2000 ms	"Live" Uplink Streaming
73		56	300 ms	10^{-8}	N/A	2000 ms	"Live" Uplink Streaming
74		56	500 ms	10^{-8}	N/A	2000 ms	"Live" Uplink Streaming
76		56	500 ms	10^{-4}	N/A	2000 ms	"Live" Uplink Streaming
5	Non-GBR	10	100 ms	10^{-6}	N/A	N/A	IMS Signaling
6		60	300 ms	10^{-6}	N/A	N/A	Video (Buffered Streaming) TCP-based (e.g., www, e-mail, chat, ftp, p2p file sharing, progressive video, etc.)
7		70	100 ms	10^{-3}	N/A	N/A	Voice, Video (Live Streaming) Interactive Gaming
8		80	300 ms	10^{-6}	N/A	N/A	Video (Buffered Streaming) TCP-based (e.g., www, e-mail, chat, ftp, p2p file sharing, progressive video, etc.)
9		90					
69 (NOTE 9, NOTE 12)		5	60 ms	10^{-6}	N/A	N/A	Mission Critical delay sensitive signaling (e.g., MC-PTT signaling)
70		55	200 ms	10^{-6}	N/A	N/A	Mission Critical Data
79		65	50 ms	10^{-2}	N/A	N/A	V2X messages

(Continued)

5QI value	Resource type	Default priority level	Packet delay budget	Packet error rate	Default maximum data burst volume	Default averaging window	Example services
80		68	10 ms	10^{-6}	N/A	N/A	Low Latency eMBB applications Augmented Reality
82	Delay Critical GBR	19	10 ms	10^{-4}	255 bytes	2000 ms	Discrete Automation
83		22	10 ms	10^{-4}	1354 bytes	2000 ms	Discrete Automation (e.g., V2X messages)
84		24	30 ms	10^{-5}	1354 bytes	2000 ms	Intelligent transport systems
85		21	5 ms	10^{-5}	255 bytes	2000 ms	Electricity Distribution— high voltage
86		18	5 ms	10^{-4}	1354 bytes	2000 ms	V2X messages (Advanced Driving: Collision Avoidance, Platooning with high LoA)

TABLE **6.3**   Standardized 5QI to QoS characteristics mapping

At the beginning, the control and charging aspects of traffic flowing through the mobile network were performed through other means, for example, Service-Based Local Policy (SBLP) and Flow-Based Charging (FBC) frameworks. These frameworks were merged when IMS came around and the need for a robust QoS policy and charging framework for addressing requirements for managing traffic growth due to multimedia applications. The PCC framework was introduced by 3GPP in 2007 in order to fill the need of a robust QoS policy and charging in IMS networks. The PCC framework has evolved since Release 7 (refer to "A review of 3GPP PCC Architectures from Release 7 to 14") [34].

As background, the 3GPP EPC PCC framework architecture of Release 14 is shown in Fig. 6.23 [33].

The 3G/4G PCC framework architecture comprises the following functions:

- Policy and Charging Rules Function (PCRF): It provides flow-based charging control and policy control. It also provides application detection and based on it gating, QoS, and charging control per application. The PCC decision-making process is carried out by the PCRF and is based on inputs from the PCEF, the BBERF if present, the TDF if present, and the SPR. In addition, if the AF is involved, based on inputs from the AF, as well as possibly PCRF's own predefined information. PCRF is the controlling engine interacting with most of the other entities in the PCC framework.

- Policy and Charging Enforcement Function (PCEF): It provides service data flow detection, policy enforcement, and flow-based charging functionalities. It enforces PCRF-provided policies to SDFs via gate enforcement (to allow SDF through only if gate is open) and QoS enforcement (to apply QoS rules on SDFs as provided by PCRF).

- Bearer Binding and Event Reporting Function (BBERF): It provides bearer binding and event report to PCRF. BBERF ensures that the resources which can be used by an authorized set of SDFs are within the "authorized resources" specified via the Gxx interface by "authorized QoS." The authorized QoS provides an upper bound on the resources that can be reserved (GBR) or allocated (MBR) for the service data flows.

- Traffic Detection Function (TDF): It performs application detection and reporting of detected application and its service data flow description to the PCRF. The TDF supports solicited application reporting and/or unsolicited application reporting. It supports usage reporting as well.

- RAN Congestion Awareness Function (RCAF): This is an added functionality in PCC framework for detecting RAN user plane and reporting to the PCRF to

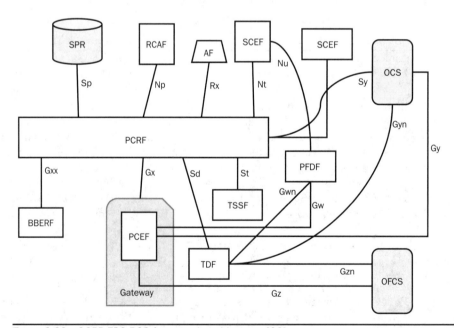

FIGURE 6.23   3GPP EPC PCC framework architecture [33]

enable the PCRF to take the RAN user-plane congestion status into account for policy decisions. The PCRF can then mitigate the congestion by taking action which will be applied by the PCEF, the TDF, or the AF.

- Service Capability Exposure Function (SCEF): This is an added functionality in Release 13 PCC framework to allow operators to interact with third-party service providers in exposing a greater set of 3GPP service capabilities as requested by third parties, using a standardized service capability exposure framework. To fill this need, Service Capability Exposure Function (SCEF) was introduced in 3GPP Release 13 for a secured exposure of 3GPP network service capability to third-party applications.

- Traffic Steering Support Function (TSSF): This is an added functionality in Release 13 to provide traffic redirecting. This redirection of traffic or traffic steering allows operators to optimize resource utilization, users' service perception and terminal or base station power consumption by directing the traffic to the Radio Access Technology (RAT) that provides the best performance, as indicated by PCRF.

- Packet Flow Description Function (PFDF): This is an added functionality to PCC framework in Release 14 to enable the management of Packet Flow Descriptions (PFDs). A PFD is an extension of an application detection filter that is preconfigured in the PCEF/TDF. It is a set of information enabling the detection of application traffic. When PFDs are provided by third-party AS via the SCEF, the PFDF stores the PFDs associated with an application identifier and transfers them to the PCEF/TDF via Gw/Gwn interface to enable the PCEF/TDF to perform accurate application detection. Provision, modification, and removal of Packet Flow Descriptions in PFDF are performed via the Nu reference point which connects the SCEF to the PFDF.

- Application Function (AF): It is an element residing application that requires dynamic behavior for policy and/or charging control over IP-CAN user plane. The AF shall communicate with the PCRF to transfer dynamic session information, required for PCRF decisions as well as to receive IPCAN-specific information and notifications about IPCAN bearer level events. One example of an AF is the PCSCF of the IMS.

- Subscription Profile Repository (SPR): It contains all subscriber/subscription-related information needed for subscription-based policies and IPCAN bearer level PCC rules by the PCRF.

- Online Charging System (OCS): It performs online credit control functions as specified in 3GPP TS 32.240 [35].

- Offline Charging System (OFCS): It performs offline credit control functions as specified in 3GPP TS 32.240 [35].

The PCC framework was developed for 3G in 2007 and enhanced further for 4G. The PCRF is renamed to Policy Control Function (PCF) in 5GC. The 5G PCF performs the same functions as the PCRF providing policy rules for control-plane functions which in turn control the traffic flowing through the user plane. This includes traffic management for network slices. The PCRF or PCF is a network element in the mobile core network which collects data in real time from different parts of the mobile network (e.g., operations support system, subscriber database, and other NFs) to determine policy rules that need to be applied for traffic handling. It is a tool that provides operators the ability to apply network and subscriber policy in real time for efficient use of network resources. In addition, it is used in applying traffic charging rules.

6.6.2   5G PCC Framework

The 5G PCC framework is an extension of 3G/4G PCC framework and applied to SBA [26]. The key NF in the PCC framework architecture is PCF which interacts (via SBI) with several other NFs in the SBA control plane in real time to assist in managing userplane network resources based on the information it gathers from other NFs. The PCF SBA is depicted in Fig. 6.24.

The following list provides the mapping of 3G/4G PCC framework entities to 5G SBA-based PCC framework (3GPP TS 23.503) [26]:

- The IP-CAN session in EPC maps to the PDU session in 5GC.
- The APN in EPC maps to DNN in 5GC.
- The IP-CAN bearer in EPC maps to the QoS flow in 5GC.
- The PCRF in EPC maps to the PCF in 5GC.
- The PCEF in EPC maps to the combination of SMF and UPF in 5GC.
- The BBFRF in EPC is located in the PCEF.
- The TDF in EPC does not apply in 5GC.

The PCF provides policy control for PDU sessions and SDFs. It interacts with AF (to gather information about type of application), UDR (for subscription information), and SMF (for requested QoS by the UE) to calculate the applicable QoS (bitrates) authorized for the SDF per access type. Depending on the type of service being provided, the PCF makes appropriate policy decisions from inputs provided by other nodes and its own predefined information. Therefore, the subscriber data handling is quite dynamic based on many factors: type of data, type of subscriber, type of scenario, amount of data traversing the network, amount of network resources available, time of day, known bottlenecks in the network, etc.

The SMF is the session manager which controls the traffic flow through the UPFs. Hence, SMF is responsible for enforcement of policy and charging decisions similar to PCEF in the 3G/4G PCC framework. The policy and charging rules calculated by the PCF are provided to SMF to apply to service data flows in the user plane. The SMF control of the UPF(s) is described in TS 23.501 [20] as well as the interaction principles between SMF and RAN and between SMF and UE. The procedures for the interaction between SMF and UPF, SMF and RAN as well as SMF and UE are described in TS 23.502 [21].

The Service Data Flow detection capability in the SMF uses the service data flow template included in a PCC Rule provided by the PCF, to send to UPF to apply these rules to SDFs. These rules also include information about the QoS to be applied to the SDFs.

6.7   Core Network Procedures

Core network procedures (Stage 2) provide the logic (i.e., interactions between the NFs) that deals with the execution of network services for timely delivery to its subscribers,

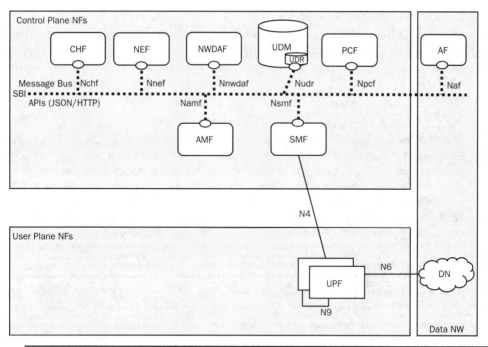

Figure 6.24   5GC PCC framework architecture (non-roaming)

such as authentication, registration, incoming/outgoing voice calls, etc. These Stage 2 procedures drive the Stage 3 protocol work. In this section, some of the basic core network procedures, executed in a mobile system, are described to give readers an idea of how NF services interact to perform different tasks.

The mobile system is designed to serve the mobile subscribers owning mobile devices (UE in 3GPP terminology). All mobile devices that require service from a mobile system have to first register on the selected home network. When a new mobile device is to be brought online on a selected home network, it has to be installed with a removable SIM or USIM card, which has a unique identity that is linked to the subscriber. This unique identity is referred to as International Mobile Subscriber Identity (IMSI), and is used to authenticate the subscriber as a valid paying customer of the selected home network, when the subscriber is roaming away from his/her home network provider and served by a visited network owned by a different mobile network provider. Initially, the home network service provider enters the SIM/USIM card information into the HLR/HSS and is stored there permanently as long as the subscriber remains as a paying customer of the service provider. The SIM/USIM card stores subscriber data, such as IMSI, Integrated Circuit Card ID (ICCID), home network ID or Service Provider Name (SPN), phone number (Mobile Country Code [MCC] + Mobile Network Code [MNC] + Mobile Subscriber Identity Number [MSIN]), personal security keys (described in UICC applications in Chap. 3), contacts list, and text messages.

6.7.1   Registration Procedure

The registration procedure of the newly bought mobile device has many variations depending on the type of mobile device and network used, such as GSM, UMTS, GPRS, EPC, or 5GC. The registration procedure called the Attach procedure is described here for the understanding of the reader, followed by its variations depending on the type of network.

6.7.1.1   Initial Attach Procedure (Release 1999)

In a GSM network, when a mobile device (called Mobile Station [MS] in GSM) is powered on for the first time, the IMSI initial attach procedure is executed. This procedure is required for the MSC and VLR to register the MS in the network. When the MS with a valid SIM card is activated (switched on) for the first time, it searches for a mobile network to connect to by communicating to the nearest cell tower, i.e., one with the strongest signal belonging to the selected home network mobile operator. Once the communication path is established between the mobile device and the radio access network, the process of initial attachment of the new mobile device with the selected home network begins, called the IMSI initial attach request procedure.

The MS transmits IMSI over the radio path which is received by the RAN. The RAN does the first check on the parameters sent by the MS and if valid forwards them to the MSC/VLR of the visited network mobile operator (referred to as VPLMN), to which the subscriber wants to be served by. Once the MS has a communication path established with the visited network, the VLR checks its database to determine whether there is an existing record of the particular subscriber. If no record is found, the VLR communicates with the subscriber's HLR/AuC to obtain necessary information so that the MS can be authenticated. If the MS passes authentication, the VLR then communicates with the home network to obtain a copy of the subscription information. This obtained information from the HLR is then stored in the database of the VLR. This is followed by sending an acknowledgment message by the serving network to the MS indicating that it is now ready to be attached to the network. At this point, the IMSI initial attach procedure is executed as shown in Fig. 6.25.

1. The newly activated MS sends a channel request message to the BSC on the radio access channel requesting a communication path establishment with the mobile network.

2. The BSC allocates a dedicated control channel to the MS for communicating with the mobile network and returns channel assigned message to the MS.

3. The MS uses the dedicated control channel and sends an Initial attach request message to the BSC, along with the IMSI and location. Since this is the first time the MS is attaching to the mobile network, the IMSI is sent as part of the Initial attach request. The IMSI is sent in clear text over the radio interface and could be vulnerable against IMSI catcher.

4. The BSC sends an acknowledgment to the MS indicating the attach request is in process.

5. The BSC forwards the Initial attach request to the MSC.

6. Since this is the initial attachment of the MS, the MSC sends the IMSI to the HLR/AuC for verification, as well as request authentication triplets.

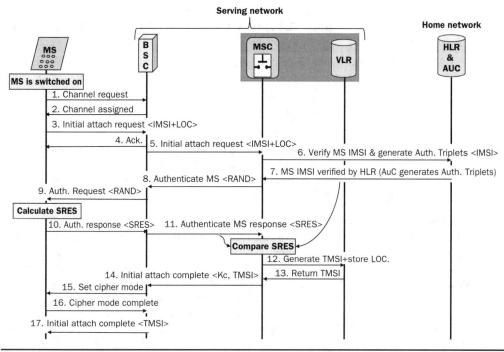

FIGURE 6.25   IMSI initial attach request procedure

7. The HLR validates the IMSI to ensure it belongs to a valid subscriber. The AuC is used to generate the authentication triplets (RAND, Kc, SRES). The HLR then forwards the IMSI and authentication triplets to the MSC.

8. The MSC stores Kc and SRES, and sends the RAND to the MS via the BSC to authenticate the MS.

9. The BSC sends authentication request message to the MS along with the RAND. The MS uses the RAND to calculate the SRES.

10. The MS sends the calculated SRES to the BSC over the dedicated control channel.

11. The BSC forwards the SRES up to the MSC. The MSC compares the SRES generated by the AuC with the SRES generated by the MS. If they match, then the authentication process of the MS is completed successfully.

12. The MSC sends a request to VLR to assign a TMSI to the MS using the MS credentials and store the location of the MS. The TMSI is used instead of IMSI to protect subscriber from being identified and also prevent radio interface eavesdropping.

13. The VLR returns the new TMSI to the MSC.

14. The MSC forwards Kc and TMSI to the BSC as part of the Initial attach complete message. The Kc and TMSI are NOT sent across the Air Interface to the MS. Instead the BSC stores the Kc and TMSI.

15. The BSC forwards the Set cipher mode command to the MS indicating the MS which encryption to use (A5/X), no other information is included.

16. The MS immediately switches to cipher mode using the A5 encryption algorithm. All transmissions are now enciphered. It sends a Cipher mode complete message to the BSC.

17. The BSC sends the TMSI in the Initial attach complete message to the MS. Since the radio Interface is now in cipher mode, the TMSI is not compromised. After the completion of the initial attachment, the MS goes into idle mode. The dedicated control channel is released.

The above is the IMSI initial attach procedure for 3GPP Release 1999 for a CS mobile network. Once the MS has gone through the initial attachment procedure, it will use Temporary Mobile Subscriber Identity (TMSI), instead of IMSI, for any subsequent correspondence with the network. This is necessary to prevent any rogue eavesdropping on the radio interface and compromising the subscriber credentials. The TMSI is randomly assigned to every MS by the VLR in the serving network, and as the subscriber roams, the TMSI is regenerated in the new serving network. The main use of the TMSI is in paging of the MS when an incoming call request comes to be delivered to it. The paging of the MS is done using the broadcast mechanism of the mobile system to locate the MS in a known given area.

To avoid the subscriber from being identified, and tracked by eavesdroppers on the radio interface, the mobile network normally will change the TMSI of the mobile frequently. This frequent changing of the TMSI makes it difficult to trace which mobile is which, except briefly, when the MS is just switched on, or when the data in the MS becomes invalid for one reason or another. In such rare situations, the global IMSI must be sent to the network.

When the MS is powered off, it loses all communication with the mobile system. However, the last assigned TMSI is saved in the MS, so when the MS is powered on again it will send the saved TMSI to the serving network, and go through the attach procedure without requiring to involve HPLMN. The attach procedure in this case is similar to as described in Fig. 6.25, in which the MS uses TMSI instead of the IMSI.

If the MS is registered and has gone through the initial attach procedure, and then changed location while it was powered off, then instead of the IMSI attach procedure it will execute location update procedure only.

6.7.1.2 Combined Attach Procedure (UMTS—Release 4)

In Release 97 the packet core was added to enable data services to be supported to the GSM network. This required that the MSs that were able to support both CS and PS services do a combined attach procedure for both parts of the core network. The combined attach was also introduced to GPRS, i.e., once the MS attached to the SGSN, the SGSN notified the VLR of the MSC about the registration state. The combined attach procedure uses the Gs interface between SGSN and MSC/VLR. The combined attach saves the radio resources by avoiding execution of two different attach procedures. The

UMTS combined attach procedure, very similar to the Release 97 GSM procedure, is shown in Fig. 6.26.

Prior to the initiation of the UMTS combined attach procedure the following preconditions exist:

- UE is powered up (switched-on) and was registered in the visited serving network.

- UE moves into a new visited serving network.

With the preconditions, the UMTS combined attached procedure is performed as follows:

1. The UE sends a channel request message to the RNC on the radio access channel requesting a communication path establishment with the mobile network.

2. The RNC allocates a dedicated control channel to the UE for communicating with the mobile network and returns channel assigned message to the UE.

3. The UE uses the dedicated control channel and sends an attach request message to the RNC, along with the TMSI, assuming that the UE has already registered once before in the system.

4. The RNC sends an acknowledgment to the UE indicating the attach request is in process.

5. The RNC forwards the initial attach request to the new SGSN with the TMSI.

6. Since UE was previously attached and registered, the new SGSN sends Identification request to the old SGSN to identify the UE's credentials. The new SGSN can work out the location of the old SGSN from the construct of the TMSI.

7. The old SGSN confirms the UE's credentials and receives the UE's associated IMSI with the TMSI.

8. The process of generating a new set of authentication triplets for UE is executed between the new SGSN and the HLR/AuC.

9. The UE is authenticated.

10. The location updated procedure is initiated by the new SGSN to update the new MSC/VLR about the new location of the UE in the current visited network.

11. The new MSC/VLR sends the location update message to the HLR.

12. The HLR sends the cancel location message to the old MSC/VLR in the previous visited network to remove UE's location as the UE has moved to the new visited network.

13. The old MSC sends an acknowledgment.

14. The HLR sends insert subscriber data message to the new MSC/VLR to store the visiting UE's information, such as IMSI, subscription information, etc.

15. The new MSC/VLR sends acknowledges to the HLR that the visiting UE's subscriber data is received.

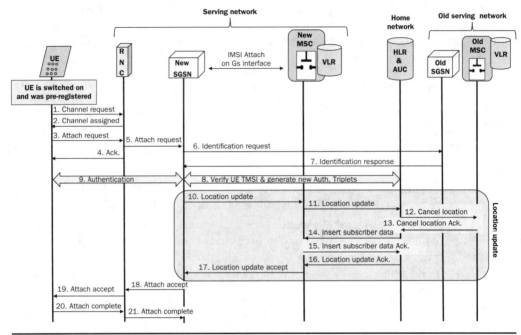

Figure 6.26   UMTS combined attach procedure

16. The HLR sends an acknowledgment to the new MSC/VLR that the location update procedure is complete.

17. The new MSC/VLR sends location update accept message to the new SGSN indicating the UE's location update is complete.

18. The new SGSN sends attach accept message to the RNC.

19. The RNC forwards attach accept message to the UE.

20. The UE immediately switches to cipher mode using the A5 encryption algorithm. All transmissions are now enciphered. It sends an attach complete message to the RNC.

21. The RNC forwards the attach complete message to the new SGSN. The UE goes into idle mode. The dedicated control channel is released.

Figure 6.26 describes a combined GPRS/IMSI attach procedure in a cell that supports GPRS. The new SGSN uses the location update procedure as soon as the new SGSN receives data related to the GPRS subscriber from the HLR. When the new SGSN receives the acceptance of IMSI attach from the MSC/VLR entity, it transmits the IMSI and GPRS attach confirmation. If the MS receives a new P-TMSI identifier or a new TMSI identifier, then it acknowledges it to end the combined attach procedure.

6.7.1.3 Attach Procedure (LTE–Release 8)
Since LTE/EPC was developed as a new system in Release 8, it was designed as an "always on" system with IP connectivity directly during the Attach procedure to the network. The simplified attach procedure is shown in Fig. 6.27.

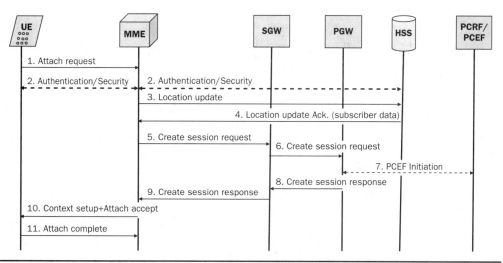

FIGURE 6.27   LTE attach procedure

1. The UE sends an attach request to the MME, which is the terminating point for the NAS protocol.

2. The MME may decide to authenticate the UE and would request authentication information from the HSS.

3. The MME sends a location update message to the HSS and requests information with respect to the UEs subscription.

4. The MME retrieves the subscriber data from the HSS in the location update acknowledgment. The subscriber data contains the PDN subscription contexts with the subscribed QoS profiles.

5. The MME selects the PGW corresponding to the default APN for default bearer activation and sends a create session request to the SGW.

6. The SGW forwards the create session request to the PGW.

7. The PGW interacts with the PCRF in case dynamic QoS is deployed and retrieves updated QoS parameters and PCC rules for the default bearer.

8. The PGW returns a create session response to the SGW with all relevant information to the session.

9. The SGW returns the create session response message to the MME.

10. The MME sends the attach accept to the UE, which is part of the initial context setup to the eNB. The eNB then initiates an RRC reconfiguration of radio bearers with the UE.

11. The UE sends an attach complete to the MME.

The UE is now attached and recognized by the network and is authorized to receive services.

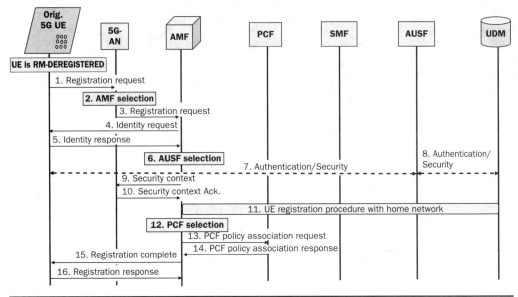

Figure 6.28 5G initial registration procedure

6.7.1.4 5G UE Registration (Release 15)

The 5G UE registration procedure in 3GPP TS 23.502, clause 4.2.2.2 [19] has many different variations depending upon UE's current state, type of UE, and type of registration requested, but the focus here will be on initial registration when a UE is in idle state (called RM-DEGISTERED state) and needs to register with the network to get authorized to receive services, to enable mobility tracking, and to enable reachability. The type of registration helps the 5GC in selection of the AMF. This initial registration is similar to the initial attach in EPC. However, in this initial registration, the UE is also requesting a network slice to be allocated. In 5G, the UEs are becoming complex due to the many different types of services/capabilities they can support (e.g., CIoT, V2X, network slicing, dual mode, etc.) along with multiple radio access types. The 5G UE initial registration procedure call flow is shown in Fig. 6.28.

It is assumed the following conditions exist prior to execution of the 5G UE initial registration procedure:

- UE is registering on 5G RAN (NR).

- Old AMF is not involved since this is initial registration.

1. UE sends registration request to NG-RAN (with parameters: registration type = initial registration, UE ID = 5G-GUTI+GUAMI or 5G-S-TMSI, requested NSSAI, UE radio capability information, UE policy container). The requested NSSAI corresponds to the slice(s) to which the UE wishes to register, in addition to the 5G-S-TMSI. The UE radio capability information contains information on RATs that the UE supports (e.g., power class, frequency bands, etc.).

2. The 5G-AN selects an AMF according to the type of UE and its capabilities. In the call flow presented here, the UE supports slicing and sends the requested NSSAI and a 5G-S-TMSI or a GUAMI in RRC connection establishment, then 5G-AN prepares to forward the request to this AMF.

3. 5G-AN sends registration request to the selected AMF, with parameters: Selected PLMN ID (or PLMN ID and NID, see TS 23.501 [20], clause 5.34), Location Information and Cell Identity related to the cell in which the UE is camping, UE Context Request which indicates that a UE context including security information needs to be set up at the 5G-AN.

4. If the SUCI is not provided by the UE, the identity request procedure is initiated by AMF sending an identity request message to the UE requesting the SUCI.

5. The UE responds with an identity response message including the SUCI. The UE derives the SUCI by using the provisioned public key of the HPLMN.

6. The UE here is being registered for the first time, hence requires authentication. Therefore, the AMF initiates UE authentication by invoking an AUSF. In that case, the AMF selects an AUSF instance, based on SUCI, which performs authentication between the UE and 5G CN in the HPLMN.

7. Upon request from the AMF, the AUSF executes authentication of the UE. The AUSF selects the HPLMN UDM and gets the authentication data from UDM.

8. Once the UE has been authenticated the AUSF provides relevant security-related information to the AMF. In case the AMF provides an SUCI to AUSF, the AUSF shall return the SUPI to AMF only after the authentication is successful.

9. The AMF provides the security context information to the 5G-AN.

10. The 5G-AN stores the security context and acknowledges to the AMF. The 5G-AN uses the security context to protect the messages exchanged with the UE.

11. The AMF performs registration procedure with the UDM in the UEs home network. This procedure will download the subscribers' profile containing their identity and services they have subscribed to (e.g., support of IMS voice over PS session).

12. To support UE roaming, the AMF initiates PCF communication. AMF decides to perform PCF discovery and selection and the AMF selects a (V)-PCF and may select an H-PCF (for roaming scenario).

13. AMF performs a UE policy association establishment sending Npcf_UEPolicyControl create request.

14. PCF sends an Npcf_UEPolicyControl create response to the new AMF.

15. The AMF sends a registration accept message to the UE indicating that the registration request has been accepted. The allowed NSSAI provided in the registration accept is valid in the registration area and it applies for all the PLMNs which have their tracking areas included in the registration area.

16. UE sends registration complete to AMF.

The UE is now registered in the 5GS and is authorized to receive services.

6.7.2 Data Bearer Establishment

When UE initiates a specific service (e.g., initial registration, voice call origination, IMS mobile originated SMS, as shown in Figs. 6.28, 6.31, and 6.50, respectively), it needs to establish a data connection with network. This is referred to as data radio bearer (DRB) establishment in 2G/3G, or PDN connection in 4G, or PDU session establishment in 5G. The DRB is constructed between the UE and the cellular base station (cell tower) to prepare UE for data transmission over its radio bearer. On the other hand, PDN connection or PDU session establishment is for identifying IP addresses of various nodes in the network over which the UE will communicate. This is explained in this section.

6.7.2.1 2G/3G Data Bearer Establishment

1. UE sends an activate PDP context request to the SGSN, and it contains an identifier of the service that the UE wants to activate. This service activation identifier is called an Access Point Name (APN) and usually in the format of label#1.label#2.label#3, e.g., webapn.mobileoperator.com. At the SGSN, the SGSN will perform a DNS query to determine which GGSN to contact.

2. The SGSN will send a create PDP context request to the GGSN containing the identity of the UE. Upon receipt of the request the GGSN will create the PDP context for the data bearer.

3. The GGSN sends a response back to the SGSN.

4. The SGSN sends a success message back to the UE.

5. The data bearer establishment is complete, as shown in Fig. 6.29.

6.7.2.2 5G PDU Session Establishment Procedure

1. The UE sends a PDU session establishment request including the S-NSSAI to the AMF.

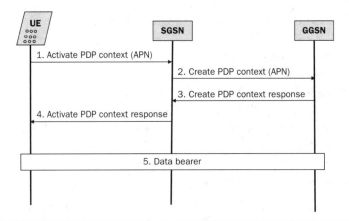

FIGURE 6.29 2G/3G establishment of data bearer

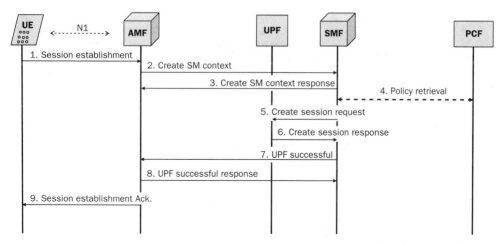

FIGURE 6.30   5G PDU session establishment procedure

2. The AMF selects an SMF and sends a create SM context request with the S-NSSAI to the SMF.

3. The SMF may retrieve session management subscription data from the UDM (not shown here) and creates an SM context and responds to the AMF with a create SM response with the SM context ID.

4. In order to get the default PCC rules for the PDU session, the SMF may perform an SM policy association establishment procedure and establish an SM policy association with the PCF.

5. The SMF selects the UPF and sends a create session establishment request for this PDU session.

6. The UPF acknowledges the session establishment with create session response.

7. The SMF reports the successful session establishment at the UPF with all relevant parameters, e.g., QoS as well as information for the RAN, to the AMF.

8. The AMF acknowledges the message from the SMF.

9. The AMF sends a session establishment acknowledgment response to the UE.

6.7.3   Mobile Voice Call Origination

6.7.3.1   Mobile Voice Call Origination in GSM

An MS attached to the GSM network can initiate a CS voice call to the remote party. The call flow for MS call origination to remote party is shown in Fig. 6.31.

MS-originated call in GSM

1. MS sends a request to BSC for the allocation of a dedicated signaling channel to perform the call setup. The BSC sends the request to MSC.

2. The dedicated signaling channel is assigned for the MS to begin direct communication with the MSC.

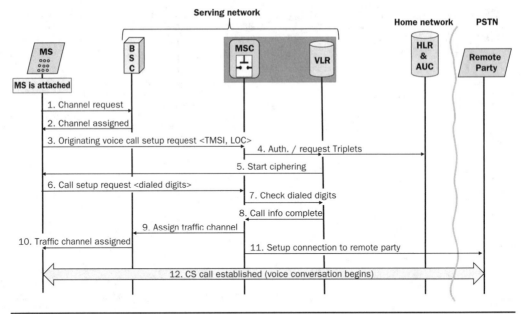

FIGURE 6.31   GSM MS CS voice call origination procedure

3. MS sends a request to MSC for originating voice call setup, and includes the TMSI and its last location.

4. MSC sends request to VLR to authenticate the requesting MS's TMSI followed by VLR sending a request to HLR to generate authentication triplets to enable ciphering between MS and MSC/VLR.

5. After VLR has authenticated the TMSI and received the triplets from HLR, the path between the MS and MSC/VLR is secure and ciphering enabled.

6. MS sends the call setup request with the dialed digits to the MSC.

7. MSC then forwards the dialed digits and subscriber call information to VLR for verification if the requested service can be enabled.

8. VLR verification is completed.

9. After VLR verification is done successfully, the MSC commands the BSC to assign a traffic channel (i.e., resources for speech data transmission) to the MS.

10. The BSC assigns a traffic channel TCH to the MS.

11. The MSC initiates connection to the remote called party in the PSTN.

12. After the remote party is successfully connected, the voice communication between originating MS and the remote called party can begin over a dedicated circuit connection.

Note: There can be several error cases in the setup of the voice call which are not shown. In addition, there can be several variations depending on where the remote party is located which are not covered. The above call flow conveys the basic idea on how the various system entities interact to complete the call origination service requested.

6.7.3.2   *Mobile Voice Call Origination in LTE/EPC*

6.7.3.2.1   General   The UE that is VoLTE capable can originate a voice call over LTE (VoLTE) either to another terminating VoLTE UE, or to a terminating UE in the CS domain. Since EPC is an IP-based system using SIP signaling, the voice call is handled by IMS based on SIP signaling. A VoLTE UE, under LTE coverage, shall automatically perform an LTE attach followed by an IMS registration for VoLTE, if the network supports VoLTE. This ensures that the VoLTE UE shall be available for VoLTE services (i.e., incoming calls, outgoing calls, and supplementary services), similar to the voice experience in CS GSM network deployments.

When IMS is used there are two phases for the call to be set up. One is the establishment of the signaling path. This will use an EPS bearer that is dedicated for IMS signaling. In addition, another EPS bearer will be assigned to carry the user plane, voice traffic.

6.7.3.2.2   Control-Plane Signaling   The VoLTE mobile originating call flow is shown in Fig. 6.32.

The originating VoLTE terminal initiating the VoLTE voice call is attached to the LTE/EPC and registered.

1. Originating VoLTE terminal sends SIP "INVITE" message to terminating VoLTE terminal via the IMS network, sending its terminal ID and codec used.

2. Terminating VoLTE terminal responds with SIP "183 session progress" sending its codec used to P-CSCF.

3. P-CSCF starts the process to establish dedicated bearer with EPC for voice media by involving PCRF.

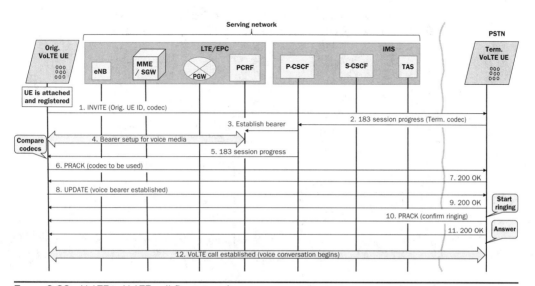

FIGURE 6.32   VoLTE-to-VoLTE call flow procedure

4. Dedicated bearer is established by PGW, SGW, eNB, and originating terminal, using QCI = 1 for voice conversation.

5. In parallel to (4), P-CSCF forwards SIP "183 session progress" message to originating terminal. The originating terminal compares the codec used by terminating terminal with its own to ensure the same codec is used on both ends.

6. Originating terminal sends SIP "PRACK" message to the terminating terminal confirming the codec to be used.

7. Terminating terminal responds with SIP "200 OK" message.

8. Originating terminal sends SIP "UPDATE" message indicating establishment of voice bearer.

9. Terminating terminal acknowledges by responding with "200 OK."

10. The terminating terminal starts ringing and informs the originating terminal by sending SIP "PRACK."

11. The call is answered by the terminating terminal and informs originating terminal by sending "200 OK."

12. The call proceeds successfully.

The originating and terminating parties are in conversation.

6.7.3.2.3   User-Plane Bearer Establishment   When P-CSCF receives a 183 containing an SDP answer it will take that information and convert it into the necessary format to send to the PCRF using Diameter protocol. The PCRF then sends that information to the SGW/PGW so that the dedicated bearer can be activated toward the UE (Fig. 6.33).

1. The P-CSCF will receive an SIP 183 that contains an SDP answer. The SDP answer contains information regarding what service is being set up and its characteristics.

2. The P-CSCF sends a Diameter message called AA Request (AAR) command (see 3GPP TS x.y.z) containing media type being requested, codec, etc.

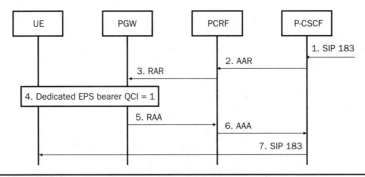

FIGURE 6.33   User-plane bearer establishment procedure

3. The PCRF will send a Re-Auth-Request (RAR) message to the PGW containing a PCC rule (e.g., QCI, GBR, ARP, precedence).

4. A dedicated bearer will be set up to the UE from the PGW. The QCI will be 1 which is used for voice call.

5. PGW sends back a Re-Auth-Answer (RAA) message with information like identity of the PGW, Radio access technology being used by the UE.

6. The PCRF sends AA Answer (AAA) to the P-CSCF.

7. The P-CSCF sends SIP 183 to the UE.

6.7.3.2.4 EPS Fallback

6.7.3.2.4.1 Mobile Voice Call Establishment in 5GC Voice of 5GC will be supported using IMS. The 5GC will not be deployed everywhere initially, so in the interim it will coexist with EPC, with support of EPC interworking possible via N26 interface. In 3GPP Release 15, voice calling will also be supported on 5GC using EPS fallback (using dual mode 5G UE) or RAT fallback. RAT fallback is where E-UTRA (LTE) radio is connected to the 5GC. For EPS fallback, the UE originating a voice call is moved from NR (5G) to LTE (4G) and the voice call is established on 5GC, i.e., the number dialing is done on 5GC but the voice call is established on EPC.

As background, in EPS (eNB+EPC), if Circuit-Switched Fallback (CSFB) is supported, the UE falls back from EPS to 2G/3G UMTS to make or receive calls. Or if the EPS supports Voice over LTE (VoLTE) with call control provided via IMS, the UE will remain on EPS to make or receive calls. Initially, 5GC will not be deployed widely. Therefore, 5G UE will make or receive calls using EPS fallback (not shown) or IMS, as shown in Fig. 6.51, sec. 6.9.8.5.2.

6.7.4 Handover

An active mobile device without any voice/data session in progress is still always communicating over the radio interface with a radio cell tower. This allows the mobile network to be aware of mobile device's location in case a service request has to be delivered to it. When this active mobile device moves to another location, it reattaches to mobile network via a different radio cell tower in that location and its new location is updated. This movement of the mobile device is called reattachment. It is not handover. The handover procedure comes into play when the mobile device which has an active voice/data session in progress is in motion. As the mobile device moves, its signal deteriorates as it moves away from the cell tower it is connected to, and reconnects to the next cell tower which is closer to the mobile device at that time with stronger signal. During this process the voice/data session in progress has to be maintained by handing over the voice call from one cell tower to another. At a high level, there are two types of handovers:

- Break before make (hard handover)
- Make before break (soft handover)

Figure 6.34 depicts the hard and soft handover. During the hard handover, the voice/data session will experience a slight interruption. For a voice session this is hardly noticeable to human ear. However, during the soft handover, the mobile has two connections ongoing (one to the old cell tower and the other to the new one). The new connection is established prior to dropping the old one. In this way, the handover of session does not experience any interruption. The mobile operators may select which type of handovers they want to support in their network.

6.7.4.1   N2-Based Handover in 5GC

This section provides details regarding the inter NG-RAN handover of active PDU sessions (for handover of both downlink and uplink user-plane data) using the N2 interface (between AMF and NG-RAN) without any involvement of the Xn interface (between two NG-RANs) [21]. If the Xn-based handover is not used (or not available due to the specific deployment) then the handover from source to target NG-RAN is totally performed under control of the 5GC. The N2-based handover is triggered from source NG-RAN to the target NG-RAN when weak radio conditions are detected at the source NG-RAN and when there is no Xn connectivity between the source NG-RAN and the target NG-RAN.

The handover procedure is done in two phases as shown in Fig. 6.35 (preparation phase) and Fig. 6.36 (execution phase). This handover can use direct forwarding path when transferring downlink user-plane data between source and target NG-RANs which requires that an IP connectivity between the source and target NG-RANs and security association(s) are in place. Otherwise, the option of indirect forwarding via Source UPF can be used (also shown).

There are many variations and options available in the N2-based handover procedure, which are not detailed here but can be found in TS 23.502 [21]. Moreover, there are other types of handover procedures (e.g., Xn-based handover) which are not detailed here.

6.7.4.1.1   N2-Based Handover—Preparation Phase   The N2-based handover is triggered by the source NG-RAN when it detects weak radio conditions and when there is no Xn connectivity between the source NG-RAN and the target NG-RAN.

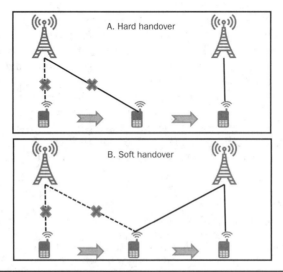

FIGURE 6.34   Voice call hard and soft handover

1. The Source NG-RAN sends "Handover Required" message to the Source AMF (with parameters: Target NG-RAN ID carried in Source to Target transparent container, Source NG-RAN information to be used by Target NG-RAN, user-plane security enforcement, QoS information, Direct Forwarding Path information, PDU session IDs, intra system handover indication). All PDU sessions handled by Source RAN (i.e., all existing PDU sessions with active UP connections) shall be included in the Handover Required message.

2. The Source AMF selects Target AMF when it can't serve the Originating 5G UE anymore.

3. The Source AMF initiates Handover resource allocation procedure by invoking the "Namf_Communication_CreateUEContext Request" toward the Target AMF, including many parameters sent in step 1.

4. The Target AMF sends "Nsmf_PDUSession_UpdateSMContext" message to the SMF (with parameters: PDU session ID, Target NG-RAN ID, Target AMF ID, N2 SM Information). PDU session ID indicates a PDU session candidate for N2 Handover. Target ID corresponds to Target ID provided by S-RAN in step 1. SM N2 Information includes the Direct Forwarding Path Availability if the direct data forwarding is available between the Source NG-RAN and the Target NG-RAN and has been inserted by the Source NG-RAN.

5. Based on the Target NG-RAN ID, SMF checks if N2-based Handover for the indicated PDU session can be accepted. The SMF checks also the UPF Selection Criteria [19].

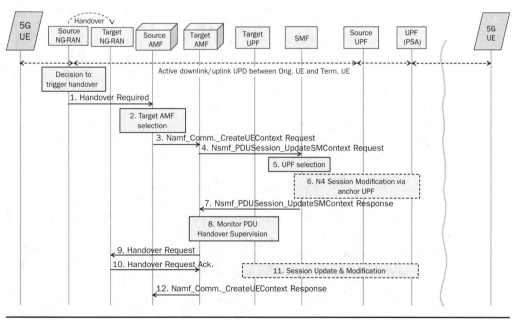

FIGURE 6.35   Inter NG-RAN node N2-based handover procedure: Preparation phase

6. This includes several steps carried out between SMF and UPF (PSA), and between SMF and Target UPF. These steps are required to prepare the UPF (PSA) and Target UPF to get ready to accept the PDU session transfer during the execution phase.

7. The SMF sends "Nsmf_PDUSession_UpdateSMContext Response message to the Target AMF with parameters: PDU session ID, N2 SM information, reason for non-acceptance.

8. The Target AMF supervises the maximum time it should wait for "Nsmf_PDUSession_UpdateSMContext Response" messages from the involved SMFs, for different active PDU sessions that are subject to handover. The lowest value of the Max delay indications for the PDU sessions that are candidates for handover gives the maximum time AMF may wait before continuing with the N2 Handover procedure.

9. The Target AMF sends "Handover Request" message to the Target NG-RAN (with parameters: Source to Target transparent container with Target NG-RAN ID, N2 MM Information, N2 SM Information list, Tracing Requirements if available, UE Radio Capability ID). Source to Target transparent container is forwarded as received from Source NG-RAN. N2 SM Information list includes N2 SM Information received from SMFs for the Target RAN in the Nsmf_PDUSession_UpdateSMContext Response messages received within allowed max delay supervised by the Target AMF mentioned in step 8.

10. The Target NG-RAN sends "Handover Request Acknowledge" message to the Target AMF (with parameters: Target to Source transparent container, List of PDU sessions to handover with N2 SM information, List of PDU sessions that failed to be established with the failure cause given in the N2 SM information element). Target to Source transparent container includes a UE container with an access stratum part and an NAS part. The UE container is sent transparently via Target AMF, Source AMF, and Source RAN to the UE.

11. This includes several steps carried from Target AMF to SMF, SMF to Target UPF, Target UPF to SMF, SMF to Source UPF, Source UPF to SMF, and SMF to Target AMF. All these steps are required to prepare Source and Target UPFs for acceptance of PDU session transfer in the execution phase.

12. The T-AMF sends "Namf_Communication_CreateUEContext Response" message to the Source AMF [with parameters: N2 information necessary for S-AMF to send Handover Command to Source RAN including Target to Source transparent container, PDU sessions failed to be set up list, N2 SM information (N3 DL forwarding Information, PCF ID)]. The Target to Source transport container is sent by the Target NG-RAN and the N2 SM Information is received from step 11. After this step the handover execution phase can proceed.

6.7.4.1.2 N2-Based Handover—Execution Phase The AMF interacts with UDM for registration purpose, which is not shown in the call flow.

1. The Source AMF sends "Handover Command" to the Source NG-RAN (with parameters: Target to Source transparent container as received, List of PDU

sessions to be handed over with N2 SM information containing information received from target NG-RAN during the handover preparation phase, List of PDU sessions failed to be set up).

2. The Source NG-RAN sends "Handover Command" to the 5G UE with the UE container parameter, which is the UE part of the Target to Source transparent container sent transparently from target NG-RAN.

3. The Source NG-RAN sends the "Uplink RAN Status Transfer" message to the source AMF.

4. The Source AMF sends "Namf_Communication_N1N2MessageTransfer" to the Target AMF. This message is needed to relocate the AMF.

5. The Target AMF sends the information to the target NG-RAN via the "Downlink RAN Status Transfer" message. At this point the downlink user-plane data path is established between the terminating UE and the source NG-RAN via anchor UPF (PSA) and source UPF. Now the uplink user-plane data path needs to be set up next.

6. The Source NG-RAN starts forwarding of downlink data toward the Target NG-RAN, via either direct (step 6a) or indirect (step 6b) forwarding based on QoS data forwarding rules applied to data packets.

7. The 5G UE synchronizes to the target cell in the target NG-RAN.

8. After the the 5G UE has successfully synchronized to the target cell, it sends a "Handover Confirm" message to the Target NG-RAN. At this point the handover is pronounced successful by the originating 5G UE, and now the Target NG-RAN starts forwarding downlink data to the UE which in turns starts sending uplink data to the terminating the 5G UE via target UPF and anchor UPF (PSA).

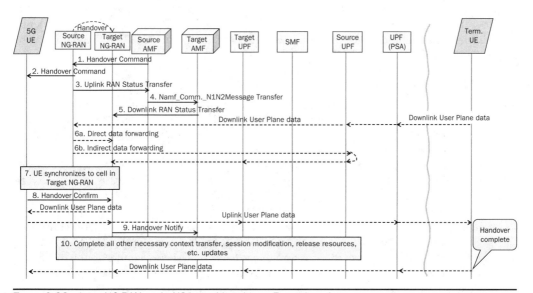

FIGURE 6.36 Inter NG-RAN node N2-based handover: Execution phase

9. The Target NG-RAN "Handover Notify" to the Target to AMF. This is to notify the Target AMF that handover has been successful.

10. Complete all other necessary context transfer, session modification, release of resources, etc. Downlink data starts flowing from terminating UE to originating 5G UE. The two-way path is established between originating the 5G UE and terminating UE via the target NG-RAN. The handover is now complete.

6.8   Network Slicing Feature

In the development of 5G standards, the network slicing feature received a lot of attention as it enables operators to manage their network resources in unique ways based on customer traffic needs. This can be done by partitioning the network resources via creation of network slices (NSs) consisting of dedicated network resources that are needed to do the job, rather than using the entire network and wasting resources. Like what was done previously when there was only a single monolithic network whose resources could not be partitioned resulting in under- or overutilization of resources. For example, NSs can be created to address different requirements on functionality (i.e., priority, charging, policy control, security, and mobility), or different requirements on performance (i.e., latency, mobility, availability, reliability, and data rates), or an NS can be created to serve only specific users (i.e., MPS users, public safety users, corporate customers, roamers, or hosting an MVNO).

In this direction, the 5G NS requirements are provided in the 3GPP Stage 1 specification TS 22.261 [36]. A network slice is a complete logically isolated network which is customized per vertical requirements, including both 3GPP (RAN, CN) and non-3GPP domains (transport network). In Chap. 5, the concept of network slicing is mentioned from the RAN perspective. In this chapter, the concept of network slicing is mentioned from the CN perspective.

The 3GPP Release 15 technical study on "Management and orchestration of network slicing for next-generation network," captured in TR 23.801, provides in-depth background behind the development of network slicing feature [37]. The other relevant 3GPP specifications are TS 28.530 [38] and TS 28.531 [39].

6.8.1   Background

The concept of network slicing has existed for a long period in computer networking, but not under the network slicing name. The virtual machines (VMs) and virtual private networks (VPNs) in computer networking have similar functionality of a network slice. In telecommunications, network slicing closely resembles VPNs, which are separate isolated networks created for use by private enterprises/companies. The VPNs were one of the key features of telecom Intelligent Networks in the early 1990s, which were initially standardized by ITU-T in Q.1200 Recommendation series [40–44]. The VPNs were created by partitioning the common telecom network resources into separately managed SDNs or virtual machines dedicated to serve specific customers. The VPNs are isolated from each other to provide a secure networking environment for enterprise users. VPNs allow company employees to create a secure connection, while away from office, to their internal company network via a secure encrypted and authenticated communication channel or tunnel between two endpoints over the internet. The

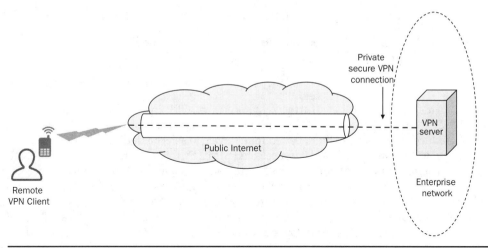

Private
secure VPN
connection

VPN
server

Public Internet

Remote
VPN Client

Enterprise
network

FIGURE 6.37   Virtual private network

authentication, encryption, and authentication of the channel or tunnel is dependent on the underlying network being used—Point-to-Point Tunneling Protocol (PPTP) or Layer 2 Tunneling Protocol/IPSec (L2TP/IPSec).

Figure 6.37 shows a typical VPN environment.

Mobile users frequently connect to their enterprise network from remote locations over unsecure Wi-Fi hotspot enterprises; therefore, it is recommended to use VPNs to prevent a data breach. VPNs disguise the IP address of the device on one end of the tunnel by "virtually" connecting it to the network on the other end of the tunnel masking the real location of the end user. Since VPNs provide a secure end-to-end channel over the internet, it circumvents political or government censorship imposed on the internet. It should be noted that the use of VPN over public internet is banned in certain countries.

Another feature developed by the 3GPP standards for EPC in Release 13 is called Dedicated Core Network (DECOR or DCN), clause 4.3.25 in 3GPP TS 23.401, clause 4.3.25 [12], in PS or CS domain, within a single PLMN to serve different types of subscribers/end devices. A DCN, as its name implies, is a network created using common mobile network resources to cater to different terminal types, based on the traffic type. The need for DCN came about when it became necessary to segregate traffic that is generated by low-end legacy devices from the traffic generated by smart devices (smartphones). The smart devices need faster response times and better quality of service; therefore, it became necessary to segregate the traffic generated from smartphones to be handled by a separate dedicated core network. The rerouting of traffic from different terminals can be done at the time of attach procedure by checking the characteristics of the device. DCN provides the capability to have a dedicated MME/SGW per UE type. To identify which UE type belongs to which DCN, 3GPP in Release 13 introduced a new subscription parameter called "UE Usage Type" which is stored in the HSS within the subscription information of UE.

Note: Each device supported by 3GPP networks can have no more than one type.

To select the correct MME/SGW, 3GPP has defined the following methods for EPC:

- Network Node Selection Function (NNSF) in eNodeB
- Local configuration on the eNodeB
- Query DNS and load balance with weighing depending on UE location

The DCN feature was further enhanced by 3GPP in Release 14, called enhanced DECOR (eDECOR) or eDCN,[6] by adding the assistance information for devices in order to reduce network signaling needed for a device to select a dedicated core network, i.e., to avoid the rerouting from a first selected MME by the RAN to the dedicated MME.

The signaling reduction in eDCN happens by avoiding the device to select the wrong dedicated core network, in the first place, and making it unnecessary to use the reroute function as previously required in the DCN feature. In eDCN, two additional parameters were added in devices to assist in the selection of the dedicated core network: (1) DCN Selection Assistance parameter identifying the device's intended type of usage, for example, MBB DCN, V2X DCN, etc. and (2) NAS type parameter used during DCN selection and serving node selection to ensure the selected node can respond to NAS messages. One NAS type may indicate the use of narrowband IoT and another NAS type may indicate mobile broadband IoT.

The Release 13 DECOR and Release 14 eDECOR features can be viewed as a precursor to network slicing. However, network slicing is much more advanced entailing the end-to-end mobile system.

6.8.2   5G Network Slicing Architecture

Network slicing specified for 5G in Release 15 is a concept that allows MNOs to slice their existing end-to-end network infrastructure (RF, Transport, and Core) into several network slice instances (NSIs) where each NSI may be fully or partly, logically and/or physically, isolated from another NSI. An NSI may be composed by none, one, or more Network Slice Subnet Instance (NSSI), and may be shared by another NSI. Similarly, the NSSI is formed of a set of NFs, which can be either virtual network functions (VNFs) or physical network functions (PNFs). With network slicing, the MNO can maximize the usage of network resources through sharing the network infrastructure across different groups of services not possible in traditional dedicated monolithic network models of previous generations. The MNO shared infrastructure means that NSIs can be leased to vertical service providers called Network Slicing as a Service (NSaaS). NSaaS can be designed with different granularities based on customer needs and managed accordingly. This will be a new multi-tenancy resource usage model used in 5G. Figure 6.38 based on ETSI GR NFV-EVE 012 [41] helps visualize the network slicing layered architecture concept for the creation and management of NSIs.

Network slicing feature in 5G provides many more capabilities than VPN, DCN, or eDCN, and is an integral part of the SDN and NFV technologies. It consists of network slices as isolated end-to-end complete logical networks, with compute and storage capabilities, running on a common underlying physical or virtual network. These network slices can be created on demand by the operators and managed separately. Each

---

[6]Whenever an "e" appears at the beginning of a 3GPP feature name it means that an existing feature has been enhanced.

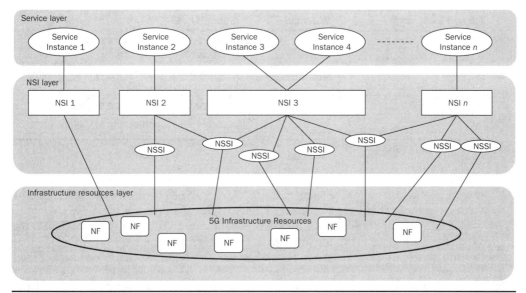

FIGURE 6.38   Network slicing management architecture

network slice is created to support diverse use cases as demanded by the customer, allowing operators to split the network resources into multiple virtual networks. Up until 3G systems, all types of services were provided through a single network. In 4G, DCN and eDCN features allowed traffic generated from different UE types to be treated separately. Now, in 5G, with network slicing capability the operators will be able to divide their entire infrastructure into different slices each with its own configuration and specific QoS requirements. This will allow efficient utilization of network resources not possible in previous generations. As mentioned previously, network slicing is an integral part of SDN and NFV, and these technologies are facilitating the operator networks toward software-based automation.

In 3GPP Spec TS 23.501 [20], the following identifiers have been defined which are used in network slicing:

- Single Network Slice Selection Assistance Information (S-NSSAI): This identifies the network slice. It is used by the UE residing in the access network of the PLMN to identify the network slice being used. It comprises the following:

  o Slice/Service Type (SST): The expected network slice behavior in terms of features and services.

  o Slice Differentiator (SD): An optional identifier that complements the SST to differentiate amongst multiple NSs of the same SST.

The SSTs which are standardized are listed in Table 6.4.

The support of all standardized SST values is not required in an hPLMN or vPLMN. The service types listed in Table 6.4 for each SST value can also be supported by means of other SSTs. What does this mean? It means that PLMNs have the liberty to assign their own nonstandard SST values for the network slicing services they offer, and they are not bound to support network slicing for roaming subscribers in a standardized manner. It

Slice/Service type	SST value	Characteristics
eMBB	1	Slice suitable for the handling of 5G-enhanced Mobile Broadband.
URLLC	2	Slice suitable for the handling of ultra-reliable low-latency communications.
MIoT	3	Slice suitable for the handling of massive IoT.
V2X	4	Slice suitable for the handling of V2X services.

Table 6.4   Standardized SST values

is worth noting that vPLMN can map the hPLMN S-NSSAI to its own vPLMN S-NSSAI value, which can be the same value or different value than the hPLMN S-NSSAI.

At the time of subscription, each UE is configured with one or more S-NSSAIs. When the UE registers with the serving network, it sends the requested S-NSSAI(s) that it wants to use in the registration request to the serving network. If none is sent then the serving network will assign a default S-NSSAI based on UE and type of services it supports. If the UE is roaming away from its home network, the VPLMN receives from HPLMN UDM the UE subscribed S-NSSAI(s). The serving network informs the UE which S-NSSAIs the UE is allowed to use for this registration.

Another parameter in the 5GC is NSSAI which is a collection of supported S-NSSAIs by the UE to allow an easier way to communicate the multiple supported S-NSSAIs to the serving network in a single request. There can be at most eight S-NSSAIs allowed and requested for simultaneous use in an NSSAI signaling message between the UE and the 5GC. The RAN may use requested NSSAI in access stratum signaling to handle the UE control-plane connection for use of AMF selection. It is worth noting that the UE may or may not send any S-NSSAI or NSSAI in the access stratum layer due to security reasons and depending on operator policy. By default, the UE usually does not send anything in the access stratum layer.

6.8.3   Operational Aspects of Network Slicing

The role of selecting network slice(s), comprised of a collection of specific NFIs, identified via S-NSSAI or NSSAI is allocated to serving AMF which is the first point of contact for the UE in 5GC. The NS selection process starts when UE initially registers with the 5GS to identify the list of configured slice IDs (S-NSSAI or NSSAI) it wishes to use, along with UE capabilities. This information, along with the configured UE SLA, is used by the 5GS to preselect one or more RAN slices and CN slices. The selected RAN and CN slice(s) are prepared to host user- and control-plane flows. The SMF is responsible for establishing one or more PDU sessions in the user plane to the required data network (DN) via the selected network slice(s). This is depicted in Fig. 6.39.

The URLLC NSI is a critical part of the 5GC required to deliver highly reliable connections for NSI(s) supporting vertical services where reliability cannot be compromised. This is the reason URLLC NSI is used in combination with the other vertical service NSIs when high-reliability communications are required (e.g., V2X).

The establishment of user-plane connectivity to a data network (AMF-UPF-DNN) via an NSI comprises two steps:

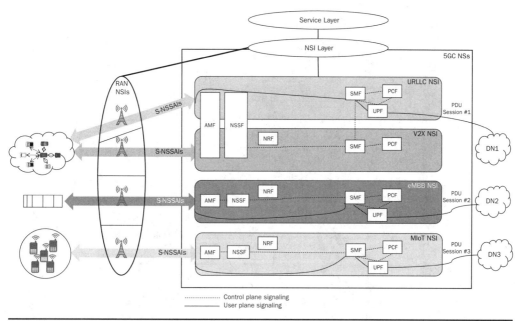

FIGURE 6.39 Example deployments of network slices

- The CN executes the Reachability Management procedure to select an AMF that supports the network slices required by the UE.

- The UE sends a PDU session establishment request to start connectivity to a Data Network via the NSI(s) managed by SMF.

The tools available to MNOs for the configuration of NSs will be numerous. These tools will assist in creation and managing the NSs based on customer needs. The NSs will be able to dynamically adjust on the use of NF and UPF resources, as traffic increases or decreases.

6.8.4 Interworking with EPC

5GC supports interworking with EPC; therefore, network slicing interworking with the EPS is supported. However, mobility between 5GC and EPC does not guarantee all active PDU session(s) can be transferred to the EPC. In order to interwork with EPC, the UE that supports both 5GC and EPC NAS can operate in single registration mode (i.e., using a single radio interface to register UE to either EPC or 5GC at any moment) or dual registration mode (i.e., using dual radio interfaces to register UE to EPC and 5GC). In addition, network slicing for UE in idle mode and connected mode is supported when UE moves from EPS-to-5GS or 5GS-to-EPS. For interworking between 5GC and EPC, the AMF and MME interact to exchange relevant information. The EPC interworking architecture is shown in Fig. 6.40.

During PDN connection establishment in the EPC, the UE allocates the PDU session ID and sends it to the PGW-C+SMF via Protocol Configuration Options (PCO), which is used in LTE as a means by which the UE can indirectly exchange information

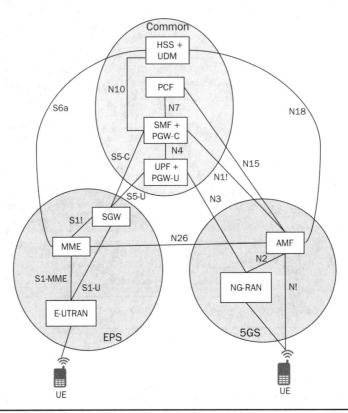

Figure 6.40 5GC/EPC interworking

with the PGW. In 5GC, an S-NSSAI associated with the PDN connection is determined based on the operator policy by the PGW-C+SMF, based on a combination of PGW-C+SMF address and APN, and sent to the UE in PCO together with a PLMN ID that the S-NSSAI relates to. For additional details on interworking between 5GC and EPC, refer to [20].

6.9 IP Multimedia System (IMS)

Traditionally, the mobile system was a closed system, i.e., the mobile services built by utilizing the features and capabilities of the mobile system were self-contained to ensure reliability (QoS and security) in delivering these services to its subscribers. The development of new services was a slow process as it required modifications in many parts of the mobile system. This paradigm had to be changed as use of the internet became widespread with proliferation of IP-based services offered to users. Mobile operators felt the need to stay abreast with this changing paradigm by adopting IP-based protocols as a means of integrating internet services within the telecom world, and deliver them to users with inbuilt Telecom reliability. This is one of the main reasons for the development of the IP-Multimedia System (IMS). In 3GPP, IMS was originally designed as part of the GPRS evolution enabling it to deliver IP-multimedia services. IMS is also a glue between fixed and mobile networks, called Fixed Mobile Convergence (FMC), allowing delivery of multimedia applications and voice services over IP transport. The IP transport allows IMS to communicate over any network that supports IP. This section provides an overview of IMS subsystem. There are several references listed [45–47] for readers who want to gain in-depth knowledge about IMS subsystem.

The 3GPP work on IMS specification began in Release 5 (2002) as part of the core network evolution from circuit-switching to packet-switching and was refined in subsequent Releases 6 and 7. Initially, IMS was an all-IP system designed to assist mobile operators to deliver next generation interactive and interoperable services, cost-effectively, over a separate architecture to enable the use of the internet. Session Initiation Protocol (SIP) was selected as the signaling mechanism for IMS, thereby allowing voice, text, and multimedia services to traverse all connected networks. 3GPP works closely with experts in the IETF to ensure maximum reusability of internet standards, preventing fragmentation of IMS standards. IMS is basically used for VoIP services, including emergency calls. IMS work has progressed in 3GPP for a long period and its use has grown within the mobile networks.

IMS is based on a number of IETF protocols. Figure 6.41 shows an example protocol stack. IMS is physical layer agnostic, meaning that it can use wireless or fixed transport layers. The basic requirement is that it uses IPv4/v6.

6.9.1   Session Initiation Protocol (SIP)

6.9.1.1   *General*

The Session Initiation Protocol (SIP) is the basis of IMS; it is used to set up, modify, and terminate a session between two logical endpoints called SIP User agents (UA). It is defined in IETF RFC 3261 [48] and is a textual description language in English, meaning that the protocol is human readable. The originator of a message will be the SIP UA client (UAC) and the receiver SIP UA server (UAS). The SIP UAS may send a response to the UAC; in this situation both entities keep their roles. However, if the UAS at some point in time needs to originate a message, it will become a UAC and the receiving entity will be the UAS. There are times when an entity that receives a message will originate a new message based on the receipt; in this situation the function is called a

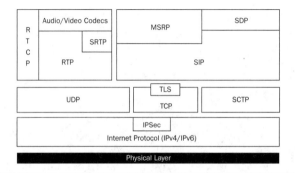

FIGURE 6.41   Example IMS protocol stack

back-to-back user agent (B2BUA) as it implements both UAS and UAC. Messages consist of requests and responses.

The protocol consists of a number of headers and body. Each header has a specific purpose for creation, modification, and termination of a session. Some contain basic information such as the origination and destination address. Messages can be sent within or outside a dialog. A dialog is a concept that is used to create a relationship between two SIP endpoints and allows for proper routing of messages and sequencing. The protocol has been expanded/extended many times in different 3GPP releases to include new functionalities.

6.9.1.2 Requests

Requests have a request line that contains a method name. There are numerous methods defined in IETF RFC 3261 [48] and other RFCs; some of them are listed below.

IETF RFC 3261 [48] methods
1. INVITE = Establishes a session.

2. ACK = Confirms an INVITE request.

3. BYE = Ends a session.

4. CANCEL = Cancels establishing of a session.

5. REGISTER = Communicates user location (hostname, IP).

6. OPTIONS = Communicates information about the capabilities of the calling and receiving SIP phones.

IETF RFC 3262 [49] method
7. PRACK = Provisional acknowledgment.

IETF RFC 6665 [50] methods
8. SUBSCRIBE = Subscribes for notification from the notifier.

9. NOTIFY = Notifies the subscriber of a new event.

IETF RFC 3903 [51] method
10. PUBLISH = Publishes an event to the server.

IETF RFC 6086 [52] method
11. INFO = Sends mid-session information.

IETF RFC 3515 [53] method
12. REFER = Asks the recipient to issue call transfer.

IETF RFC 3428 [54] method
13. MESSAGE = Transports instant messages.

IETF RFC 3311 [55] method
14. UPDATE = Modifies the state of a session.

6.9.1.3 Responses

A response is a message after a request message. They fall into different types, some meaning the session is proceeding, some errors, and some successes. In order to differentiate the response message, a response will have a status line. The status line contains a status code. The codes can be split into six groups:

1xx: Provisional—request received, continuing to process the request;

2xx: Success—the action was successfully received, understood, and accepted;

3xx: Redirection—further action needs to be taken in order to complete the request;

4xx: Client error—the request contains bad syntax or cannot be fulfilled at this server;

5xx: Server error—the server failed to fulfill an apparently valid request;

6xx: Global failure—the request cannot be fulfilled at any server.

These responses have some familiarity with Hyper Text Markup Language (HTML) codes.

6.9.1.4   Basic Headers

General Headers: These are headers that are used in both directions (requests/responses); some example headers are To:, From:, Call-ID: Contact.
Request Headers: Headers that only go in requests: Subject:, Priority:
Response Headers: Headers that only go in responses: Unsupported

6.9.2   Session Description Language (SDP)

6.9.2.1   General

SDP is a textual description language that describes a session to be set up. It is transport agnostic and can be used with Session Announcement Protocol (SAP), Real Time Streaming Protocol (RTSP), and, in the context of this chapter, Session Initiation Protocol (SIP). It does not carry any media either. The structure is that it consists of one to many lines of description, each description is represented by a letter that corresponds to some data—"c" is for connection information and is represented "c=." When mentioned it might be called the "c line." In another form the general syntax is: <type>=<value>.

SDP generally consists of the following:

1. Media (e.g., video, audio, H.261, MPEG, etc.) and transport information (e.g., RTP/UDP/IP, IP addresses, port information)

2. Time description that either bounds or does not bound the session in time.

IETF RFC 4566 [56] contains a list of the lines, some of which are provided in Table 6.5. Furthermore, each line might have numerous attributes. For example, m line has the following attributes:

m=<media> <port>/<number of ports> <proto> <fmt> This field is used to advertise properties of the media stream, such as the port it will be using for transmitting, the protocol used for streaming, and the format or codec.

<media> It is used to specify media type, e.g., audio, video, text, etc.

<port> The port to which the media stream will be sent. Multiple ports can also be specified if more than one port is being used.

<proto> The transport protocol used for streaming, e.g., RTP (real-time protocol).

<fmt> The format of the media being sent, in which codec is the media encoded, e.g., PCMU, GSM, etc.

Session description
    v= (protocol version)
    o= (originator and session identifier)
    s= (session name)
    i=* (session information)
    u=* (URI of description)
    e=* (e-mail address)
    p=* (phone number)
    c=* (connection information—not required if included in
        all media)
    b=* (zero or more bandwidth information lines)
    One or more time descriptions ("t=" and "r=" lines; see below)
    z=* (time zone adjustments)
    k=* (encryption key)
    a=* (zero or more session attribute lines)
    Zero or more media descriptions

Time description
    t= (time the session is active)
    r=* (zero or more repeat times)

Media description, if present
    m= (media name and transport address)
    i=* (media title)
    c=* (connection information—optional if included at
        session level)
    b=* (zero or more bandwidth information lines)
    k=* (encryption key)
    a=* (zero or more media attribute lines)

TABLE 6.5   SDP lines from RFC 4566 [56]

An SDP communication consists of an OFFER and an ANSWER. The OFFER, as the word implies, contains a description of the session; in case of a voice call, voice being offer, e.g., codec, QoS, etc. There are a number of alternatives—codecs can be contained in the OFFER, and the other end then has the ability to determine what to accept and provides a response back in the ANSWER. An OFFER or ANSWER typically looks like that shown below where there is Session Level Information and can contain two to many Media descriptions.

Because of the parameters in the OFFER and ANSWER, there may be a number of messages sent back and forth until an agreement can be reached.

6.9.2.2 Preconditions

Preconditions were initially defined in IETF RFC 3312 [57] and extended in IETF RFC 5027 [58] and IETF RFC 5898 [59]. It can be seen as an extension of SDP, where media attribute lines are defined, e.g., "a line." Preconditions set a number of conditions

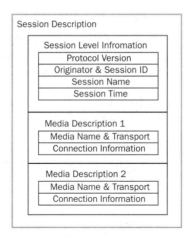

FIGURE 6.42   Session description

that need to be met for the session to be established; in the case of a voice call media resources need to be allocated so that events like "ring tone" and "voice communication" can take place. The basic syntax from IETF RFC 3312 [57] is as follows:

```
current-status   = «a=curr:» precondition-type
             SP status-type SP direction-tag
desired-status   = «a=des:» precondition-type
             SP strength-tag SP status-type
             SP direction-tag
confirm-status   = «a=conf:» precondition-type
             SP status-type SP direction-tag
precondition-type = «qos» | token
strength-tag     = («mandatory» | «optional» | «none»
             = | «failure» | «unknown»)
status-type      = («e2e» | «local» | «remote»)
direction-tag    = («none» | «send» | «recv» | «sendrecv»)
```

There are three types of status—current, desired, and confirmed. Then, there is a precondition type required, only to indicate QoS. Then, there is status type for either end to end (e2e), local, or remote, and it is "send" only, "receive" only, or both:

a=curr:qos local none

a=curr:qos remote none

a=des:qos mandatory local sendrecv

a=des:qos mandatory remote sendrecv

Here, current status at the local (originating point) and remote (terminating point) is none or unknown for QoS (or bearer status) and the desired goal for the local and remote ends should be "send and receive". The mandatory implies this goal SHALL be met. In the SDP Offer Answer exchange end-points will provide updates until the desired status is met.

6.9.3   Message Session Relay Protocol (MSRP)

MSRP is defined in IETF RFC 4975 [60]. It is a messaging protocol that is textual and is used to transport textual information. It can use SIP and SDP to set up the MSRP session which has similar format to SIP, and it has a SEND message and number of response codes. In addition to SEND message and response codes, MSRP also caters for the need to send reports. Entities that receive a message in a SEND may need to send that message on or do something with it. It does not mean that the intended target will successfully receive the message so REPORTs can be created to inform a sender if a message was successful or a failure occurred.

An example message exchange using MSRP is shown in Fig. 6.43. It is seen that SIP sets up the session, and then MSRP session takes place followed by SIP terminating the session.

6.9.4   Real Time Protocol (RTP)

RTP is defined in IETF RFC 3550 [61]. It is used to transport real-time media, e.g., video and audio. It provides a sequence number and timestamp to each packet so that a receiving end can reconstruct the incoming data stream and detect packet losses quickly. It is a binary-based protocol unlike SIP and SDP. Figure 6.44 provides an overview of RTP packet format, in which the first 12 bytes are present in every RTP packet.

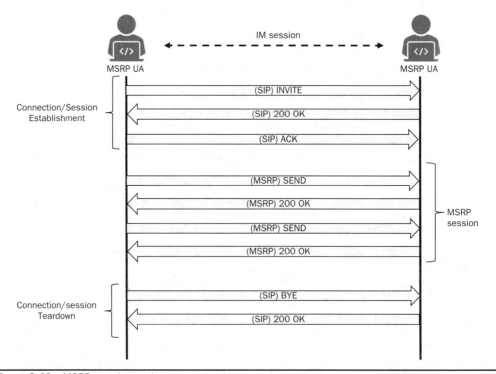

Figure 6.43   MSRP session exchange

As well as timestamp and sequence number there is a Synchronization Source Identifier (SSRC). The SSRC identifies the synchronization source, which is the entity that provided the timestamp in the RTP packet. RTP also supports multiplexing of media streams, by using different destination transport addresses.

If point A is communicating with point B in a teleconference that consists of video and audio encoded in separate RTP streams then each destination address should be different. Destination address is a combination of IP address and port, so in the example described the IP address could be the same and the port number different.

RTP allows for intermediate nodes to participate in the communications. There are following two forms:

- Mixers: As the word means multiple RTP sources are received and then a single RTP source is generated. A mixer can have a new SSRC.

- Translator: This is more of a transparent entity; a new SSRC is not generated; however, the transport address will be that of the translator.

6.9.5   Secure RTP (SRTP)

This is a secure version of RTP. It is an extension of RTP using RTP extension capability. It contains a Master Key Identifier (MKI) and an Authentication tag.

SRTP also has a corresponding control protocol part, SRTCP.

6.9.6   Real Time Control Protocol (RTCP)

RTCP is the second part of the RTP specification. It provides the control aspects required for RTP. Each participant in the session will send RTCP information to all the other participants. It provides feedback on the quality of the data being distributed, hence how well is each participant communicating with all other participants. The data sent is called a report containing:

- Receiver reception report (RR): Reception report from participants that are not active senders.

- Sender report (SR): Transmission and reception from participants that are active senders.

- Source description report (SDES): It includes the CNAME, which is a unique identification of an endpoint. It allows for SSRC to change at an endpoint.

Reports contain information concerning number of RTP packers sent, number lost, and inter-arrival jitter.

Other RTCP messages include:

BYE: End of the participant session

APP: Application specific information

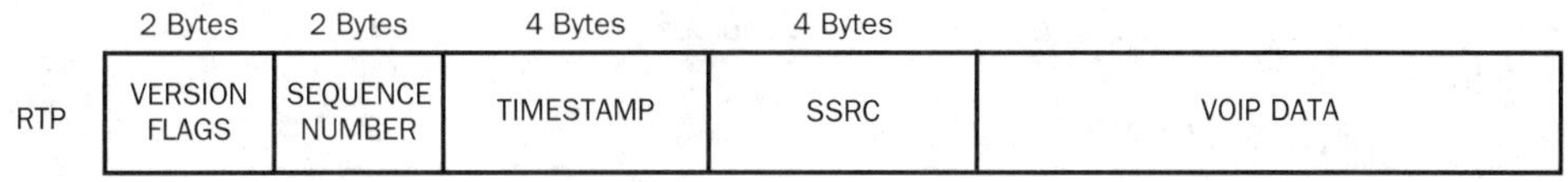

	2 Bytes	2 Bytes	4 Bytes	4 Bytes	
RTP	VERSION FLAGS	SEQUENCE NUMBER	TIMESTAMP	SSRC	VOIP DATA

Figure 6.44   RTP packet format

6.9.7 Stream Control Transmission Protocol (SCTP)

SCTP, defined in IETF RFC 4960 [62], can be seen as a hybrid of User Datagram Protocol (UDP) and Transmission Control Protocol (TCP). It is message-orientated protocol like UDP but has congestion control like TCP. Unlike the other two protocols it can provide multi-homing and redundant paths to increase the resilience of the communication path. Multi-homing is the ability of an endpoint to have multiple IP addresses that can be used for the same communication. These IP addresses could be from the different or the same networks (Table 6.6).

The SCTP protocol frame structure consists of source and destination ports followed by data. The data is structured into "chunks" that are structured as: tag, flag, length, and data. The tag defines what type of message or data is being transported in the "chunk." The length indicates the length of the data field and finally the data field contains the data. In addition, the tag top 2 bits define the action that the receiving end should perform if it does not recognize the "chunk" type.

ID value	Chunk type
0	Payload Data (DATA)
1	Initiation (INIT)
2	Initiation Acknowledgment (INIT ACK)
3	Selective Acknowledgment (SACK)
4	Heartbeat Request (HEARTBEAT)
5	Heartbeat Acknowledgment (HEARTBEAT ACK)
6	Abort (ABORT)
7	Shutdown (SHUTDOWN)
8	Shutdown Acknowledgment (SHUTDOWN ACK)
9	Operation Error (ERROR)
10	State Cookie (COOKIE ECHO)
11	Cookie Acknowledgment (COOKIE ACK)
12	Reserved for Explicit Congestion Notification Echo (ECNE)
13	Reserved for Congestion Window Reduced (CWR)
14	Shutdown Complete (SHUTDOWN COMPLETE)
15 to 62	Available
63	Reserved for IETF-defined Chunk Extensions
64 to 126	Available
127	Reserved for IETF-defined Chunk Extensions
128 to 190	Available
191	Reserved for IETF-defined Chunk Extensions
192 to 254	Available
255	Reserved for IETF-defined Chunk Extensions

TABLE 6.6 Extract from IETF RFC 4960 [62]

6.9.8   IMS Architecture

Figure 6.45 is a simplified architecture diagram of the IMS subsystem. For detailed IMS architecture, refer to 3GPP TS 23.228 [45].

6.9.8.1   *Call Session Control Function (CSCF)*

There are three call control functions that form the heart of the IMS network:

- P-CSCF Proxy Call Session Control Function is the first point of contact with the IMS subsystem from the UE. Maintains security association with the UE and performs policing of messages and other policy functions.
- S-CSCF Serving CSCF maintains session state, performs the Registrar functions defined in RFC 3261 [48]. Determines which Application Servers (ASs) to involve in a session.
- I-CSCF Interrogating CSCF. It is the contact point for terminating connections to a user. It queries HSS to see which S-CSCF the user is hosted on. For UE registration it is involved in assigning S-CSCF for the user.

6.9.8.2   *Application Servers (ASs)*

SIP AS is the function that hosts services. It can act as either SIP UA or a B2BUA. It can communicate with other functions in the IMS network. The most common ones are S-CSCF and HSS. The S-CSCF routes SIP messages to and from the AS where the service is executed. The HSS provides subscriber data necessary to execute the service. Some common ASs that have been defined in the 3GPP system are the following:

- Service Continuity Controller (SCC) AS: This AS supports a number of services that were designed for IMS. These were IMS Centralized Services, Service Continuity, and Inter UE transfer.

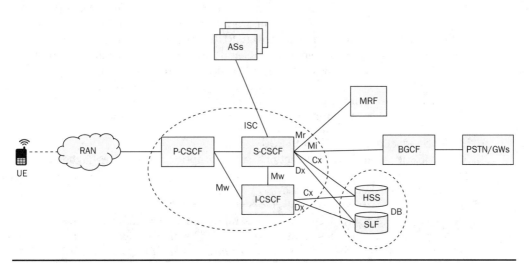

FIGURE 6.45   IMS architecture (based on 3GPP TS 23.228 [45])

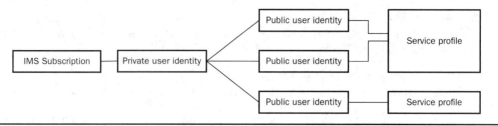

FIGURE 6.46   Relationship between private and public user identities (see 3GPP TS 23.228 [45])

- Mobile Telephony Application Server MMTEL AS or Telephony Application Server (TAS): Initially specified to host legacy GSM/ISDN type services like call waiting, call forwarding, and SMSoIP.

6.9.8.3   Identity Management

IMS has the following two identities:

- Pubic User Identity: This identity is known to the user and is given to other parties so that the user can be contacted. A user may have many public user identities and they can be of the form of a Tel URI or SIP URI. A Tel URI is a phone number at a domain (e.g., +155581212@operator.com) where as an SIP URI is somewhat similar to an e-mail address (e.g., Fred.blogs@operator.com).
- Private User Identity: This identity is typically not known by the user and is used to identify the subscription. It may have one to many public user identities associated with it. It may be used for authentication, authorization, and accounting purposes. It can have a format similar to the SIP URI.

Figure 6.46 shows relationship between private user identities and public user identities.

Given a public user identity can be shared with many private user identities there is a concept of appending the equipment identifier to the public user identity; this is called a Globally Routable User agent URI (GRUU). There might be a situation as user has many devices (e.g., watch, smartphone, and headset) and the origination party of the session knows they want to have a video call with the terminating party so they want to reach that person on the smartphone. This allows UE to call a specific public user ID at a certain device. Figure 6.47 shows this relationship.

The instance ID is an identity that identifies the terminal—it could be an IMEI.

6.9.8.4   Support of IMS Entities in 5GC

In 5GC, IMS entities and their interactions can be supported using SBA via the SBI message bus, as shown in Fig. 6.48.

Service-based interfaces to support IMS:

Npcf:	Service-based interface exhibited by PCF.
Nhss_ims:	Service-based interface exhibited by an SBI-capable HSS.
Npcscf:	Service-based interface exhibited by an SBI-capable P-CSCF.
Nicscf:	Service-based interface exhibited by an SBI-capable I-CSCF.
Nscscf:	Service-based interface exhibited by an SBI-capable S-CSCF.

These SBI services provide equivalent functionality to the Diameter Rx and Cx/Sh reference points.

Figure 6.49 shows the p2p-based 5GC architecture to support interactions between IMS entities using the reference point representation.

P2P Reference points to support IMS:

Following reference points are realized by service-based interfaces in IMS:

N5:   Reference point between the PCF and an AF.

(Note: P-CSCF acts as an AF from PCF point of view. N5 Reference point is defined in TS 23.501 [20].)

N70:   Reference point between SBI-capable I/S-CSCF and HSS.

N71:   Reference point between SBI-capable IMS AS and HSS.

To support coexistence of IMS nodes supporting SBA services and IMS nodes not supporting SBA services, SBI-enabled IMS nodes may support both SBI and non-SBI interfaces.

6.9.8.5   *IMS Procedures (Voice Call and SMS)*

For making or receiving voice calls in 5GS and even SMS, IMS will continue as the core voice network as it is today for VoLTE on EPC. The need for IMS remains for voice

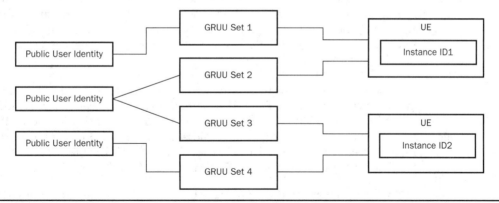

FIGURE 6.47   Relationship between public user identity and UE/ME (see 3GPP TS 23.228 [45])

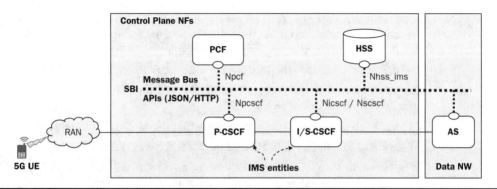

FIGURE 6.48   5GC to support IMS entities in SBA representation

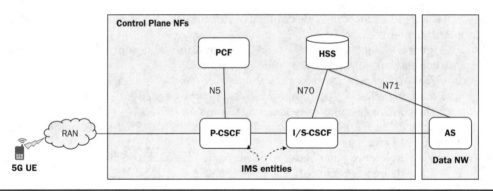

FIGURE 6.49 5GC to support IMS entities in reference point representation

calling and SMS for both EPC and 5GS, as both systems will coexist for some time to come. The operators who have IMS already deployed for use in EPC will continue to use it for 5GC, but start virtualizing it, if not already done. SMS may be supported using native SMS procedures if IMS is not supported.

6.9.8.5.1 IMS Registration In order to use the IMS system a UE first needs to REGISTER with the IMS and be authenticated. Once authenticated the UE can use the services provided by the IMS. The IMS registration procedure is shown in Fig. 6.50.

Before a UE can send the REGISTER message it first must obtain data connectivity; see Sec. 6.7.2. During the data connectivity procedure the UE will be provided with the address of the P-CSCF via PCO information element.

The basic resgiration flow starts with is the UE sending an unprotected REGISTER message; this means that from the UE to IMS system the message will not be encrypted in anyway. Once the REGISTER message has been received in the IMS network it will challenge the UE with authentication vectors. More detailed description of how IMS authentication works can be found in Chap. 7. The subsequent REGISTER message that contains the response to the authentication challenge (step 10) will be encrypted.

6.9.8.5.2 IMS Session Origination and Termination This section describes an information flow for both an IMS Session origination and Session Termination (see FIg. 6.51).

1. A UE will send an SIP INVITE message containing an SDP offer indicating preconditions are required. As described in Sec. 6.9.2, the SDP will describe the media. In this instance we will assume it is voice.

   (Note: It is possible that the INVITE contains no SDP offer.)

2. The terminating UE will receive the SIP INVITE message containing the SDP offer and examine its contents.

3. IMS network sends 100 Trying to indicate that the message was received.

4. UE sends 100 Trying to indicate that the message was received.

5. In response to step 2, the UE will determine how it will respond to the SDP Offer it received. It will send back an SDP Answer in a 183 message containing preconditions that local and remote ends are unknown.

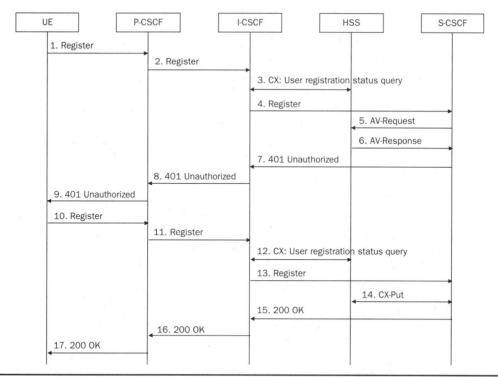

FIGURE 6.50 IMS registration procedure

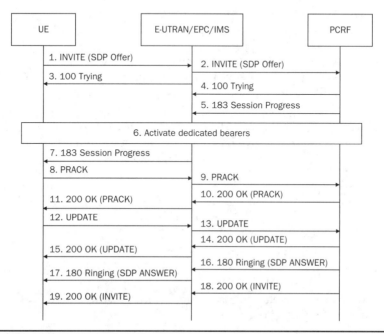

FIGURE 6.51 IMS mobile originated and terminated voice call procedure

a=curr:qos local none

a=curr:qos remote none

a=des:qos mandatory local sendrecv

a=des:qos mandatory remote sendrecv

Optionally, the terminating UE can include that it wants to receive back an acknowledgment to the provisional 183 message.

6. The network will start the process of activating the bearers that will be used to transport the media (see Sec. 6.7.3.2.3). This could be for both UEs.

7. The originating UE will receive the 183 with the SDP Answer and current status of the preconditions.

8. If the terminating UE wanted confirmation that the 183 was received, the originating UE sends a PRACK message.

9. The PRACK is received by the terminating UE.

10. The terminating UE responds to the PRACK with 200 OK.

11. PRACK is forwarded to the UE.

12. When the originating UE has a bearer available for voice media it will send an SIP UPDATE message updating the preconditions to state current state that it is able to send and receive data.

a=curr:qos local sendrecv

a=curr:qos remote none

a=des:qos mandatory local sendrecv

a=des:qos mandatory remote sendrecv

13. UPDATE is sent to the terminating UE. The UE will receive this and determine from a=curr:qos local sendrecv that the originating end has met the necessary preconditions for the session.

14. When the terminating UE is aware that its dedicated bearer has been allocated it will send the 200 OK including SDP Answer that contains preconditions that it can send and receive media locally. In addition the preconditions will also highlight that the remote end can send and receive.

a=curr:qos local sendrecv

a=curr:qos remote sendrecv

a=des:qos mandatory local sendrecv

a=des:qos mandatory remote sendrecv

15. 200 OK is received by the UE.

16. 180 Ringing is returned given that dedicated bearers for both UEs have been successfully assigned.

17. 180 Ringing is sent to the UE and ring tone is played.

18. The called UE answers the session and sends a 200 OK.

19. The 200OK is received by the UE and the session is established between both UE's.

6.9.8.5.3 SMS   The short message service has also been adapted to run over IMS. It uses the SIP METHOD Message to transport a 3GPP SMS Protocol Data Unit (PDU) as defined in 3GPP TS 23.040 [63]. This ensures that the originating entity can use the legacy SMS protocol stack and adapt insert it into the available transport protocol, this being SIP.

The architecture defines an IP-SM-GW. The IP Short Message Gateway performs a number of functions:

a. Transport interworking: This is the process of taking an SIP-based SMS and interworking it to legacy SMS procedures that have been used in GSM. 3G and 4G network.

b. Service-level interworking: As defined in Sec. 6.9.3, there is the MSRP protocol. This is used for SIP-based messaging services. Service-level interworking is taking the SMS service and interworking it with SIP "Chat"-based services.

An IMS short message flow in described in Fig.6.52.

1. The UE registers with the IMS system.

2. The UE sends a short message. **In terms of protocol description** it is an SMS-SUBMIT (defined in 3GPP 23.040 [63]) contained in the RP-User-Data protocol data unit (defined in 3GPP TS 24.011[69]). This envelope of data **(SMS-SUBMIT contained in the RP-User-Data), called the SMS,** is encapsulated into the SIP message **which is called SIP MESSAGE.** The SIP MESSAGE contains a

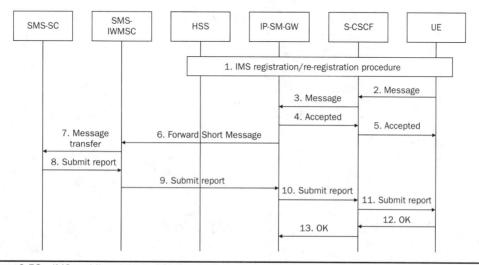

Figure 6.52   IMS mobile originated SMS over IP procedure [64]

parameter/header call Content Type (defined in RFC 3261 [48]) that identifies that it contains **the** SMS. The SIP MESSAGE is sent to the P-CSCF.

3. The S-CSCF evaluates Initial Filter Criteria to determine if the UE has the ability to use the SMS service. If so, the SIP MESSAGE is sent to the IP-SM GW.

4. The IP-SM-GW acknowledges the receipt of the SIP MESSAGE.

5. The acknowledgment is sent to the UE.

6. The IP-SM-GW will extract the SMS-SUBMIT from RP-User-Data from the SIP MESSAGE and insert the SMS-SUBMIT into a MAP Forward short message. This message will be sent to the SMS-Interworking MSC (SMS-IWMSC).

7. The SMS-IWMSC sends the SMS-SUBMIT to the Short Message Service Centre (SMS-SC).

8. Upon successful receipt the SMS-SC will generate an SMS-SUBMIT report and send it to the SMS-IWMSC.

9. The SMS-IWMSC sends the SMS-SUBMIT report in MAP message to the IP-SM-GW.

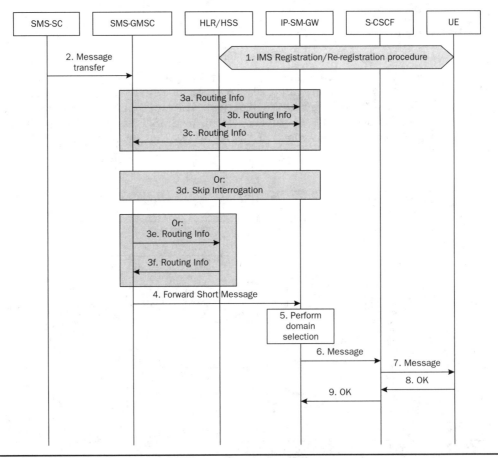

Figure 6.53   IMS mobile terminated SMS over IP procedure [64]

10. The IP-SM-GW will take the SMS-SUBMIT report and put it into an RP-ACK which is then encapsulated into an SIP MESSAGE with content Type identifying the SIP MESSAGE contains an SMS.

11. The S-CSCF sends the SIP MESSAGE to the UE.

12. UE acknowledges receipt of the message to the IP-SM-GW.

13. S-CSCF sends the acknowledgment to the IP-SM-GW.

Fig. 6.53 describes an IMS information flow for an IMS mobile terminated SMS over IP procedure.

1. UE registers with the IMS network.

2. SMS-SC sends the short message in SMS-SUBMIT-DELIVER to the SMS-GMSC.

3. The steps here determine that the SMS-SUBMIT-DELIVER needs to be sent to the IP-SM-GW.

4. SMS-GMSC sends the SMS-SUBMIT-DELIVER in MAP Forward short message to the IP-SM-GW.

5. IP-SM-GW determines that the UE is available over IMS and can receive SMS.

6. IP-SM-GW sends an SIP MESSAGE containing SMS-SUBMIT DELIVER in an RP-USER-DATA PDU to the S-CSCF.

7. S-CSCF sends the SIP MESSAGE to the UE.

8. When the UE successful receives the SIP MESSAGE it will respond with a 200 OK to indicate that the SIP MESSAGE transaction was received.

9. S-CSCF sends the 200 OK to the IP-SM-GW.

10. When the UE has unpackaged the SMS-SUBMIT DELIVER it will send an acknowledgment at the SMS later by sending an SIP MESSAGE, in the SIP MESSAGE is an SMS-DELIVER-REPORT within an RP-ACK.

11. S-CSCF sends the SIP MESSAGE to the IP-SM-GW.

6.10   Vertical Services

6.10.1   General

The 3GPP 4G EPC architecture was initially designed for generic data services; however, over the years vertical industries have become interested in using LTE/EPC to provide wireless capabilities. As such the architecture has been expanded to include new functions. This section will look at two specific vertical segments that have already been discussed elsewhere and what has been added to the architecture to support them.

6.10.2   V2X

V2X was initially developed for 3GPP Release 14. The architecture of the V2X system, designed by 3GPP to support these requirements, adds nodes and interfaces for V2X communication to the Evolved Packet Core (EPC), a core network of existing mobile network systems. Figure 6.54 shows the V2X system structure developed by 3GPP.

MME S/P-GW and HSS are core nodes that handle subscriber authentication, location management, and data traffic, while E-UTRAN is an LTE wireless network, including base stations that were present in existing mobile networks. Vehicle communication networks add the following functions to these existing mobile networks:

- Interface (PC5) for Direct Communication (V2V) between vehicles (or pedestrians).
- Interface (LTE-Uu) for base station communication (V2N) between vehicle and base station.
- Control function for V2X service management.
- Application server for V2X services.
- V2X support features such as terminal-based OBU structural support and region-based broadcasting.

V2X application servers, V2X control servers, and interfaces between them have been added and vehicle communication functions have been added to existing nodes such as base stations to provide V2X services in the V2X application of the OBU (UE, User Equipment).

- V2X Control Function: A logical function that is used for network related actions required to support V2X. The function is used to provide necessary parameters and specific information to the UE for V2X communication.
- V2X User Equipment: OBU communicates with V2X Control Function through the V3 interface and receives ID, radio parameters, application server address settings, and broadcasting service settings for V2X communication.
- V2X Application Server: This is a server that provides V2X services, which may be within the network of operators depending on the type of service, or outside

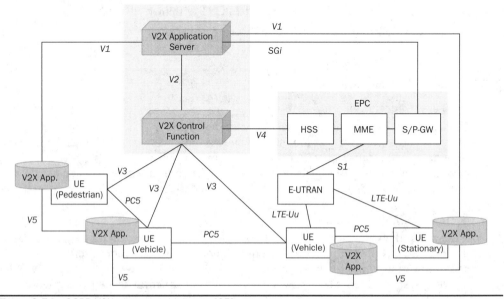

FIGURE 6.54   3GPP V2X system architecture [65]

the network of operators such as road management agency (roadwork or local government), police agency, and may require interconnection between different operators. The application server is tightly coupled with the mobile network to receive information from the unit's V2X application to provide services and to manage vehicle location and provide broadcast services.

6.10.3 MTC

Figure 6.55 shows a high-level architecture for service capability exposure which enables the 3GPP network to securely expose its services and capabilities provided by 3GPP network interface to external third-party service provider SCS/AS hosting an application(s). For example, if there exists an IoT service provider for Smart City, the platform can use some of exposed capabilities (e.g., group message delivery and 3GPP IoT device triggering) to deliver their IoT services to their IoT applications.

Figure 6.56 depicts the specified machine-type communications (MTC) architecture for the non-roaming scenario for simplicity with the different control-plane and user-plane connectivity options of the MTC UE to the network, of course not all need to be supported by the UE. Two different models are supported for IoT-related traffic as follows:

- Direct Model: The AS connects directly to the operator network in order to perform direct user-plane communications with the UE without the use of any external SCS.

- Indirect Model: The AS connects indirectly to the operator network through the services of an SCS in order to utilize additional value-added services for MTC (e.g., control-plane device triggering).

In the architecture the SCS represents a service provider which connects to the 3GPP network to exchange required service message with UEs used for IoT service. A UE can

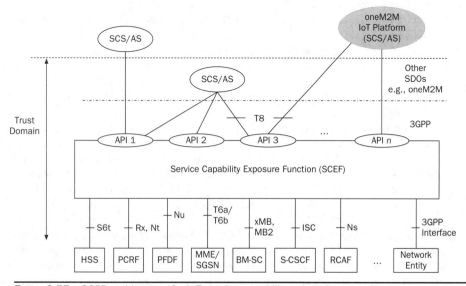

FIGURE 6.55 3GPP architecture for IoT service capability exposure

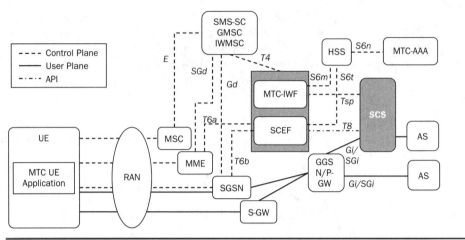

FIGURE 6.56 3GPP MTC architecture

host one or multiple IoT/MTC applications even from different service providers. Most IoT/MTC applications use Tsp and T8 interfaces for their services:

- Tsp is a 3GPP standardized interface to facilitate value-added services motivated by MTC (e.g., control-plane device triggering) and provided by an SCS.

- T8 is the interface between the SCEF and the SCS/AS. SCEF-exposed network services can be accessed by SCS/AS through APIs over T8 interface. In the indirect model, the SCS and the Application Server hosting Application(s) can be collocated.

The Service Capability Exposure Function (SCEF) is the key entity within the 3GPP architecture for service capability exposure that provides a means to securely expose the services and capabilities provided by 3GPP network interfaces. In standalone MTC-Interworking Function (IWF) deployment, MTC-IWF functionality (e.g., T4 triggering) is made available to the SCS/AS via the Tsp interface. In certain deployments, the MTC-IWF may be co-located with the SCEF in which case MTC-IWF functionality is exposed to the SCS/AS via T8 interface (i.e., API). In deployments where MTC-IWF is not co-located with SCEF, interactions between MTC-IWF and SCEF are left up to the implementation.

The gray highlighted functional entities were introduced in the context of MTC in 3GPP and are described below. The UE can be connected to the mobile network via different 3GPP RAN technologies, i.e., Global System for Mobile Communications (GSM), General Packet Radio Service (GPRS), UMTS, or LTE to the mobile core network. LTE further offers radio optimization for low power (LTE-M) or as a different radio technology (RAT), a narrowband version (NB-LTE) for achieving battery lifetimes up to 10 years. MTC offers different transport possibilities for different use cases either via control or user plane. A UE can be triggered via SMS for a certain action and could transmit data via SMS or via IP or Non-IP data. SMS can be sent via the Mobile Switching Center (MSC) or the Mobility Management Entity (MME) to the SMS-Service Center (SMS-SC) and then to the interworking function (MTC-IWF/SCEF) to the Application Server

(AS), which may belong to a third party. The MTC-IWF was first specified in Release 10 and evolved to the SCEF in later releases for exposing 3GPP network functionality to the MTC service provider/AS. IP data can be transferred via the gateway nodes Serving GPRS Support Node (SGSN) and Gateway GPRS Support Node (GGSN) in UMTS/ GPRS or via Serving Gateway (S-GW) and Packet Gateway (P-GW) in EPS. Non-IP data is encapsulated in NAS signaling between UE and MME and can be then further transported via the SCEF to the SCS/AS or via SGW/PGW (reference point not shown in Fig. 6.56). Non-IP Data delivery within a p2p tunnel to the gateway is supported in EPC to the PGW but also already supported generally in 5G to the UPF.

The latest version of the MTC specification 3GPP TS 23.682 [66] has an extensive set of features as follows:

- Device Triggering is the means by which an SCS sends information to the UE via the 3GPP network to trigger the UE to perform application-specific actions that include initiating communication with the SCS for the indirect model or an AS in the network for the hybrid model. Device Triggering is required when an IP address for the UE is not available or reachable by the SCS/AS.

- Packet-Switched (PS)-only Service Provision is providing a UE with all subscribed services via PS domain. PS-only service provision implies a subscription that allows only for services exclusively provided by the PS domain, i.e., packet bearer services and SMS services.

- Core Network-assisted RAN parameters tuning aids the RAN in optimizing the setting of RAN parameters. See 3GPP TS 23.401 [12] for details.

- UE Power Saving Mode (PSM) is for reducing UE's power consumption. In this mode the UE remains registered with the network and there is no need to reattach or re-establish PDN connections.

- Group Message Delivery allows an SCS/AS to deliver a message to a group of UEs. This can be done via either MBMS which is intended to efficiently distribute the same content to the members of a group or unicast MT NIDD for UEs which are part of the same External Group Identifier.

- Monitoring Events allows an SCS/AS to monitor specific events in 3GPP system. Such monitoring events can be accessed via the SCEF. Configuration and reporting of the monitoring events include monitoring the association of the UE and UICC, UE reachability, location of the UE, loss of connectivity, communication failure, roaming status, number of UEs present in a geographical area, availability after DDN failure, and PDN connectivity status.

- High-latency communication is a function to handle MT communication with UEs being unreachable while using power saving functions, such as UE Power Saving Mode or extended idle mode Discontinuous Reception (DRX) depending on operator configuration.

- Support of informing about potential network issues allows the SCS/AS to request the SCEF for being notified about the network status in a geographical area. The SCS/AS can request for a one-time reporting of network status or a continuous reporting of network status changes.

- Resource management of background data transfer enables the SCS/AS to request a time window and related conditions from the SCEF for background data transfer to a set of UEs via the Nt interface. The SCS/AS request shall contain the SCS/AS identifier, SCS/AS Reference ID, the volume of data expected to be transferred per UE, the expected amount of UEs, the desired time window, and optionally network area information.

- Evolved UMTS Terrestrial Radio Access (E-UTRAN) network resource optimizations based on communication patterns provided to the Mobility Management Entity (MME) allows the SCS/AS to provide predictable communication patterns of a UE to the SCEF in order to enable network resource optimizations for such UE(s).

- Support of setting up an Application Server (AS) session with required Quality of Service (QoS) allows allows the third party SCS/AS to request a data session to a UE is set up with a specific QoS (e.g., low latency or jitter) and priority handling. This functionality is exposed via the SCEF toward the SCS/AS.

- Change the chargeable party at session setup or during the session allows the SCS/AS to request the SCEF to start or stop sponsoring a data session for a UE that is served by the third-party service provider (AS session), i.e., to realize that either the third-party service provider is charged for the traffic (start) or not (stop).

- Extended idle mode Discontinuous Reception (DRX) is a function that supports a negotiation between the UE and the network over NAS signaling the use of extended idle mode DRX for reducing its power consumption, while being available for mobile terminating data and/or network-originated procedures within a certain delay dependent on the DRX cycle value.

- Non-IP Data Delivery (NIDD) is a function to be used to handle MO and MT communication with UEs, where the data used for the communication is considered unstructured from the EPS standpoint (which is also referred to as Non-IP).

- Support of Packet Flow Description (PFD) management via Service Capability Exposure Function (SCEF) is a function that the PFDs can be managed by the third-party SCS/AS via the SCEF, which ensures the secure access to the operator's network even from the third-party SCS/AS in untrusted domain.

- Mobile Subscriber Integrated Services Digital Network Number (MSISDN)-less MO-SMS via T4 is a subscription-based function. The subscription provides the information whether a UE is allowed to originate MSISDN-less MO-SMS.

- Enhanced Coverage Restriction Control via SCEF is a function that enables third-party service providers to query status of enhanced coverage restriction or enable/disable enhanced coverage restriction per individual UE(s).

- Multimedia Broadcast Multicast Service (MBMS) user service for UEs using power saving functions is a function that identifies the time intervals the UE stays awake to receive MBMS user service or to discover if there is any MBMS user service scheduled for delivery; should not necessarily be the same as the reachable intervals negotiated for extended idle mode DRX or PSM.

- Enhancements to Location Services for CIoT are functions to support Location Services for Cellular Internet of Things (CIoT) UEs. The enhancements include deferred location for the UE availability event, indication of UE RAT type and/

or coverage level to Evolved Serving Mobile Location Center (E-SMLC), support of UE positioning measurements in idle mode, addition of periodic and triggered location for EPC, and support of Last Known Location for a UE that are unreachable for long periods of times.

- MBMS user service for Narrowband (NB)-LTE or LTE-M (Machine) UE categories is a function for UEs of such categories to be able to receive MBMS service. For this function, E-UTRAN needs to be able to determine the UE category that applies to the specific service indicated by the TMGI.

- Network Parameter Configuration via SCEF allows the SCS/AS to issue network parameter configuration requests to the network, via the SCEF, to suggest parameter values that may be used for maximum latency, maximum response time, and suggested number of downlink packets.

6.10.4 CIoT Concepts in 5G

The 3GPP 5G architecture is specified in 3GPP TS 23.501 [20] and its procedures in 3GPP TS 23.502 [21]. Obviously, the major changes to the EPC as specified in 3GPP TS 23.401 [12] are the concept of network slicing and the service-based architecture (SBA). A network slice is defined in [20] as "a logical network that provides specific network capabilities and network characteristics." Further, SBA paves the way to the virtualization of network functions, easier extensibility, modularized services of each network function, and openness for exposure of network information to third parties.

The 5G architecture, as shown in Fig. 6.57 (also see Fig. 6.12 for the detailed architecture), has a clear split between control-plane and user-plane functionalities. The RAN node or also called gNB is the 5G base station for 3GPP access. Note that for

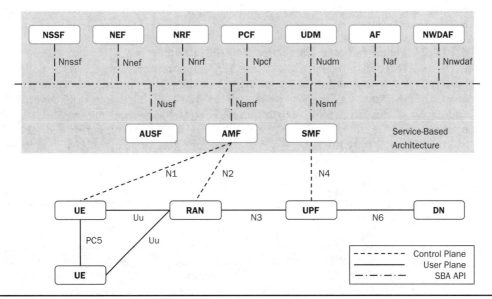

FIGURE 6.57 5GC architecture in SBA representation

non-3GPP access, e.g., WLAN access, the current architecture can be applied as well with an additional interworking function for untrusted access. A UE can now send NAS signaling not only via 3GPP access but also via non-3GPP access. The UPF is the gateway node to the Data Network (DN), which could be operator services like the IP Multimedia Subsystem (IMS), internet access, or third-party services. The 5GC AMF and the SMF nodes provide the services similar to the MME node in EPC, e.g., terminate the NAS signaling of the UE via N1, idle or connected mode mobility, or protocol data unit (PDU) connectivity service. The Authentication Server Function (AUSF) takes care of the authentication of the UE and Unified Data Management (UDM) is responsible for subscription-related information. The Network Function (NF) Repository Function (NRF) supports the service discovery of NFs and maintains the NF profiles of the available NF instances. The Network Slice Selection Function (NSSF) selects the set of Network Slice Instances serving a particular UE. The Policy Control Function (PCF) provides policy rules to CP functions for enforcement, and the Application Function (AF) provides application-related information to the policy framework for policy control. The Network Exposure Function (NEF) is somehow the 5G version of the SCEF in EPC and is responsible for exposure of capabilities and events, secure provision of information from external application to 3GPP network, and translation of internal–external information.

6.10.4.1   Non-IP Transport in 5G

From the beginning, different transport modes were considered in the 5G architecture [20]. A UE can request the establishment of a Packet Data Unit (PDU) connectivity service via PDU sessions to exchange of PDUs between a UE and a data network identified by a Data Network Name (DNN). A PDU session can support one of the following PDU session types: IPv4, IPv6, IPv4v6, Ethernet, and unstructured. The unstructured PDU session type is corresponding to the Non-IP transport mechanism and is achieved via different point-to-point (p2p) tunneling techniques based on UDP/IP encapsulation to transmit unstructured data to the destination (e.g., application server) in the DN via the N6 reference point.

6.10.4.2   Mobile Initiated Connection Only (MICO) Mode

A UE may register to the network with a MICO mode indication; the AMF will then perform the relevant authorization of the use of the feature based on several parameters. The AMF serving area may be the whole PLMN (Public Land Mobile Network) and it may set the "all PLMN" registration area in order to prevent any mobility-related re-registrations to the same PLMN. Periodic re-registration is still performed. The UE in MICO mode is only reachable when it is in Connected mode in the AMF and it does not listen to any paging messages.

6.10.4.3   Network Exposure

Capabilities of network functions can be exposed to external application servers via the NEF. There are three types of service exposure:

- Monitoring capability: The 5GS network functions are responsible for monitoring specific events of each UE. This capability provides identification of the suitable 5G network function responsible for monitoring the UE event, its configuration, detection, and reporting to the authorized external party.

- Provisioning capability: UE behavioral information can be provisioned via the NEF to an NF. The capability consists of authorization of the provisioning external third party, and receiving, storing, and distributing the information to relevant NFs.

- Policy/charging capability: It is based on the request from authorized external party to provide applicable UE policy and charging. The 5GS uses the applicable QoS and charging policy information for the UE, e.g., session and charging policy, QoS policy to be enforced, applicable accounting functionality, and provides to authorized external party.

6.10.4.4   Secondary Authentication

A secondary authentication/authorization can be performed during the establishment of a PDU session to a DN and may perform only DN authorization without DN authentication, as shown in Fig. 6.58 [21]. SMF plays the role as the EAP authenticator and relies on a DN server to authenticate and authorize the UE's request to establish a PDU session. As a precondition the UE sends a PDU session establishment request to AMF. The UE is pre-configured with the credentials of the DN-AAA (Authentication Authorization Accounting) for the secondary authentication. The SMF then triggers the N4 session establishment between SMF and UPF and sends the Authentication Request to the DN-AAA server via the UPF (step 1). The Authentication/Authorization messages from/to the DN-AAA server are sent to the UE via AMF, SMF, and UPF (steps 2a–2c). Between UE and SMF, the messages are transported in NAS messages via AMF. The SMF then proxies the messages to and from the UPF via an IP connection (steps 2d–2e). The DN-AAA server confirms the successful authentication/authorization of the PDU session (step 3). The PDU session establishment continues and completes (step 4). Lastly, the SMF notifies the DN-AAA with the IP address allocated to the UE together with the Generic Public Subscription Identifier (GPSI), if requested or configured in previous steps (step 5).

6.10.4.5   IoT-Related 3GPP Items

3GPP TS 23.724 [67] defines three main objectives of CIoT. The first objective is to enable CIoT/MTC functionalities in 5G Core Network (CN). This means that the existing CIoT/MTC functionalities from the pre-5G networks are enabled in 5G CN with potential connectivity to (Wideband) WB-EUTRA (eMTC) and/or NB-IoT for 5GS capable devices. For example, selected key functionalities are monitoring, non-IP data delivery, small data transmission, additional power saving, overload control (as relevant in 5G CN), high-latency communication, reliable communication, and equivalent to group communication and messaging. 5G Phase 1 specification (3GPP TS 23.501 [20]) supports already the network attach/registration without PDN connection as well as non-IP PDN Connection type. In addition, regulatory requirements have to be fulfilled at the same level as in EPC.

The second objective is to support the coexistence and migration from EPC-based eMTC/NB-IoT to 5G CN. Solutions are studied where the same service is offered to some UEs connected to EPC and some UEs connected to 5G CN. Only solutions are considered that support 5G-NAS signaling to access the 5G CN, i.e., no EPC NAS for 5G-CN is allowed, which rules out legacy IoT device support in 5G.

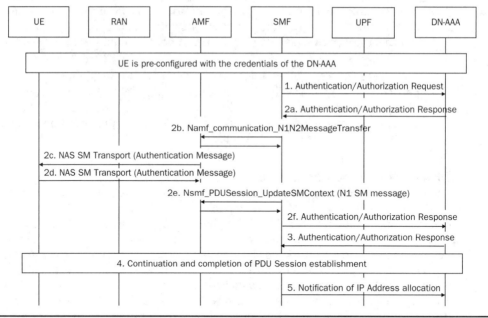

FIGURE 6.58 Secondary authentication call flow

The last objective is to enhance 5G System to address 5G service requirements (see 3GPP TS 22.261 [36] and 3GPP TR 38.913 [68]). Two requirements are identified with respect to those:

- The association between subscription and address/number of an IoT device should be changeable.
- Restricted Registration procedure should be supported to allow IoT device provisioning for eSIM profiles.

References

1. 3GPP TS 23.101, "General Universal Mobile Telecommunications System (UMTS) Architecture," v3.1.0, December 2000.
2. 3GPP TS 11.11, "Specification of the Subscriber Identity Module—Mobile Equipment (SIM-ME) Interface," Release 1999, v8.14.0, June 2007.
3. 3GPP TS 31.101, "UICC-Terminal Interface; Physical and Logical Characteristics," v15.2.0, September 2019.
4. 3GPP TS 23.205, "Bearer-Independent Circuit-Switched Core Network," Stage 2, v15.0.0, June 2018.
5. 3GPP TS 24.008, "Mobile Radio Interface Layer 3 Specification; Core Network Protocols," Stage 3, v16.1.0, June 2019.
6. 3GPP TS 29.002, "Mobile Application Part (MAP) Specification," v15.5.0, June 2019.
7. ITU-T Recommendation H.248.1, "Gateway Control Protocol," Version 3, March 2013.

8. 3GPP TS 29.232, "Media Gateway Controller (MGC)—Media Gateway (MGW) Interface," Stage 3, v15.0.0, June 2018.

9. 3GPP TS 29.205, "Application of ITU-T Q.1900 Series to Bearer Independent Circuit Switched (CS) Core Network Architecture," Stage 3, v15.0.0, June 2018.

10. 3GPP TS 29.414, "Core Network Nb Data Transport and Signaling," v15.0.0, June 2018.

11. ITU-T Recommendation Q.2630.1, "AAL Type 2 Signaling Protocol—Capability Set 1," December 1999.

12. 3GPP TS 23.401, "General Packet Radio Service (GPRS) Enhancements for Evolved Universal Terrestrial Radio Access Network (E-UTRAN) Access," v16.3.0, June 2019.

13. ITU-R Recommendation M.2012-4, "Detailed Specifications of the Terrestrial Radio Interfaces of International Mobile Telecommunications-Advanced (IMT-Advanced)," 2019.

14. 3GPP TS 23.214, "Architecture Enhancements for Control and User Plane Separation of EPC Nodes," v16.0.0, June 2019.

15. 3GPP TS 29.244, "Interface between the Control Plane and the User Plane Nodes," v16.2.0, December 2019.

16. GSMA, "Road to 5G: Introduction and Migration," April 2018.

17. C. Abrams, R. W. Schulte, Report from Gartner Research, "Service-Oriented Architecture and Guide to SOA Research," January 2008.

18. B. Kanagwa, E. K. Mugisa, "A Comparison of Service Oriented Architecture with Other Advances in Software Architectures."

19. P. Di Francesco, G. Sasso, "Architecting Microservices," 2017 IEEE International Conference on Software Architecture Workshops.

20. 3GPP TS 23.501, "System Architecture for the 5G System," Stage 2, v16.1.0, June 2019.

21. 3GPP TS 23.502, "Procedures for the 5G System," Stage 2, v16.1.1, June 2019.

22. IETF RFC 7540, "Hypertext Transfer Protocol Version 2 (HTTP/2)," May 2015.

23. IETF RFC 7159, "The JavaScript Object Notation (JSON) Data Interchange Format," March 2014.

24. TCP IETF RFC 793, "Transmission Control Protocol," September 1981.

25. IETF Internet Draft, "A UDP-Based Multiplexed and Secure Transport (QUIC)," September 2019.

26. 3GPP TS 23.503, "Policy and Charging Control Framework for the 5GS," Stage 2, v16.3.0, December 2019.

27. 3GPP TS 29.281, "General Packet Radio System (GPRS) Tunneling Protocol User Plane (GTPv1-U)," v16.0.0, December 2019.

28. 3GPP TS 36.300, "Evolved Universal Terrestrial Radio Access (E-UTRA) and Evolved Universal Terrestrial Radio Access Network (E-UTRAN) Overall Description," Stage 2, v16.0.0, December 2019.

29. 3GPP TS 38.300, "NR and NG-RAN Overall Description," Stage 2, v16.0.0, December 2019.

30. ITU-T Recommendation E.800, "Definition of Terms Related to Quality of Service," September 2008.

31. 3GPP TS 23.107, "Quality of Service (QoS) Concept and Architecture," v15.0.0, June 2018.

32. IETF RFC 7657, "Differentiated Services (Diffserv) and Real-Time Communication," November 2015.

33. 3GPP TS 23.203, "Policy and Charging Control Architecture," v16.1.0, June 2019.

34. M. Dossou, S. Chede, et al., "A Review on 3GPP PCC Architecture Evolution from Releases 7 to 14," International Journal of Advanced Research, ISSN 2320-5407.

35. 3GPP TS 32.240, "Charging Architecture and Principles," v16.1.0, December 2019.

36. 3GPP TS 22.261, "Service Requirements for the 5G System," v16.10.0, December 2019.

37. 3GPP TR 28.801, "Release 15 Study on Management and Orchestration of Network Slicing for Next-Generation Network."

38. 3GPP TS 28.530, "Services and System Aspects; Management and Orchestration; Concepts, Use Cases, and Requirements," v16.1.0, December 2019.

39. 3GPP TS 28.531, "Services and System Aspects; Management and Orchestration; Provisioning," v16.4.0, December 2019.

40. ITU-T Recommendation Q.1201, "Principles of Intelligent Network architecture."

41. ITU-T Recommendation Q.1202, "Intelligent Network—Service Plane Architecture."

42. ITU-T Recommendation Q.1203, "Intelligent Network—Global Functional Plane Architecture."

43. ITU-T Recommendation Q.1204, "Intelligent Network—Distributed Functional Plane Architecture."

44. ITU-T Recommendation Q.1205, "Intelligent Network—Physical Plane Architecture."

45. 3GPP TS 23.228, "IP Multimedia Subsystem (IMS)," Stage 2, v16.2.0, September 2019.

46. 3GPP TS 24.228, "Signaling Flow for the IP Multimedia Call Control based on Session Initiation Protocol (SIP) and Session Description Protocol (SDP)," Stage 3, v5.15.0, September 2006.

47. 3GPP TS 28.706, "IP Multimedia Subsystem (IMS) Network Resource Model (NRM) Integration Reference Point (IRP); Solution Set (SS) definitions," v15.0.0, June 2018.

48. IETF RFC 3261, "Session Initiation Protocol," June 2002.

49. IETF RFC 3262, "Reliability of Provisional Responses in SIP," June 2002.

50. IETF RFC 6665, "SIP-Specific Event Notification," July 2012.

51. IETF RFC 3903, "SIP Extension for Event State Publication," October 2004.

52. IETF RFC 6086, "SIP INFO Method and Package Framework," January 2011.

53. IETF RFC 3515, "SIP Refer Method," April 2003.

54. IETF RFC 3428, "SIP Extensions for Instant Messaging," December 2002.

55. IETF RFC 3311, "SIP UPDATE Method," September 2002.

56. IETF RFC 4566, "Session Description Protocol (SDP)," July 2006.

57. IETF RFC 3312, "Integration of Resource Management and SIP," October 2002.

58. IETF RFC 5027, "Security Preconditions for SDP Media Streams," October 2007.

59. IETF RFC 5898, "Connectivity Preconditions for SDP Media Streams," July 2010.

60. IETF RFC 4975, "The Message Session Relay Protocol (MSRP)," September 2007.

61. IETF RFC 3350, "RTP: A Transport Protocol for Real-Time Applications," July 2003.

62. IETF RFC 4960, "Stream Control Transmission Protocol (SCTP)," September 2007.

63. 3GPP TS 23.040, "Technical Realization of the Short Message Service (SMS)," v15.3.0, March 2019.

64. 3GPP TS 23.204, "Support of Short Message Service (SMS) over Generic 3GPP Internet Protocol (IP) Access," Stage 2, v16.0.0, December 2019.

65. 3GPP TS 23.285, "Architecture Enhancements for V2X Services," v16.2.0, December 2019.

66. 3GPP TS 23.682, "Architecture Enhancements to Facilitate Communications with Packet Data Networks and Applications," v16.5.0, December 2019.

67. 3GPP TR 23.724, "Study on Cellular Internet of Things (CIoT) Support and Evolution for the 5G System (5GS)," v16.1.0, June 2019.

68. 3GPP TR 38.913, "Study on Scenarios and Requirements for Next Generation Access Technologies," v15.0.0, June 2018.
69. 3GPP TS 24.011, "Point-to-Point (P2P) Short Message Service (SMS) Support on Mobile Radio Interface."

Definitions

Domain: The highest-level group of physical entities. Reference points are defined between domains.

Network Function: In 3GPP, this term is used for the network element (e.g., AMF, SMF, etc.) in the CN architecture which has a defined functional behavior. A network function can be implemented either as a network element on a dedicated hardware, as a software instance running on a dedicated hardware, or as a virtualized function instantiated on a cloud infrastructure.

Stratum: Grouping of protocols related to one aspect of the services provided by one or several domains.

Hypervisor: Responsible for extracting physical resources into virtual resources.

CN main functions: Subscriber management, switching network fabric, location and mobility management, network management, and policy and charging management.

Service-Oriented Architecture: SOA is a software development platform for designing services that can be used by applications deployed across different software platforms through standard communication protocol.

Service-Level Agreement: A contract between a service provider and an end user that stipulates a specified level of service, support option, a guaranteed level of system performance as relates to downtime or uptime.

Useful Technical Terms in Mobile Communications

Access Stratum (AS) is a functional layer in wireless communication protocol stacks between the radio access network and mobile device (UE). This layer is used to manage radio resources for establishing connection between UE and the radio access network.

Non-Access Stratum (NAS) is a functional layer in wireless communication between the core network and mobile device (UE). This layer is used to manage the establishment of communication sessions and for maintaining continuous communications with the user equipment as it moves. From a protocol stack perspective, the NAS is the highest stratum of the control plane. NAS is responsible for the mobile device mobility, connection, reachability, and session management.

Mobility Management (MM) includes procedures related to the UE's mobility over the radio access. It is also responsible for the authentication and security of the mobile device.

Registration Management (RegM) refers to the registration of UE to the network in order to be authorized to receive services.

Connection Management (CM) provides functions to support the connection of the UE to the core network (e.g., service request initiated by mobile device to start the establishment of NAS signaling, paging request initiated by the core network when a downlink NAS message is received to locate the UE to deliver an incoming service request, etc.).

Reachability Management (ReaM) is responsible for detecting whether the UE is reachable and providing UE location (i.e., access node) in the RAN to reach the UE. This is done by paging the UE in the previously known UE tracking area obtained during location tracking or registration area update.

Session Management (SM) supports the establishment and handling of user data in the NAS.

Location tracking deals with the mobile network keeping track of the geographic location of a UE based on its tracking area where UE has registered. The UE location is known by the network on a Tracking Area List granularity. The UE location tracking includes both UE registration area tracking.

When the UE is in CM-IDLE state, it does periodic registration to allow the network to keep track of the UE location.

Service-Level Agreement is an agreement between the mobile operator and a subscriber which specifies the QoS in traffic delivery based on both subscriber needs and operator system capability. There is a minimum level of guaranteed QoS in traffic delivery for each subscriber but higher QoS can be purchased by subscriber, if desired.

Problems/Exercise Questions

Background-related questions:

1. Why are standards critical for telecom systems?
2. What are the three stages used in development of standards?
3. Where is 3GPP standards organization headquartered?
4. What is the major difference between circuit switching and packet switching?

General CN architecture-related questions:

1. Which entity in the UMTS architecture stores all the subscriber-related information and what type of subscriber information is stored?
2. Are HLR and HSS entities the same?
3. What is UDM in the 5GC architecture?
4. Which call processing procedure is executed to activate a newly acquired device?
5. What is network virtualization?
6. What is cloud computing?

5GC capability-related questions:

1. What is the major difference in the design methodology used in the 5GC architecture than its predecessors?

2. What is CUPS feature and what are its major advantages?

3. How do the network functions in the 5GC control plane communicate?

4. What is the major function of AMF in the 5GC control plane?

5. Which network function in the 5GC architecture provides the policy control and enforcement?

6. URLLC is a key requirement of the 5G System. What capabilities are provided in the 5GC in support of URLLC?

7. Is network slicing a new concept in mobile communications?

8. What are the key benefits of network slicing?

9. How does a device indicate to the network which slice it wants to use?

10. What is NSSAI?

11. Which entity in the network is used to manage the traffic flow in the network to ensure adequate QoS?

12. What is the major difference in QoS concept applied to EPC versus 5GC?

13. What is QCI in EPC and 5QI in 5GC?

14. Are data bearer establishment and PDU session establishment the same?

15. What are the key differences between 3GPP and non-3GPP access technologies?

IMS-related questions:

1. List eight SIP request methods.

2. For each request method describe its purpose.

3. Identity three SIP response types and what they are used for.

4. What is SDP used for?

Security

7.0 Introduction

As vertical industries are thriving, e.g., V2X, IoT, AR/VR, etc., the rise of new business, architecture, and technologies in 5G will present new challenges to security and privacy protection. To better capture how security has evolved toward 5G systems, this chapter aims to present an overview of mobile communications' security fundamentals, starting from traditional security architectures in 4G systems, and explicitly describing the key enabling features in 5G era, as specified in 3GPP. The chapter is organized as follows: Initially, the 4G architecture and challenges are presented in Sec. 7.1. In Secs. 7.2 and 7.3 the 3GPP organization and timeline related to security are provided, and the key features till Release 15 are highlighted, respectively. Finally, in Sec. 7.4, IMS security is discussed.

7.1 4G Security

7.1.1 4G Security Architecture

The 4G security requirements and architecture are described in the 3GPP technical specification TS 33.401 [1]. It specifies the security architecture which is divided into five security relationships between different domains of the system, where four of them are depicted in Fig. 7.1:

1. The network access security which takes care of the UE security associations for the non-access stratum (NAS) [7] layer connection between UE and mobility management entity (MME), in addition to the access stratum (AS) layer for the radio resource control (RRC) [8] connection between UE and eNB.

2. The network domain security (NDS) which ensures that the communication between the network elements is secured, e.g., with an IPsec tunnel.

3. The user domain security which provides the security features in order to access the mobile phone, e.g., storage and access of the PIN code in the universal subscriber identity module (USIM). The mobile phone or user equipment (UE) consists of two parts: the mobile equipment (ME), which is terminating all protocols, and the USIM, which is storing and providing all relevant security keys and data.

4. The application domain security, not standardized in 3GPP, which provides a secure message exchange of applications between the user domain and the provider domain.

5. This is the domain known as visibility and configurability of security, not shown in Fig. 7.1, which provides an indication to the user on the mobile device's screen whether a security feature is used or not. It is also called "ciphering indicator," but in reality, there are no known implementations of this in current mobile phones.

The 5G security architecture is, in principle, the same as the 4G security architecture [2], and also includes the service-based architecture (SBA) domain security [2], which is described later.

4G compared to 3G provides a few new features related to security, e.g., the UE and the serving network, which may be the home network or a visited network where the UE is roaming in, needs to perform mutual authentication with the Evolved Packet System (EPS) Authentication Key Agreement (AKA) procedure. The serving network (as identified by the public land mobile network [PLMN] code, consisting of a mobile country code [MCC] and a mobile network code [MNC]) is included in the derivation of the key, which is sent to the MME in the serving network to perform the authentication. This is done in order to make sure that the UE is actually located in the serving network. Another new feature is the integrity protection for the control plane messages both on Radio RRC and on NAS layer. Confidentiality protection is optional but available for control and user plane. There is no integrity protection for the user-plane data, due to the increased processing power required in the UE and the eNB. The only exemption is for the so-called relay node, acting as a UE toward the eNB and as an eNB toward a UE, which can perform user-plane integrity protection between relay node and eNB.

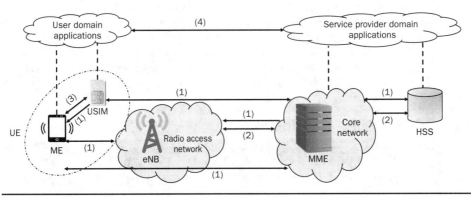

FIGURE 7.1   4G security architecture

4G also provides enhanced privacy by using temporary identities, which are refreshed by the MME at certain events. Only at the initial attach to a network, the international mobile subscriber identifier (IMSI) is sent in clear text over the air interface because the UE does not have any security context in the network.

7.1.2   4G Potential Threats

4G has some security issues and risks, e.g., the IMSI is sent unencrypted over the air interface during the initial attach procedure and could be subject to the so-called IMSI catching by the IMSI catcher. IMSI catching is the threat where a false eNB pretends to be a genuine network and performs an identity request to receive the IMSI, which may be then used for further attacks e.g., user tracking. 4G already provides temporary identities; but unfortunately the standard offers some freedom on how often they are refreshed and configured. To this end, some researchers [3] already found out that in some networks the implementation does not refresh the temporary identifier at all so that it would be possible to track a subscriber. Furthermore, 4G does not provide user-plane (UP) integrity protection between UE and eNB, where other researchers [4] were able to modify the domain name server (DNS) request with a man-in-the-middle attack and were able to change the IP address of the requested website with another one (DNS hijacking). This un-intended redirected website might be a malicious one that potentially attacks the UE with other means, e.g., on application layer with various web techniques.

Another example of attacks on 4G might be a "bidding down" attack to GSM where the security is much easier to break. "Bidding down" attacks are pretending that capabilities of a higher security system are not supported, such that a system with weaker capabilities is chosen instead. This may result in easier compromise of the system.

The interconnection protocols SS7 and Diameter were exploited by attackers, e.g., it was possible to pretend to be a roaming network and send an any time interrogation request to the home location register (HLR)/home subscriber server (HSS) and request the current location of a subscriber simply based on the mobile subscriber integrated services digital network number (MSISDN). Many mobile operators introduced Diameter firewalls, in order to prevent such a fraud and only allow requests from PLMNs where the UE is attached. For many years after the deployment of 4G, no security risk was disclosed. On the one hand, this was due to the fact that in the beginning of 4G the equipment was relatively expensive and many core network elements implemented in hardware were difficult to purchase for individuals in addition to the high price. Nevertheless, more recently it was proved that it is very easy to build up a 4G false base station and perform different attacks on the system, which explains the recent increase in detected security issues in 4G, e.g., [5,6]. With the development of 5G, many shortcomings of 4G have been improved, but under the consideration that legal interception is always to be allowed in the serving network.

7.2   3GPP Release Timeline

The 3GPP security working group "SA3" is a Stage 2 working group that takes care of selected Stage 3 protocol aspects (see Chap. 1), which is probably one of the very few working groups in 3GPP that does not 100% belong to a specific stage.

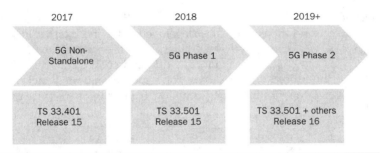

FIGURE **7.2**   SA3 specification timeline

Figure 7.2 shows the 5G security specification timeline: first, the 5G non-standalone security was defined in the 4G security specification 3GPP TS 33.401 [1], since the security is based on 4G and the 5G gNB acts like a secondary cell of the eNB. The 4G NAS protocol is used between UE and 4G core network; thus, all 4G procedures were carried out. For 5G Phase 1, a new TS was created in 33.501 [2], which is the baseline of this section. At the time of writing this textbook, the 5G Phase 2, i.e., the Release 16 of the security work, was not completed.

7.3   5G Security Features in Release 15

7.3.1   Overview of 5G Security Features

Figure 7.3 shows the new security features that are introduced in 5G in comparison to 4G and that are described in detail in the subsequent sections of this chapter. 5G offers increased home control, i.e., the Home PLMN (HPLMN) is able to verify whether a UE is actually residing in a particular (roaming) PLMN or Visited PLMN (VPLMN). The radio access network (RAN) security for the central unit (CU) and distributed unit (DU) split is a new RAN feature in Release 15 and handled by the NDS. The interworking security between 5GS and 4G EPS is handling the connected mode and idle mode security, when the UE is moving between the different systems as well as how to map the different security contexts. The primary authentication got extended to support EAP-AKA, in addition to 5G-AKA, an enhancement of EPS-AKA. Subscriber's privacy is improved, since now no unique subscription identity is transferred in the clear (i.e., unencrypted) over-the-air interface at any time, except when the feature is not used. The 5G core network is using web service techniques, supported by the newly introduced service-based architecture (SBA), so as to allow for easier functional virtualization, hence requiring special security means. Non-standalone security refers to the deployment model where the 5G gNB is attached to a 4G eNB, as a secondary cell. This assumes that the UE and evolved packet core (EPC) still communicate with 4G NAS protocol.

In 5G, there are enhanced possibilities to provide a finer granularity of the visibility and configurability of security features toward the end user, i.e., to display on the phone screen whether a security feature is used or not. For the first time, 5G supports more than one NAS connection, including to different PLMNs for the case that non-3GPP access, e.g., WLAN, is used. The user-plane security now can perform integrity protection, which was not possible in any 3GPP technology before. For interconnect scenarios, new solutions had to be developed since the Internet Protocol (IP) Packet eXchange (IPX) providers between different mobile operator networks perform mediation services and messages cannot be end-to-end encrypted. The secondary authentication was introduced to grant access for a special requested packet data unit (PDU) session, after authentication is performed with a third-party service provider.

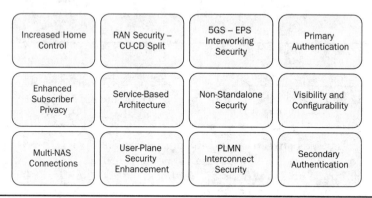

Figure 7.3   New 5G security features

7.3.2   5G Security Architecture

The 5G security architecture in terms of security domains is similar to the 4G security architecture, as shown in Sec. 7.1.1 with an additional SBA security domain, which considers only the security between SBA network functions as described in the following section. 5G architecture design requires a few new functional entities and interfaces (see Fig. 7.4), along with the description of its key functional components.

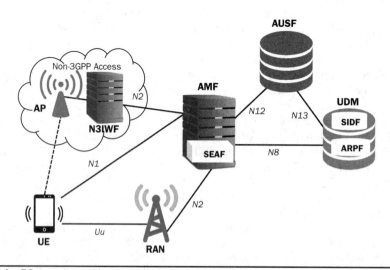

Figure 7.4   5G security architecture

The unified data management (UDM) performs subscription management and access authorization. It also holds two functionalities: the subscription identifier de-concealing function (SIDF) and the authentication credential repository and processing function (ARPF) for authentication credentials generation. The authentication server function (AUSF) uses those credentials and performs authentication of the UE via 3GPP access and untrusted non-3GPP access.

The access and mobility management function (AMF) terminates the NAS signaling and takes care of many aspects like registration management, connection management, reachability management, mobility management, etc. The AMF further is collocated with the security anchor function (SEAF), which may change, in future, 3GPP Releases depending on feature evolution and whether there is a reason that a UE is served by multiple AMFs in the same PLMN at the same time. The non-3GPP interworking function (N3IWF) is only used for untrusted non-3GPP access, which is mostly WLAN but also refers to fixed-line access.

7.3.3 Service-Based Architecture

SBA adopts the service-oriented architecture (SOA) and utilizes technologies already established in the domain of web services, such as REST [9] and HTTP/JSON [10,11]. A network function (NF) of SBA as shown in Fig. 7.5 can act as service producer and/ or service consumer and has to support transport layer security (TLS) following the profile of clause 6.2 of TS 33.210 [12] which should be compliant with the profile given by HTTP/2 in RFC 7540 [10]. Furthermore, all NFs shall support both server-side and client-side certificates. In case TLS is not used, then NDS shall be in place to protect the service-based interfaces (SBI); nevertheless, NDS (TS 33.210 [12] and TS 33.310 [13]) can also be used in addition to TLS.

At the event of discovery, registration, and access token request, the NF and the network repository function (NRF) shall authenticate each other. If TLS is used for transport layer protection then also the authentication provided by TLS shall be used

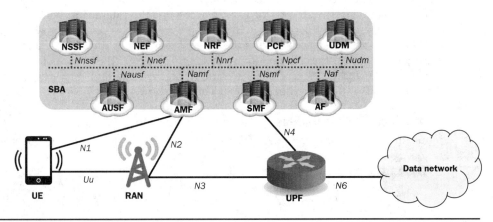

FIGURE 7.5   Service-based architecture and the service-based interfaces

for mutual authentication of the NRF and NF, else in case of NDS the authentication is implicit.

If an NF requests a service, then this has to be authorized by the NRF and granted with an access token. OAuth 2.0 (RFC 6749 [14]) is used for the service authorization procedure where the access tokens shall be JSON web tokens (RFC 7519 [15]) which are secured with digital signatures or message authentication codes (MAC) based on JSON web signature (JWS) (RFC 7515 [16]). The following roles are assigned to the NRF and NFs based on OAuth 2.0:

a. The NRF acts as OAuth 2.0 authorization server.

b. The NF service consumer acts as OAuth 2.0 client.

c. The NF service producer acts as OAuth 2.0 resource server.

After successful mutual authentication, the NF service consumer can send an access token request to the NRF for an NF type of the expected NF producer instance as well as the expected NF service name(s). The NRF may authorize the request from the NF and then generate the access token which is sent back to the NF including an expiring time and scope of the token. The token is signed by the NRF based on a shared secret or private key as described in RFC 7515 [16]. Then the NF service consumer can send a service request including the access token to the NF service producer. Both NFs perform mutual authentication explicit based on TLS or implicit based on NDS. The NF service producer verifies the access token and executes the requested service including sending a response back to the NF service consumer.

The following NFs are also part of the SBA and were not mentioned up to now: the session management function (SMF), whose main task is to take care of session establishment, session modification and release, as well as tunnel management between user-plane function (UPF) and access network (AN) node, e.g., gNB. The policy control function (PCF) provides policy rules to control-plane function(s) like the SMF to enforce them, e.g., in the UPF or in the RAN. The network exposure function (NEF) interfaces with external entities for exposing capabilities and events, provisioning of information from external applications to the 3GPP network, and further providing translation of internal-external information. The application function (AF) may communicate directly with other NFs or indirectly via the NEF for the provisioning of services like application-influenced traffic routing, NEF access, or the interaction with the policy framework. The network slice selection function (NSSF) is responsible for selecting the set of network slice instances serving the UE and determines the allowed and configured NSSAI as well as the corresponding AMF set. The UPF is the gateway to other data networks (DN) and takes care of user-plane packet routing and forwarding besides other tasks. UPF and DN are not part of SBA but are shown in Fig. 7.5 for completeness.

7.3.4   Security Associations

Figure 7.6 shows the different security associations on top of the SBA architecture in Fig. 7.5.

There are five different security associations defined in 5G:

1. Primary Authentication: The UE authenticates with the AUSF via the AMF and with credentials from the UDM.

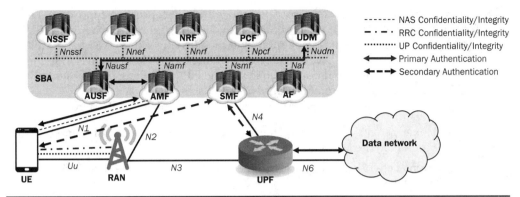

FIGURE 7.6 Security associations

2. NAS Confidentiality/Integrity: Once the UE is authenticated, NAS protocol confidentiality and integrity are enabled with the NAS security mode command (SMC) between UE and AMF.

3. RRC Confidentiality/Integrity: After NAS security is set up, the RRC security between UE and gNB is enabled with the access stratum (AS) SMC procedure between UE and gNB.

4. UP Confidentiality/Integrity: Optionally based on the security policy provided by the SMF, the UE and gNB may perform UP confidentiality and integrity.

5. Secondary Authentication: The secondary authentication is performed between UE and (external) AAA server via the SMF and UPF according to the user subscription and local policies to grant PDU session establishment for the subscribed service in the 5G core network.

7.3.5   Identities and Enhanced Subscriber Privacy

In all previous technologies, the international mobile subscriber ID (IMSI), which uniquely identifies a mobile subscriber, or more precisely the USIM in the UE, is transferred in the clear (i.e., unencrypted) over-the-radio interface, e.g., during initial attach/identity request procedures before NAS SMC and security are activated.

The IMSI is a number with maximum of 15 digits, where the first 3 digits represent the MCC and next 2 to 3 digits the MNC, i.e., they point to a specific mobile operator in a specific country, followed by 9 to 10 digits of the mobile subscriber identification number (MSIN).

As mentioned before, a so-called IMSI catcher can passively track individual UEs, and further lead to other active attacks, e.g., man-in-the-middle attack where the information sent over-the-air interface is intercepted and potentially modified. The IMSI as such got replaced in 5G with the subscriber permanent ID (SUPI) that is intended not to be transferred in the clear over-the-radio interface. The SUPI can be represented in two different formats: one is still the traditional IMSI format and the other one the network access identifier (NAI) format as username@realm as specified in IETF RFC 7542 [17]. The USIM or ME calculates the subscription concealed identifier (SUCI), the encrypted SUPI, and only the subscription identifier de-concealing function (SIDF), a service offered by the UDM in the HPLMN can de-conceal the SUCI back to the SUPI.

In order to route the message to the right mobile network, the MCC and MNC in IMSI representation as well as the realm (e.g., @mobile-operator.com) in NAI representation are not encrypted and point uniquely to the HPLMN. Figure 7.7 shows how the SUPI concealment is performed for the two formats.

The following parameters are required in order to calculate the SUCI from the SUPI:

- Home network public key, the key used for encryption configured by the HPLMN in the USIM.

- Home network public key identifier, in case of multiple keys identifies the key used for SUPI protection.

- Protection scheme profile, elliptic curve integrated encryption scheme (ECIES) parameters [18,19,20].

- Protection scheme identifier, so far three values are specified: NULL scheme, profile A, and profile B.

In addition to the above protection schemes, other proprietary protection schemes for concealing the SUPI can be used depending on the mobile operator. The NULL scheme, however, provides no protection and the SUCI is therefore identical to the SUPI. The two ECIES schemes, with profile A and B, differ mostly in the elliptic curve (EC) domain parameters (A: [18], B: [20]) and EC Diffie-Hellman primitive (A: [18], B: [19]).

The SUCI length is limited to 3000 octets plus the size of the username used in case of NAI format or the size of the MSIN in case of IMSI.

The SUCI has then the following format as shown in Fig. 7.7:

- SUPI type is either an IMSI format or an NAI format. There are still bits left for other formats in the future.

- The routing indicator is assigned by the home network operator and provisioned in the USIM in order to find the right UDM instance within one UDM, which is capable of serving the subscriber. It is set to the value 0 if no routing indicator is configured on the USIM or if no home network public key is provisioned.

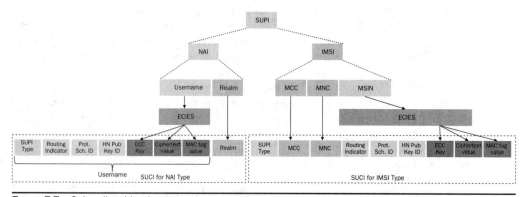

FIGURE 7.7   Subscriber identity concealment

- The protection scheme identifier identifies which scheme is used, i.e., the NULL scheme or a non-NULL scheme (profile A or B) or a protection scheme specified by the HPLMN.

- The home network public key identifier points to the public key used to protect the SUPI. If the NULL scheme is used, then this data field is set to the value 0.

- The scheme output is the encrypted username or MSIN and the length depends on the used protection scheme. It also contains the ECC ephemeral public key value (ECC Key) for freshness and the MAC tag value for integrity verification at the SIDF.

The UE can send the SUCI only in the following three NAS messages, else it should send a temporary identity:

1. Initial registration request to a PLMN where the UE does not have a 5G-GUTI.

2. Response to an identity request message from a PLMN.

3. De-registration request, in case the UE did not receive a registration accept message during the initial registration procedure.

On the other hand, the UE should not use the NULL scheme, i.e., the NULL scheme usage is restricted only to the following scenarios:

- If the UE does not have a 5G-GUTI and is making an unauthenticated emergency session.

- If the home network has configured "null-scheme" to be used.

- If the home network has not provisioned the public key needed to generate a SUCI, this may be also the case if the USIM is not capable of storing the required security parameters.

7.3.6 Visibility and Configurability

The visibility of the status of the security in use of a UE was already introduced in GSM [21] and onward as the so-called ciphering indicator. It was a mandatory feature for the mobile station, but it was not widely implemented by the operating systems [22]. In addition, the home network operator could simply disable the ciphering indicator feature in the SIM/USIM. In practice, the ciphering indicator did not exist, but the idea was revisited for 5G, in order to have a finer granularity of the indication. In this context, the following security features can now be distinguished:

- AS confidentiality: AS confidentiality, confidentiality algorithm, bearer information

- AS integrity: AS integrity, integrity algorithm, bearer information

- NAS confidentiality: NAS confidentiality, confidentiality algorithm

- NAS integrity: NAS integrity, integrity algorithm

However, the problem now still exists: there is no procedure on how the security visibility can be applied; only one requirement was specified:

1. The UE shall provide above security information to the applications in the UE (e.g., via APIs), on a per PDU session granularity.

This means that the operating system could provide APIs to the UE to query this information and provide it to the correspondent application, but if the operating system does not support it, the feature will not be used again.

Another visibility requirement is that the serving network identifier shall be available for applications in the UE.

7.3.7   Key Hierarchy

The key hierarchy describes what keys are derived in which entities, based on the procedures described in the following clauses. Each of the security procedures, i.e., primary authentication, NAS security mode command, and AS security mode command (see clause 7.3.8 for the procedures overview), use a defined set of security keys for the protection of the respective messages on the different layers. The key hierarchy is organized like a tree (see Fig. 7.8), which is growing from the root (key K) to different "branches" with derived keys for different protocols and functions. Here only the dependency and storage of the different keys are described; their usage is explained within the respective procedures. The long-term key or root key K is the shared secret and stored in the USIM and in the ARPF and can be either 128 bits or 256 bits long. When primary authentication is requested, the ARPF derives the subsequent keys CK, IK (for 5G-AKA authentication) or CK′, IK′ (for EAP-AKA′ authentication) and then

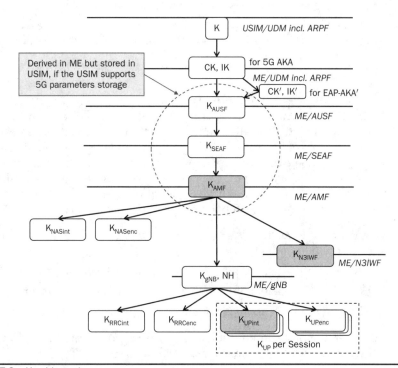

FIGURE 7.8   Key hierarchy

derives the K_{AUSF} which is sent with the authentication vector (AV) to the AUSF to perform the challenge/response authentication with the UE. The ME derives the K_{AUSF} in the same way as the APRF, depending on the authentication scheme (see Fig. 7.8). The K_{AUSF} can be stored in the ARPF until the next authentication is performed and a new K_{AUSF} is generated. The storage of the K_{AUSF} in the UE is optional, but if the USIM supports storing the 5G parameters then it should be stored there, else in the non-volatile memory of the ME. The K_{AUSF} is required for two features, "steering of roaming" and "UE parameters update," which do not rely on the keys derived in the serving network, i.e., the K_{AUSF} is used to integrity protect the message from the HPLMN to the UE in case the serving network wants to change the message content, e.g., change priority order of PLMNs. Besides the K_{AUSF}, the APRF also stores the home network private key that is used by the SIDF in the UDM to de-conceal the SUPI from the SUCI, sent by the UE.

From the K_{AUSF}, the ME and the AUSF derive the anchor key called K_{SEAF} at the time of primary authentication, which is stored in the USIM (if supported) or otherwise in the MEs' non-volatile memory, and the AUSF provides the K_{SEAF} to the SEAF in the serving network. The derivation function also includes the serving network name in the key derivation in order to make sure the UE is actually located in this particular serving network, i.e., in case of an authentication relay attack [23], where an attacker makes the victim UE camping at a false base station and then tunnels all signaling to a (far) away UE of the attacker, which is then pretending to be the victim UE. In that case, the attacking UE could be in a different country and would cause a billing fraud to the victim UE, but with the serving network name used in the K_{SEAF} derivation, the NAS SMC would fail since the keys in the UE and in the PLMN, where the attacking UE is located, are different. In 5G Phase 1, the SEAF and the AMF are co-located, but both were defined as separate entities for future evolution of the security architecture in case a specific feature requires a split between two, or, one SEAF is serving more than one AMF as a security anchor.

The key in the AMF, the K_{AMF} is then derived from the K_{SEAF} in the SEAF and the UE including the SUPI and anti-bidding down between architectures (ABBA) parameter, which is a new parameter that represents the security features of the serving network. By first including it in the K_{AMF} and then sending it between network and UE, an attacker in case of man-in-the-middle attack cannot change the parameter (e.g., cannot downgrade any security features), since the K_{AMF} would then be different, which would also lead to a failure of the NAS SMC procedure.

The K_{AMF} is also stored in the USIM (if supported) or in the MEs' non-volatile memory otherwise, similar to K_{SEAF} and K_{AUSF}. The K_{AMF} is created during primary authentication, NAS key re-keying, NAS key refresh or interworking procedures with EPS. A 5G key set identifier (ngKSI) is associated with the K_{AMF} and the K_{SEAF} and is stored together with the 5G globally unique temporary identity (5G-GUTI), which points to the serving AMF of the UE. The ngKSI has the type "native" or the type "mapped" and the purpose is to identify a native security context in the UE and in the AMF without invoking the authentication procedure; that is, during subsequent connection setups, the native security context can be reused. A mapped ngKSI is generated in the UE and in the AMF when the UE is moving from EPS to 5G and the K_{AMF} is derived from the K_{ASME}. The mapped ngKSI and the mapped K_{AMF} are also stored together.

The keys for the NAS encryption K_{NASenc} and integrity protection K_{NASint} are derived from the K_{AMF} at the time of the NAS SMC procedure and used in the AMF and the UE to protect the NAS signaling between both of them.

From the K_{AMF} two keys are derived in the AMF and the UE, depending on the access the UE is using. For non-3GPP access (e.g., WLAN), the K_{N3IWF} is derived from the K_{AMF}, and for 3GPP access, the K_{gNB} is derived in a similar way from the K_{AMF}, with the difference that uplink NAS COUNT and access-type distinguisher flag is used as input to the key derivation function (KDF) to ensure that K_{gNB} and K_{N3IWF} are different in case they are derived from the same K_{AMF}. The K_{N3IWF} is then used in the N3IWF for securing the IPsec tunnels between UE and N3IWF for untrusted non-3GPP access. The K_{gNB} is used in the gNB and UE for deriving the keys for integrity and ciphering of the RRC control signaling (K_{RRCint}, K_{RRCenc}) and for the user plane (K_{UPint}, K_{UPenc}) on AS layer between ME and gNB.

There are some conditions when the ME has to remove the stored keys K_{AUSF} and K_{SEAF} of the 5G security context:

1. The ME is in power on state and the USIM is removed from the ME.
2. The ME powers up, but the USIM gets changed (SIM swapping) when powered off.
3. The ME powers up and the USIM gets removed when powered off.

7.3.8   Overview of the Security Procedures

Figure 7.9 shows the relationship and sequence on how the security procedures are performed; some are within the context of the (initial) registration to the mobile network; some are executed afterward.

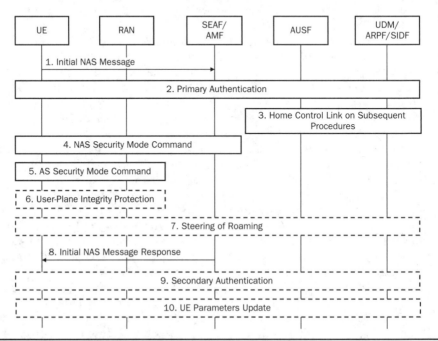

FIGURE 7.9   Security procedures overview

The UE first has to transit from IDLE state and send a registration request to the network. The UE might never have been registered to the network before, e.g., a subscriber just arrived at the airport in a country and powers on the phone. In that case, the initial NAS message protection would be performed, i.e., the UE just sends a very limited number of cleartext (i.e., unencrypted) parameters, just enough to perform the primary authentication. After the successful authentication, the AUSF provisions the result to the UDM, which then authorizes subsequent requests for this UE from the serving network, which might be a roaming network.

The AMF then sets up the NAS security with the NAS SMC and provides the gNB key to the gNB, where the AS SMC is performed for AS security on RRC layer between UE and gNB. The user plane might be protected as well, depending on the UE capabilities, operator requirements, and gNB capabilities, i.e., besides ciphering, integrity protection also may be applied. Optionally, the so-called steering of roaming is also performed during or after the registration procedure; that is, the home operator can provision a list of preferred PLMN/access technology combinations. The AMF then closes the registration procedure with a NAS response to the initial NAS message. If the UE wants to access a specific DN where the subscription profile requires another authentication, then the secondary authentication is triggered between the UE and an AAA server in the DN. Furthermore, the operator may trigger UE parameters update which is so far the same procedure as steering of roaming and basically updates the routing ID in order to find the right UDM instance in the home network.

7.3.9 Primary Authentication and Key Agreement

For 5G, there are two authentication procedures specified: EAP-AKA′ as specified in RFC 5448 [27] and 5G-AKA as specified in TS 33.501 [2]. Which procedure is finally used in the network is based on the decision of the mobile operator. The main difference between the two procedures is that EAP-AKA′ is based on the EAP framework and evaluates the response from the UE in the AUSF, i.e., always in the home network and not in the visited network as, e.g., in 4G.

Figure 7.10 shows the functional entities involved for the primary authentication and key agreement for both procedures. The same procedure was applied also to the 5G-AKA as an extension of the EPS-AKA, specified for 4G, i.e., the result of the

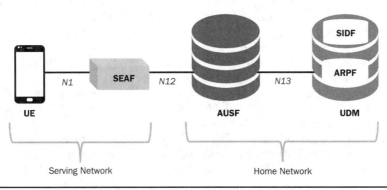

FIGURE 7.10 Authentication architecture

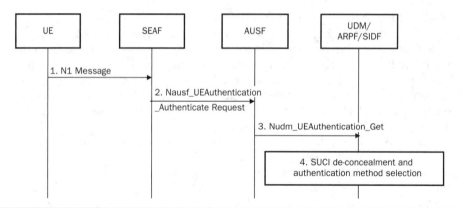

Figure 7.11 Primary authentication initiation

challenge, computed by the UE, is verified finally in the AUSF in the home network but with the difference that the SEAF in the visited network performs the first check already.

In terms of the number of messages, there is no difference between EAP-AKA' and 5G-AKA. The main differences are more in the way the authentication vector is constructed, the results are computed, and how the keys for the visited network are derived. Note that 5G-AKA is only envisioned to use NAS messages as transport, while EAP-AKA' can use other protocols as transport as well, e.g., for non-3GPP access (see Sec. 7.3.18). Potentially EAP-AKA' might have additional EAP messages compared to the minimum number of steps, as shown in Fig. 7.11.

The primary authentication is triggered by an N1 request message from the UE, e.g., an initial registration to the network or mobility re-registration events. The call flow is shown in Fig. 7.11 and applies to trigger one of the two (5G-AKA or EAP-AKA') authentication methods.

The UE sends N1 message request and includes the SUCI in case of initial registration, i.e., the very first registration to the network or the 5G-GUTI in case of any other registration message. The SEAF in the serving network can decide based on operator policy when to re-authenticate the UE if it already successfully preformed a registration before and the UE has a valid 5G-GUTI. The SEAF then sends in step 2 a Nausf_UEAuthentication_Authenticate request message to the AUSF and includes the SUCI (initial registration) or the SUPI (SEAF triggered re-authentication, UE was sending a 5G-GUTI in step 1) and in addition the serving network (SN) name of its mobile operator network (the SEAF is in the serving network, i.e., it could be located in the visited network when the UE is roaming). When the AUSF receives in step 3 the request, it authorizes the request from the SEAF with an SN name check. The AUSF sends in step 3 a Nudm_UEAuthentication_Get request to UDM, including also either SUCI or SUPI, depending on the received information from the SEAF as well as the SN name. If the SUCI is sent (initial registration), then the SIDF de-conceals the SUCI to a SUPI using the home network private key. The SUPI is then used in the UDM/ARPF to point to the subscription data in order to choose the authentication method.

7.3.9.1 5G-AKA

5G-AKA is an enhancement of EPS-AKA [1] for 4G with additional home control for the successful authentication, i.e., the home network receives a proof from the visited network that the UE is successfully authenticated, and is specified in [2]. In comparison to the EPS-AKA, only one authentication vector is sent from the UDM/ARPF. The authentication procedure is usually triggered by the AMF/SEAF in the serving network based on certain events initiated by the UE, e.g., initial registration, mobility registration, etc., as shown in Fig. 7.11. Figure 7.12 depicts the call flow for 5G-AKA from the starting point where the UDM/ARPF generates the authentication vector, i.e., the call flow from Fig. 7.12 was already performed as a prerequisite and 5G-AKA is selected as an authentication method.

The 5G-AKA procedure is explained here with the steps from Fig. 7.12:

1. When the UDM/ARPF receives a Nudm_Authenticate_Get request (see Fig. 7.11, step 3), then the UDM/ARPF creates a 5G home environment authentication vector (5G HE AV), consisting of random challenge (RAND), authentication token (AUTN), expected response (XRES*), and the AUSF key

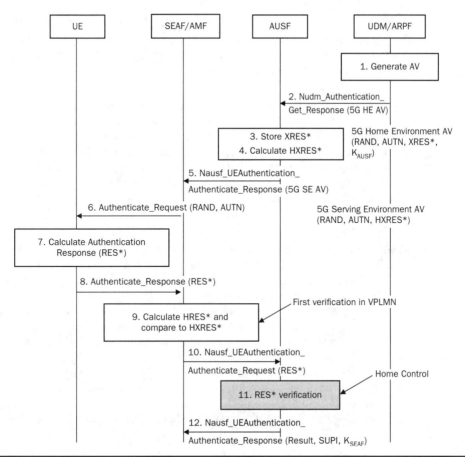

Figure 7.12 Authentication procedure for 5G-AKA [2]

K_{AUSF}. The UDM/ARPF sets the authentication management field (AMF) "separation bit" set to "1" indicating 5G-RAN access to 5G system. The AMF field is a part of the AUTN and the "separation bit" is bit 0 of the AMF field of AUTN. The AUTN consists of the concealed sequence number SQN with an anonymity key AK (SQN $\oplus$ AK), the AMF field, and the MAC of the both: AUTN = SQN $\oplus$ AK $\|$ AMF $\|$ MAC. Note that the overloading of the abbreviation AMF is an unlucky coincidence, which was pointed out already in the study phase of 5G, but the system architecture group was not able to change the name for the access and mobility management function anymore. Furthermore, the UDM/ARPF derives the AUSF key K_{AUSF} and calculates the expected result XRES*.

2. The UDM sends the 5G HE AV to the AUSF in a Nudm_UEAuthentication_Get response. The response also includes an indication that the AV is an 5G-AKA authentication vector and includes the SUPI if the request (Fig. 7.11, step 3) included a SUCI.

3. The AUSF stores the XRES* temporarily together with the received SUPI.

4. The AUSF generates the 5G AV from the 5G HE AV, i.e., it computes the Hash eXpected RESponse (HXRES*) from XRES* and the SEAF key K_{SEAF} from K_{AUSF}.

5. The AUSF creates the 5G serving environment authentication vector (5G SE AV) consisting of RAND, AUTN, HXRES*, and sends it to the SEAF in a Nausf_UEAuthentication_Authenticate response.

6. The SEAF sends the authentication challenge with RAND, AUTN to the UE in a NAS message authentication request including the ngKSI which is used by the UE and AMF to identify the K_{AMF} and after successful authentication also the new partial native security context. In addition, the anti-bidding down between architectures (ABBA) parameter also is included to indicate a set of security features and is used for the AMF key K_{AMF} derivation, i.e., a bidding down attack to a weaker security set would result in a K_{AMF} mismatch between UE and AMF. For Release 15 only the initial 5G security set is defined but may be extended in future releases. When the UE receives the challenge from the SEAF, the ME forwards the RAND and AUTN to the USIM.

7. The USIM verifies the freshness of the 5G AV by checking the AUTN: first the USIM calculates the sequence number SQN from the RAND and the first part (the concealed SQN) of the AUTN, then it calculates with the SQN and the AMF field an expected message authentication code XMAC and compares it with the MAC of the AUTN. The AUTN is acceptable if the XMAC and MAC are identical and when the sequence number SQN is in the correct range. The USIM then computes a response RES and returns it together with the keys CK, IK to the ME. The ME then shall compute RES* from RES, RAND, and the serving network name. The ME then calculates the AUSF and SEAF keys subsequently, i.e., K_{AUSF} from CK $\|$ IK, serving network name and concealed SQN (SQN $\oplus$ AK) of the AUTN and K_{SEAF} from K_{AUSF} and serving network name. The ME accessing 5G checks during the authentication whether the "separation bit" in the AMF field of AUTN is set to 1.

8. The UE sends a NAS authentication response message with the RES* to the SEAF.

9. The SEAF computes the hash HRES* from RES* and compares the HRES* with the HXRES* received from the AUSF in step 5. The SEAF considers the authentication successful from serving network point of view when both are equal.

10. The SEAF then sends a Nausf_UEAuthentication_Authenticate request message with the RES* from the UE to the AUSF.

11. AUSF compares the received RES* with the stored XRES* and the AUSF considers the authentication as successful from the home network point of view if both are equal. AUSF informs the UDM about the authentication result and stores the K_{AUSF}.

12. The AUSF sends a Nausf_UEAuthentication_Authenticate response to the SEAF with an indication that the authentication was successful, including the K_{SEAF} and the SUPI, in case SUCI was sent by the UE initially (Fig. 7.12). The SEAF derives the AMF key K_{AMF} from the K_{SEAF}, the ABBA parameter and the SUPI and sends the ngKSI and the K_{AMF} to the AMF. The SEAF will not provide any communication services to the UE until it knows the SUPI.

7.3.9.2 *EAP-AKA′*

As shown in Fig. 7.13, EAP-AKA′ as specified in RFC 5448 [27] and the 3GPP 5G profile for EAP-AKA′ is specified in the TS 33.501 [2] with respect to subscriber privacy and subscriber identity and key derivation.

1. The UDM/ARPF generates an authentication vector with authentication management field (AMF) separation bit = 1 and computes the keys CK′ and IK′ with the serving network name as the value of AN identity. CK and IK are then replaced by CK′ and IK′. The "network name" is a concept from RFC 5448 [27] and is used for the key derivation in EAP-AKA′. The value of <network name> parameter is defined for 5G in TS 33.501 [2] as "serving network name."

2. The UDM sends the transformed authentication vector AV′ (RAND, AUTN, XRES, CK′, IK′) to the AUSF and indicates that the AV′ is for EAP-AKA′ and includes the SUPI if the request (Fig. 7.11, step 3) included a SUCI. The RAND, AUTN, and XRES are similar to the 5G-AKA parameters; see Sec. 7.3.9.1, step 1.

3. The AUSF sends the Nausf_UEAuthentication_Authenticate response message with the EAP-request/AKA′-challenge message to the SEAF.

4. The SEAF sends a NAS authentication request message to the UE containing the EAP-request/AKA′-challenge message and the ngKSI and ABBA parameter similar to 5G-AKA (Fig. 7.12, step 6). When the UE receives the message, the ME forwards the RAND, AUTN ngKSI, and ABBA parameter to the USIM. Once authentication is successful, UE and AMF create the partial native security context which is identified with ngKSI.

5. The USIM verifies the freshness of the AV′ similar to 5G-AKA (Fig. 7.12, step 7) and computes a response RES. The USIM then returns RES, CK, IK to the ME

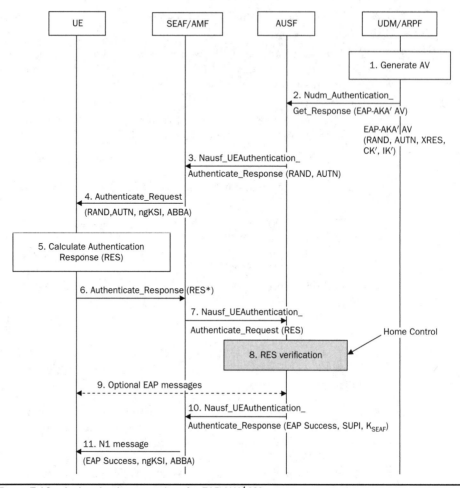

FIGURE 7.13 Authentication procedure for EAP-AKA′ [2]

which derives CK′ and IK′ using the serving network name as value of the AN identity.

6. The UE sends a NAS authentication response message containing the EAP-response/AKA′-challenge message with the computed RES.

7. The SEAF forwards the EAP-response/AKA′-challenge message transparently to the AUSF in a Nausf_UEAuthentication_Authenticate request message.

8. The AUSF verifies the message and informs the UDM about the authentication result.

9. EAP notifications (RFC 3748 [34]) as well as EAP-AKA′ notifications (RFC 4187 [35]) may be sent at any time in the EAP-AKA′ exchange between the AUSF and the UE and forwarded transparently by the SEAF.

10. The AUSF derives the extended master session key (EMSK) from CK′ and IK′ according to RFC 5448 [27]. The most significant 256 bits of EMSK are then used as AUSF key K_{AUSF} and for the calculation of K_{SEAF}. The authentication is successfully completed when the AUSF sends the EAP-success message to the SEAF together with the K_{SEAF} and the SUPI (in case the UE sent initially a SUCI) within a Nausf_UEAuthentication_Authenticate response, which is further sent to the UE.

11. The EAP-success message from the SEAF to the UE includes the ngKSI and the ABBA parameter. The SEAF then derives the AMF key K_{AMF} from the K_{SEAF}, ABBA parameter, and SUPI and sends it to the AMF. When the UE receives the EAP-success message, it starts to derive all keys on the UE side, i.e., EMSK from CK′ and IK′, taking the most significant 256 bits of the EMSK as the K_{AUSF}, calculates K_{SEAF} and derives the K_{AMF}.

The SEAF finally provides the ngKSI and the K_{AMF} to the AMF.

7.3.9.3 *Linking Increased Home Control to Subsequent Procedures*

In comparison to the EPS-AKA in 4G, the new 5G authentication and key agreement protocols provide increased home control. In order to prevent frauds with subsequent procedures, e.g., fraudulent Nudm_UECM_Registration request, the authentication result is reported to the UDM and needs to be linked to those subsequent procedures (see Fig. 7.14).

1. The UE authentication with EAP-AKA′ or 5G-AKA is performed according to the procedures described in the previous sections.

2. Once the authentication is completed, the AUSF informs the UDM by sending a Nudm_UEAuthentication_ResultConfirmation request including all relevant information such as SUPI, timestamp of the authentication, the authentication type (e.g., EAP method or 5G-AKA), and the serving network name.

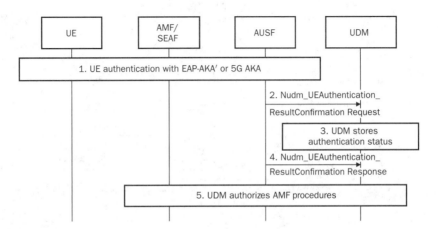

FIGURE 7.14 Linking increased home control to subsequent procedures

3. The UDM stores the authentication information of the UE, received from the AUSF.

4. UDM acknowledges the request from the AUSF with a Nudm_ UEAuthentication_ResultConfirmation response.

5. Now the UDM authorizes requests from the AMF for this UE such as Nudm_ UECM_Registration_Request.

7.3.10   Non-Access Stratum Security

7.3.10.1   Initial NAS Message Protection

Initial NAS message is sent after the UE transitions from the idle state. At the very first time, e.g., the first message when roaming to another PLMN or the first activation in the HPLMN, the UE does not have a NAS security context and this was exploited by so-called IMSI catchers in previous generations to retrieve the subscriber identity in cleartext (i.e., unencrypted) over-the-air interface. The difference to the previous generations is that if the UE has no NAS security context, then it sends only a limited set of information elements (IEs), called the cleartext IEs. Once the security context is established, the UE sends the complete initial message in the NAS security mode complete message in a NAS container, which is protected. The following cleartext IEs are only allowed to be sent without protection: subscription identifiers (e.g., SUCI or GUTIs), UE security capabilities, ngKSI, indication that the UE is moving from EPC, additional GUTI, and IE containing the TAU request in the case of idle mobility from LTE.

If the UE already has a NAS security context then the UE uses it to cipher the initial NAS message without the cleartext IEs in a NAS container and to integrity protect the whole message including cleartext IEs.

The following description only highlights the case when the UE has no security context (see Fig. 7.15).

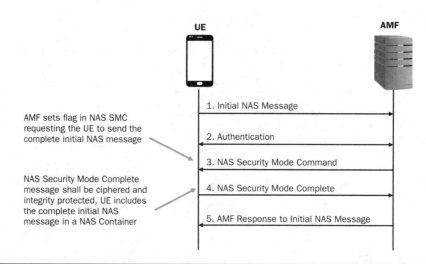

FIGURE 7.15   Initial NAS message protection

1. The UE sends the initial NAS message to the AMF and if the UE has no NAS security context, it contains only the cleartext IEs.

2. The AMF initiates an authentication procedure (EAP-AKA′ or 5G-AKA) with the UE.

3. The AMF sends the NAS SMC message if the authentication of the UE is successful. The procedure of the NAS SMC is described in Sec. 7.3.10.2. If the initial NAS message was sent protected by the UE but the AMF does not have the security context, then the AMF includes a flag to request the UE to send the complete initial NAS message in the NAS security mode complete message.

4. The UE sends the ciphered and integrity protected NAS security mode complete message to the AMF including the complete initial NAS message in a NAS container if the UE sent the initial NAS message unprotected, or the AMF set the flag in step 3.

5. The AMF sends its ciphered and integrity protected response to the initial NAS message.

7.3.10.2   NAS Security Mode Command

In order to set up the NAS security context in the UE and the AMF, the SMC procedure is carried out, as shown in Fig. 7.16. It can be triggered with (re-registration) requests from the UE or when the AMF's and UE's security contexts are not synchronized, which is an error scenario and might happen due to implementation issues, e.g., at intersystem handovers and/or AMF changes.

In Fig. 7.16, the two keys for NAS integrity protection K_{NASint} and ciphering K_{NASenc} are derived in the UE and the AMF from the K_{AMF} key, indicated by the ngKSI, during the registration procedure. Afterward, the security context is identified by the ngKSI.

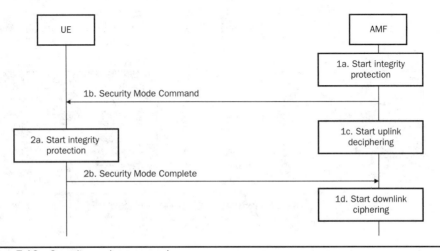

FIGURE 7.16   Security mode command

1a. The AMF starts the NAS integrity protection before it generates the NAS SMC message.

1b. The AMF assembles the NAS SMC message with the following parameters: the replayed UE security capabilities, the selected NAS algorithms, and the ngKSI for identifying the K_{AMF}. Optionally the following parameters are included: K_AMF_change_flag (indication that a new K_{AMF} is calculated), a flag to request the complete initial NAS message (see Fig. 7.15, step 3), ABBA parameter. The AMF may request the UE to send the permanent equipment identifier (PEI) in the protected response message. The AMF sends the NAS SMC message integrity protected but not ciphered to the UE.

1c. The AMF activates NAS uplink deciphering after sending the NAS SMC message.

2a. The UE verifies the integrity protection of the message and whether the received UE capabilities are the same as it previously sent to the AMF. If the K_AMF_change_flag is included, the UE derives a new K_{AMF} and sets the NAS COUNT to zero. The UE selects the NAS security context indicated by ngKSI and starts NAS integrity protection and ciphering/deciphering with the corresponding NAS keys.

2b. The UE sends now all NAS messages ciphered and integrity protected. If the AMF requested the PEI and/or the complete initial NAS message in the SMC message, then the NAS security mode complete message includes them accordingly.

1d. The AMF verifies the integrity of the received NAS security mode complete message and starts NAS downlink ciphering.

7.3.10.3 Multiple NAS Connections

5G is bringing a new principle to accommodate simultaneous non-3GPP access in parallel to 3GPP access, i.e., multiple NAS connections are now possible at the same time via 3GPP access and non-3GPP access. The rationale behind is that the same authenticationprocedures "EAP-AKA" and "5G-AKA" for primary authentication can be reused via the non-3GPP access also. Figure 7.17 shows the scenario where the UE is connected

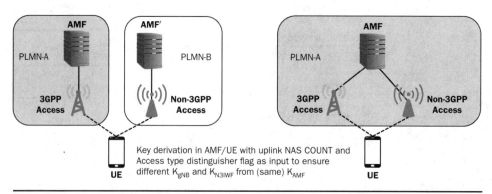

Key derivation in AMF/UE with uplink NAS COUNT and Access type distinguisher flag as input to ensure different K_{gNB} and K_{N3IWF} from (same) K_{AMF}

Figure 7.17 Multiple NAS connections

via 3GPP and non-3GPP access at the same time, where both ANs belong to the same operator, i.e., the NAS signaling is terminated at the same AMF and the same security context is reused, i.e., the same NAS keys for integrity protection and ciphering. The distinguishing of the security keys takes place at lower layer for deriving the key K_{gNB} for 3GPP access and the key K_{N3IWF} for non-3GPP access.

The scenario that the non-3GPP access and the 3GPP access belong to different operators is also supported, but then the two NAS connections are independent, i.e., authentication is performed per access with the credentials of the respective operator.

Key derivation in AMF/UE with uplink NAS COUNT and access-type distinguisher flag as input ensure different K_{gNB} and K_{N3IWF} from the (same) K_{AMF}.

7.3.11 Access Stratum Security

After the security is set up on the NAS layer, the AMF generates the gNB key K_{gNB} and sends it to the gNB together with the UE security policy for UP traffic as described in Sec. 7.3.12. The K_{gNB} is then used as the basis for the subsequent keys for RRC signaling protection (K_{RRCint}, K_{RRCenc}) as well as user-plane traffic protection (K_{UPint}, K_{UPenc}). Similar to the NAS security context setup, an access stratum security mode command (AS SMC) procedure, as shown in Fig. 7.18, is executed in order to negotiate the RRC and UP security algorithms and activate the RRC security. The AS SMC procedure is only executed for an initial context setup between the UE and the gNB. For each RRC_IDLE to RRC_CONNECTED state transition, a new K_{gNB} is generated in order to ensure a fresh K_{gNB} for the AS SMC.

The gNB starts RRC integrity protection and sends the AS SMC message including the selected RRC and UP encryption and integrity algorithms to the UE. The gNB already integrity protects the message with a MAC generated with the key K_{RRCint}. When the gNB sends the AS SMC it also starts with the downlink ciphering.

When the UE receives the AS SMC and successfully verifies the MAC, it starts RRC integrity protection and RRC downlink deciphering. The UE answers to the AS SMC with an AS security mode complete message, which is also integrity protected with a

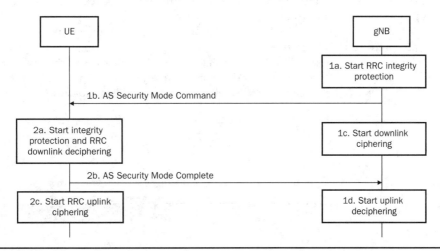

Figure 7.18 AS security mode command

MAC but not ciphered. The UE starts the uplink ciphering after sending the AS security mode complete message.

After RRC security setup, the UP security can be activated according to the UP security policy as described in Sec. 7.3.12.

7.3.12   User-Plane Security Enhancements

In 4G, only integrity protection of the RRC and NAS signaling was mandatory and confidentiality protection for signaling and user-plane data was optional. There was integrity protection for user-plane (UP) data but the usage was restricted only between relay node and donor eNB due to the required computational power for UP integrity protection.

For 5G the UP integrity protection can be now configured by the UP security policy, which is provisioned by the SMF in the following way; see Fig. 7.19.

When the UE requests a PDU session then it sends its UE integrity protection maximum data rate parameter, i.e., the maximum data the UE hardware is able to provide integrity protection for user-plane data. For Release 15 and Release 16, i.e., Phase 1 and 2 of 5G, there are only two values defined: 64 kbps and full rate, i.e., either the UE indicates a maximum of 64 kbps or it indicates with full rate that it supports integrity protection at all data rates the UE can handle. The minimum value of 64 kbps is mandatory to support.

There are three values defined in the UP security policy:

- Required: always enabled
- Preferred: can be changed by the gNB during handover
- None needed: always disabled

At the time of the PDU session establishment, the SMF checks whether a UP security policy should be applied for the requested PDU Session, i.e., whether UP confidentiality and/or UP integrity should be activated (required or preferred). The policy is then provisioned from the SMF to the gNB and is valid for all data radio bearers (DRB) that belong to the same PDU session. The UE and the gNB enforce the policy with ciphering and/or integrity protection in the packet data convergence protocol (PDCP) layer.

If the policy is "Required" and the gNB cannot activate the requested UP confidentiality and/or UP integrity for the DRBs of a PDU session, e.g., due to hardware constraints, then the gNB will reject the establishment of resources in the gNB and inform the SMF about it. The same situation would happen in case of a handover scenario via the Xn interface between two gNBs, when the target gNB cannot fulfill the UP security

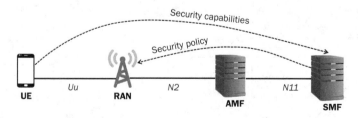

FIGURE 7.19   User-plane security enhancements

policy set to "Required" of the source gNB for a PDU session. If the policy is "Preferred" then changes in activation and de-activation of the UP integrity protection may happen.

The UP security is activated by the gNB with an RRC connection reconfiguration message to the UE which contains an indication for each DRB whether UP integrity and UP ciphering is activated. The UE generates a key K_{UPint} and/or a key K_{UPenc}, respectively, for those DRBs where UP integrity protection and/or UP ciphering is activated. The UE then sends an RRC connection reconfiguration complete message as a confirmation back to the gNB.

In case neither integrity nor ciphering is activated by the gNB, then the UE shall not include the unnecessary MAC for integrity in order to avoid unnecessary padding bits set to 0 in order to minimize the packet size.

7.3.13   Security during Mobility Procedures

In principle, there are two different types of handover: via a direct interface between the gNBs called Xn, or via the AMF, called N2. There are also different variants further possible, e.g., if several cells (DUs) belong to the same CU, then the UE could perform an Intra-gNB-CU handover with changing the cell but without changing the CU. For the N2 handover there could be also the possibility that the AMF is changed in addition to the gNB.

In RRC_IDLE, the UE and the AMF only store the keys K_{AMF}, K_{NASint}, and K_{NASenc} and have no AS-related keys. When an initial AS security context is established, e.g., when the UE performs the transition from RRC_IDLE to RRC_CONNECTED, the UE and AMF derive the K_{gNB}, the next hop (NH) parameter, which is in fact an intermediate key, and a next hop chaining count (NCC) which is initialized with NCC = 0, but the first NH value is associated with NCC = 1. If the maximum NCC value is 7, then a new initial AS security context setup has to take place.

The AMF provides the K_{gNB}, NH, and NCC parameters in the AS security context to the gNB, which is then deriving the security keys for user-plane traffic and RRC signaling messages. The UE and the gNB maintain the UE 5G AS security context in RRC_INACTIVE state. At every handover and at every transition from RRC_INACTIVE to RRC_CONNECTED states, the K_{gNB} is refreshed with either vertical key derivation, i.e., a fresh {NCC, NH} pair is available at the gNB, or horizontal key derivation, i.e., NCC associated with current K_{gNB} is used and no fresh NH is available at the gNB.

When a UE transits from RRC_INACTIVE to RRC_CONNECTED it protects the message with a MAC computed with the stored K_{RRCint} and includes a resumeIdentity that helps locate the anchor gNB address, i.e., the address of the gNB that suspended the RRC connection and therefore where the current gNB sends the connection resume request. Furthermore, in case of gNB change the old gNB will provide the full AS security context to the new gNB, consisting of newly derived $K_{NG-RAN*}$ and the associated NCC, the UE 5G security capabilities, UP security policy, the UP security activation status, as well as the ciphering and integrity algorithms used by the UE with the old gNB.

Figure 7.20 shows the differences between horizontal and vertical key derivation. If a gNB has no fresh {NH, NCC} pair available, e.g., the UE is camping at the same cell and sending data from time to time so it only moves between RRC_INACTIVE and RRC_CONNECTED mode, then the gNB has used the provisioned NH from the AMF already in a previous key derivation. For horizontal key derivation, the K_{gNB} is stored in the UE and the gNB for transition into RRC_INACTIVE mode and is not deleted as

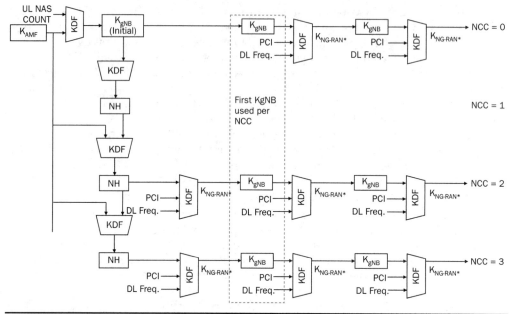

Figure 7.20   Horizontal and vertical K_{gNB} key derivation

for the vertical key derivation. For vertical key derivation, the provisioned NH value is used to derive with K_{AMF} a new NH value as an interim key for computing the K_{gNB}. During transition into RRC_INACTIVE mode, all keys besides the K_{RRCint} and NCC also are deleted in gNB and UE for vertical key derivation.

7.3.14   Security of EPC Interworking

The 5G network can interwork with EPC, taking into account that the UE has two different modes of operation available (single registration or dual registration mode), depending on the deployment scenario as also described in 5G core network in Chap. 6. For the dual registration mode, the UE is registered in the two core networks at the same time and thus can independently use and maintain two different security contexts. If the UE switches between EPC and 5G, it uses either the EPC security context or the 5G security context. Therefore, from the security point of view there is no impact on the procedures and either the 5G or the EPC procedures apply. For the single registration mode, the UE can be only in one of the networks at one point in time. There are also two subcases depending on whether the AMF and the MME support a direct interface (N26 interface) between each other.

Figure 7.21 shows the handover from 5GS to EPS and vice versa, both over N26 reference point. If the handover is performed without N26 support, then simply the stored security contexts in EPS or 5GS are reused independently.

In case of handover from 5GS to EPS over N26, the AMF receives the handover required message from the gNB and generates the mapped EPS security context, where the MME key K_{ASME}' is derived from the K_{AMF} and associated with the EPS KSI (eKSI). The AMF further derives the eNB key K_{eNB} with the uplink NAS COUNT as input to the KDF set to a value far above the maximum value in order to avoid the possibility to derive the same K_{eNB} again later on in EPS. The AMF sets NCC = 2 and derives NH two times. The UE security context is then sent from the AMF to the MME in a forward relocation request message with the following parameters: K_{ASME}', eKSI, {NH, NCC} pair, uplink and downlink EPS NAS COUNTs, UE EPS security capabilities, selected EPS NAS algorithms identifiers.

In case of handover from EPS to 5GS over N26, the MME receives the handover required message from the eNB and provides the mapped EPS security context (parameters as above) including MME ASME (Access Security Management Entity) key K_{ASME} to the AMF. The AMF derives the K_{AMF} from K_{ASME} and NH of the EPS security context and associates the mapped 5G security context with ngKSI. The AMF does not use the NH, NCC pair from the MME but derives a temporary gNB key as NH from the mapped K_{AMF}.

7.3.15   Non-Standalone (NSA) Security

When the UE is simultaneously connected to an eNB which serves as master cell, and a gNB serving as secondary cell in E-UTRA-NR dual connectivity (EN-DC), then the security depends on the type of core network, i.e., EPC or 5GC. If EPC is the serving core network, then EPS procedures are carried out; if 5GC is the serving core network then 5G procedures are performed. The master eNB (MeNB) establishes the AS security between secondary gNB (SgNB) and UE by generating the key $S-K_{gNB}$ in a similar way as it would generate the K_{eNB} for a secondary LTE cell. A countervalue is used to guarantee the freshness of the $S-K_{gNB}$ and that there is no reuse of the same key. The counter is provided to the UE via LTE RRC signaling so that the UE also can generate the same key. The UE generates a security context for the secondary cell with the RRC keys for encryption and integrity protection, as shown in Fig. 7.22. Note that even if the SgNB supports user-plane integrity protection, only ciphering can be used for the user-plane data, similar to an eNB.

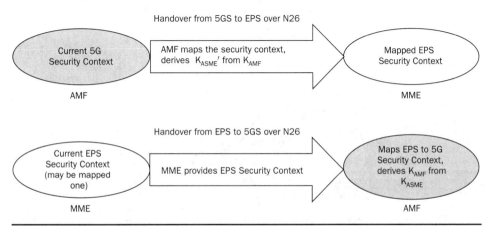

FIGURE 7.21 Security context mapping over N26

7.3.16   Secondary Authentication

The secondary authentication, as the name implies, is carried out after the successful primary authentication and occurs between the UE and the DN outside the mobile operator domain. In the pre-5G mobile networks, the DNs were independent in terms of access control without the support of mobile operator. The 5G system allows mobile operators to delegate the authentication and authorization to a third-party hosting the DN. This procedure is executed during the establishment of user-plane connection toward a specific DN name based on configuration in the subscription profile.

The secondary authentication uses the EAP framework, which is widely used and enables various credential types and authentication methods used by different application service providers, i.e., the service provider in the DN selects the authentication method, which has to be known by the UE as well.

The prerequisite of secondary authentication is the successful completion of primary authentication based on the UE's 5G network access credentials, and the establishment of NAS security context between the UE and the AMF. Additionally, the UE needs to be provisioned with the external credentials used for the authentication between itself and corresponding data network authentication authorization and accounting (DN-AAA) server.

During this procedure, the SMF in the 5G core network performs the role of EAP authenticator while the DN-AAA server performs the role of EAP authentication server, as depicted in Fig. 7.23.

When the UE requests to establish a new PDU session, the PDU session establishment request message including a DN-specific identity, the data network name (DNN), is further provided to the SMF. The SMF then determines that authentication/authorization of the PDU session establishment is required and triggers the secondary authentication procedure with the UE. First, if the UE did not already provide its

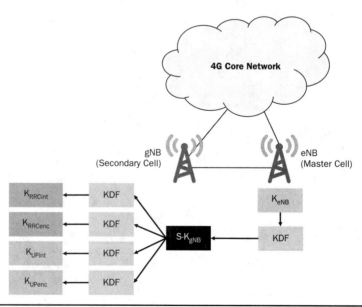

FIGURE 7.22   Non-standalone (NSA) security

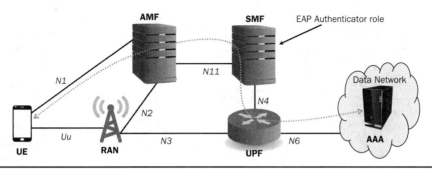

Figure 7.23 Secondary authentication

EAP identity in the PDU session establishment request, the SMF requests the EAP identity from the UE. The SMF then forwards the EAP response/identity message to the DN-AAA server via the UPF as user-plane data. Then, based on the selected EAP method, the DN-AAA server and the UE exchange corresponding EAP messages. Once the authentication is successfully completed, the DN-AAA server sends EAP-success message to the SMF via the UPF and the reception of the EAP-success in the SMF completes the EAP authentication procedure. The SMF stores the result for the specific UE and the DN for upcoming PDU session establishment requests in order to skip the secondary authentication, once already authenticated. The PDU session establishment procedure is then completed.

The DN-AAA server or the SMF may initiate the re-authentication with the UE. If the re-authentication is initiated by the DN-AAA server, the UE is addressed by generic public subscription identifier (GPSI), which is notified to the DN-AAA server during the secondary authentication.

7.3.17 Interconnect Security

For the interconnection between different operator networks, e.g., between visited PLMN and home PLMN, the security edge protection proxy (SEPP) was introduced in 5G (see Fig. 7.24). There are two interface variants of N32 specified between two SEPPs:

- N32-c (control) interface for key agreement, parameters exchange, and error handling between two SEPPs via established TLS connection.

- N32-f interface for exchanging messages via IP eXchange (IPX) providers using JSON Web Encryption (JWE, RFC 7516 [36]).

The reason of the two variants is that the traffic between two operators is traversing IP eXchange (IPX) service providers, which provide interconnection services including traffic normalization or analytics. This is why there is no IPsec or TLS on transport layer directly between two SEPPs (except for configuration on N32-c), because many IPX business cases require certain information elements within the roaming signaling to be accessible and potentially modifiable.

The SEPP is responsible for several functionalities:

- Applying security to messages on the interconnection
- Message filtering and policing

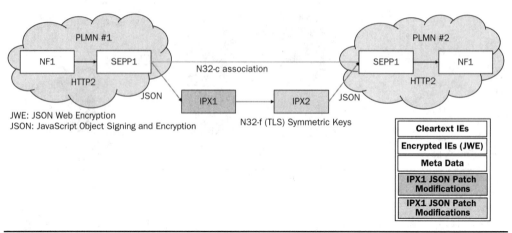

Figure 7.24   Interconnect security

- Enforcement of anti-spoofing measures
- Rate limiting
- Topology hiding

When the SEPP receives HTTP/2 request/response messages from a network function, it converts it into a JWE object which is then encapsulated into an HTTP/2 message (as the body of the message) and sent over the N32-f interface to the other SEPP via IPX service providers. The IPX service providers are allowed to perform authorized message modifications based on certificates or raw public keys, i.e., when the IPX service provider receives the message, it extracts the cleartext part of the HTTP message from the JWE object. After the modification according to the IPX policy and service, the IPX provider calculates an "operations" JSON patch object, which is appended to the received message and then sent it to the next hop, e.g., another IPX provider.

7.3.18   Steering of Roaming Security/UE Parameters Update

As shown in Fig. 7.25, the procedures for steering of roaming (SoR) and the UE parameters update (UPU) are in principle the same; therefore, they are described here together. There is a difference in the use case and thus also in the content of the messages as well as their names. The purpose of the two procedures is very different: the SoR procedure is envisioned to update the list of preferred PLMN/access technology combinations while the UPU procedure updates the routing ID for addressing the UDM in the HPLMN. The SoR procedure can be executed already at the time of registration after NAS SMC, i.e., after the generation of the K_{AUSF} and the setup of the security context, but is not limited to this. The UPU procedure is only executed after the registration took place at any time.

The commonality is that the UDM wants to send a protected message to the UE which should not be changed by the serving network, e.g., VPLMN. For this reason, the UDM is first sending the message to the AUSF, which is integrity protecting the message with a MAC and also includes a MAC for the expected acknowledgment of

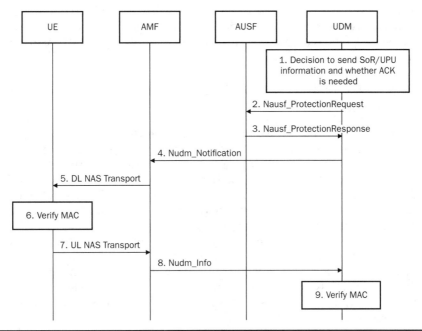

Figure 7.25 Procedure for steering of roaming security/UE parameters update

the UE (XRES) and a counter for freshness. The AUSF provides everything back to the UDM and the UDM sends the protected message with the counter to the UE. The UE is checking the integrity of the message and provides a protect acknowledgment back to the UDM, which can be verified by the UDM since it received already the expected result from the AUSF.

If the features are supported in the UE and the network, then the UE and the AUSF should be able to store the K_{AUSF} after the primary authentication. The AUSF in the home network uses the key K_{AUSF} to integrity protect the SoR list and the UE parameters by sending the corresponding MAC to UE and this prevents the serving network from altering the SoR list and UE parameters delivered to the UE. The reason why K_{AUSF} is used is that it is the lowest key in the key hierarchy known only both by the UE and the AUSF in the home network and it is not known by the serving network, i.e., potential roaming network. The SEAF/AMF in the serving network does not know the K_{AUSF} and cannot modify the message without possibility that the UE would not recognize it because the verification of the integrity protection in the UE would fail.

7.3.19 Non-3GPP Access Security

7.3.19.1 Untrusted Non-3GPP Access System Architecture

WLAN was the first non-3GPP access considered for interworking with the 3GPP system [24]. This interworking was specified already in the 3GPP Release 6 in 2006. In 3GPP Release 8 the EPC [25] and the new radio access technology LTE (long-term evolution) were introduced and other non-3GPP technologies like CDMA2000 and WiMAX were included in the procedures for interworking with the EPC [26]. For non-3GPP access, WLAN was still remaining the main focus and got integrated even tighter in later releases with the 3GPP radio access with the features LTE-WLAN aggregation (LWA), LTE-WLAN radio-level integration with IPsec tunnel (LWIP), and RAN-controlled LTE-WLAN interworking (RCLWI).

The integration of the non-3GPP access to the 3GPP core network is based on two access principles:

Untrusted Access: The access point that the user equipment (UE), i.e., the device, is connected to is not trusted by the mobile operator. This access point could be the WLAN at home or a free WLAN from a coffee shop. Also the encryption on the radio link is out of scope of the mobile operator. Since there is no trust, the mobile operator does not depend on the security mechanisms on the WLAN side and the UE tunnels all data traffic to a gateway in the network, which is trusted by the mobile operator. This is a most commonly deployed non-3GPP access for EPC (e.g., voice-over WiFi feature) and is standardized in Release 15 for 5G.

Trusted Access: As the name indicates, the mobile operator trusts and operates the access points, i.e., this includes that the encryption of the radio link is also controlled by the operator and the credentials are derived from the security context in the UE and the network. The trusted access is agreed to be standardized in Release 16 and all agreements for normative changes in the specification are currently captured in a temporary document [41].

An overview of non-3GPP access security for untrusted and trusted access to 5GC can be found in [37].

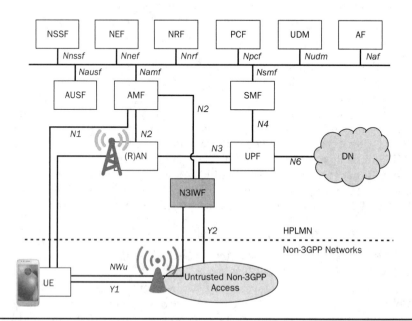

FIGURE 7.26 Untrusted non-3GPP access system architecture

Figure 7.26 shows the 5G architecture for untrusted non-3GPP access via Y1 and Y2 reference points, including the 3GPP access via N1 and N2 reference points.

For the untrusted non-3GPP access the following additional network function, the N3IWF, is defined: the UE connects first to the N3IWF and then to AMF and UPF, respectively, as for the 3GPP access.

7.3.19.2   *Untrusted Non-3GPP Access Authentication*

Untrusted non-3GPP access is the most commonly deployed access for WLAN due to the easy deployment of the N3IWF and that ability for offering additional services like voice-over WiFi, which is very useful, e.g., in locations where the mobile network radio coverage is not sufficient. The mobile operator has no influence on the WLAN access point and its deployment, e.g., in a shopping mall or at the subscriber's home. Since the operator does not trust the access point, the UE communicates with the N3IWF which is a node of trust and the termination point of the IPsec tunnel between UE and N3IWF. Before the establishment of the IPsec tunnel, the UE must be authenticated by the 5G home network, using one of the primary authentication methods. After successful authentication the key for the IPsec tunnel establishment is provisioned to the N3IWF. A new transport protocol EAP-5G was introduced in order to keep the IKEv2 between UE and the N3IWF open with EAP request/response pairs. This was required since an IKE_AUTH exchange without EAP is designed to support only one request/response between the UE and the network. EAP-AKA' or 5G-AKA in the NAS messages requires more than one exchange and would result in an IKEv2 failure. The details of EAP-5G are explained in Sec. 7.3.19.3.

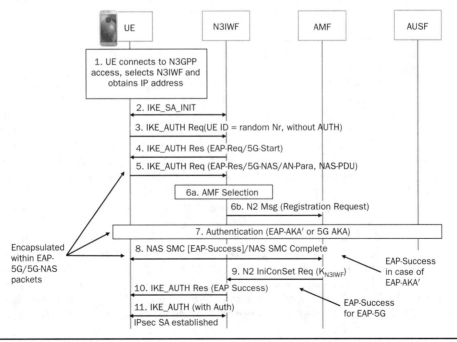

FIGURE 7.27   Untrusted non-3GPP access registration and authentication

As shown in Fig. 7.27, the UE connects to an untrusted non-3GPP AN, i.e., the UE connects to a WLAN and can reach the internet. The UE then starts with the establishment of an IPsec security association (SA) with the selected N3IWF by initiating an IKE initial exchange according to RFC 7296 [31]. All subsequent IKE messages are encrypted, and integrity protected by using the IKE SA after step 2. The UE then sends an IKE_AUTH request message without the AUTH payload. This indicates that the IKE_AUTH exchange shall use EAP signaling (in this case EAP-5G signaling). The UE uses a random number in the UE Id field in this message and shall not use its GUTI/SUCI/SUPI as the user Id. Now the EAP-5G exchange is started with the EAP-Request/5G-Start packet in step 4 to inform the UE to start sending NAS messages encapsulated within EAP-5G packets. The UE sends in step 5 a NAS registration request encapsulated in an IKE_AUTH request and tunneled in EAP-5G to the N3IWF. The N3IWF then forwards the NAS message to the AMF after the selection process. Normal primary authentication (i.e., 5G-AKA or EAP-AKA′) is performed (step 7). The final authentication message from the AUSF contains the anchor key K_{SEAF} derived from K_{AUSF}. The SEAF (not shown in the call flow but is collocated with the AMF in Release 15) derives the K_{AMF} from K_{SEAF} and sends it to the AMF. The AMF derives the NAS security keys and a security key for N3IWF (K_{N3IWF}) from the K_{AMF}. The N3IWF key is used later by the UE and N3IWF for establishing the IPsec security association (in step 11). The AMF shall send an SMC in step 8 to the UE. The NAS SMC activates the NAS security and may contain the EAP-success in case of EAP-AKA′. The UE completes the authentication and creates a NAS security context and an N3IWF key K_{N3IWF}. The EAP-5G encapsulation is no longer required and the UE sends an EAP-Response/5G-Complete packet, which triggers the N3IWF to send an EAP-success for the EAP-5G to the UE in step 10. The common N3IWF key K_{N3IWF} is now used to create the IPsec SA in step 11. All subsequent NAS messages between the UE and the N3IWF shall be sent over the established IPsec SA.

7.3.19.3 EAP-5G Protocol

The "EAP-5G" method is used only for encapsulating NAS messages between the UE and the N3IWF. EAP-5G is specified within 3GPP under its existing 3GPP Vendor-Id in [7] and [30] as a vendor-specific EAP method [27], utilizing the "Expanded" EAP type that is registered with IANA under the SMI Private Enterprise Code registry [28].

The EAP-5G protocol runs only between the UE and N3IWF and its primary purpose is to transparently relay NAS messages between the UE (over NWu) and the AMF (over N2). In addition, it enables AMF selection by the N3IWF (see Fig. 7.27).

All EAP-5G messages, except for the EAP-5G Start and Stop messages, contain a NAS message that is forwarded by the EAP-5G layer to the NAS layer. If the UE receives a NAS registration reject message, then the UE shall terminate the EAP-5G session by sending an EAP-5G Stop packet. In this case the N3IWF sends an EAP-failure message to the UE and completes the EAP-5G session.

The EAP-5G session between the UE and N3IWF is successfully performed when the EAP-5G layer in the UE receives the N3IWF key from the NAS layer and the EAP-5G layer in the N3IWF receives the N3IWF key from AMF. The UE then receives an EAP-success message from the N3IWF. After that, the EAP-5G layer in the UE and the EAP-5G layer in the N3IWF forward the common N3IWF key to the lower layer (IKEv2), which is further used for establishing an IPsec security association.

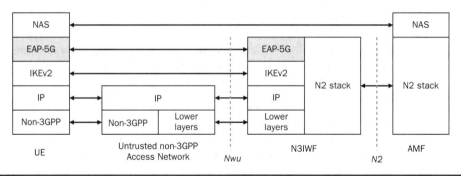

FIGURE 7.28 EAP-5G protocol stack

EAP-5G needs to support the following information elements:

- An EAP-5G Type field
- Three message identifiers for 5G-Start, 5G-NAS, and 5G-Stop
- Two attributes were defined: AN-parameters and NAS-PDU

7.3.19.3.1 EAP-5G Type Field The vendor-type field is specified in 3GPP TS 33.402 [33] annex C and set to EAP-5G method identifier of 3 (decimal) in all EAP-5G messages.

7.3.19.3.2 EAP-5G Flags The following message identifiers are considered:
5G-Start: This flag is only sent by the N3IWF to signal the initiation of an EAP-5G session. An EAP-5G Start message does not include any other information (i.e., it contains no attributes).
5G-Stop: This flag is only sent by the UE to signal the completion of an EAP-5G session due to a NAS registration failure.
5G-NAS: This flag is sent to indicate the encapsulated NAS message in the payload and is send between UE and N3IWF.

7.3.19.3.3 EAP-5G Attributes The following attributes are considered:
AN-Parameters attribute: This is included in an EAP-5G packet when the UE wants to send access network parameters (AN parameters) to N3IWF to be used for AMF selection. The use of AN parameters during a 5G registration is specified in TS 23.502 [29].
NAS-PDU attribute: This is included in an EAP-5G packet to encapsulate a NAS message.
An EAP-5G packet can include:

- No attributes: This is the case when the 5G-Start or 5G-Stop message is sent.
- Both the AN-Parameters attribute and the NAS-PDU attribute: This is the case where the UE needs to send the first NAS message and the associated AN parameters. Those AN parameters are used by the N3IWF to perform AMF selection.
- Only a NAS-PDU attribute: This is the most common case where the EAP-5G packet carries only a NAS message that should be transparently relayed by N3IWF to the AMF.

7.3.19.4 *Release 16 Enhancements*

Two new features, "Trusted Non-3GPP access" and "5GC access from WLAN UEs that do not support NAS," with respect to WLAN access were concluded to be included in the normative specifications in the ending Release 16 study on the security of the wireless and wireline convergence for the 5G system architecture [32,41].

7.3.19.4.1 Trusted Non-3GPP Access The trusted access is the corresponding deployment option of the untrusted access with the difference that the operator also trusts and controls the access point. For this reason, three new definitions were introduced:

Trusted Non-3GPP Access Network (TNAN): The TNAN consists of the trusted non-3GPP access point (TNAP) and the trusted non-3GPP gateway function (TNGF). The TNAN can connect to the 5GC by exposing north-bound interfaces compliant with N2/N3.

Trusted Non-3GPP Access Point (TNAP): The TNAP enables UEs to access the TNAN by using a non-3GPP wireless or wired access technology and corresponds to a WLAN access point.

Trusted Non-3GPP Gateway Function (TNGF): The TNGF exposes the N2/N3 interfaces and enables the UE to connect to 5GC over a non-3GPP access technology (TNAP).

The registration procedure and the authentication for trusted non-3GPP access are shown in Fig. 7.29.

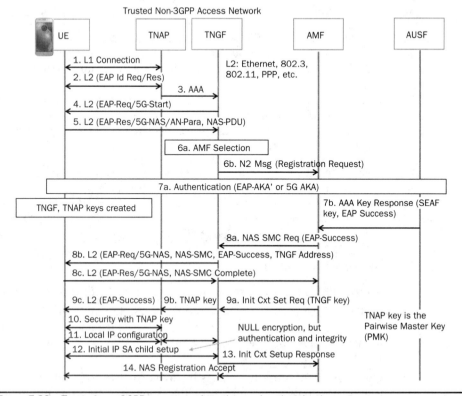

FIGURE 7.29 Trusted non-3GPP access registration and authentication

The UE registers to 5GC and authenticates with the TNAN at the same time by using the EAP-based procedure, which is essentially the same with the registration procedure for untrusted non-3GPP access. The interface between the TNAP and control-plane part of the TNGF is an AAA interface. The TNGF terminates the EAP-5G signaling and behaves as authenticator when the UE attempts to register to 5GC via the TNAN.

Similar to the untrusted non-3GPP access EAP-5G is used to encapsulate the NAS messages, starting in step 4 of Fig. 7.29. The UE derives the TNAP and TNGF keys after the primary authentication. The AMF sends a NAS SMC to the TNGF with the EAP-success flag of the primary authentication in step 6a. The TNGF includes its own IP address in the NAS SMC and forwards it to the UE. The AMF provides the TNGF key to the TNGF when it receives the NAS SMC complete message from the UE. The TNGF derives the TNAP key, which corresponds to the pairwise master key (PMK) and provides it to the TNAP. Now the EAP-5G encapsulation can be terminated and the UE can use the TNAP key for encryption over the air interface.

The security relies on layer-2 security between UE and TNAP, i.e., the encryption of the WLAN radio link. The TNAP is a trusted entity so that no IPsec encryption would be necessary between UE and TNGF, i.e., NULL encryption is sufficient for the user plane and signaling.

Separate IPsec SAs may be used for NAS transport and PDU sessions. At the end of the UE's registration to 5GC, an IPsec SA (NWt-cp) is established between the UE and TNGF (step 12) to protect NAS messages between the UE and TNGF. When the UE initiates a PDU session establishment at a later time, the TNGF initiates the establishment of one or more IPsec child SAs per PDU session. This results in additional IPsec SAs (NWt-up) to be set up between the UE and TNGF, which are then for user-plane transport between the two.

The main advantage of using IKEv2/IPsec is that the procedure for trusted non-3GPP access is almost identical with the procedure for untrusted non-3GPP access specified in TS 33.501 [2]; see also clause 7.3.19.2. Thus, the UE can use the same protocols and procedures for both trusted and untrusted non-3GPP access, and the TNGF can become very similar to N3IWF.

7.3.19.4.2   5GC Access from WLAN UEs That Do Not Support NAS   Another use case is targeting the 5G core network access from WLAN UEs that do not support NAS, i.e., those UEs would not be able to connect to the 5G core network via trusted access as described in the previous section or via untrusted access (see Sec. 7.3.19), since 5G NAS protocol support is required. A WLAN UE that does not support NAS has the following capabilities:

- The UE is capable to register to 5GC and to establish 5GC connectivity via a trusted WLAN AN.

- The UE is not capable of operating as 5G UE over a WLAN AN, i.e., it does not support normal trusted or untrusted access which requires 5G NAS support.

- The UE may be without any NAS capability at all, i.e., it may not support even 4G NAS protocol.

- The UE has 3GPP credentials, i.e., a USIM for EAP-AKA' authentication. 5G-AKA cannot be used, since it is transported in the NAS protocol.

Two new functional entities were introduced in the system:

Trusted WLAN Access Point (TWAP): The trusted WLAN access point the UE is connected to.

Trusted WLAN Interworking Function (TWIF): The interworking functionality that enables connectivity of the UE with the 5GC. The TWIF supports the NAS protocol stack and exchanges NAS messages with the AMF on behalf of the UE.

Figure 7.30 shows the registration and authentication procedure for a WLAN UE that does not support NAS protocol.

A single EAP-AKA' authentication procedure is executed for connecting the UE both to the trusted WLAN access network and to the 5G core network. The UE selects a TWAP, which initiates in step 1 the EAP-AKA' authentication with an EAP identity request that the UE answers with its SUCI. The TWAP then selects a TWIF for the interworking between UE and 5G core network. The TWIF creates a NAS registration request, including the SUCI, and after AMF selection sends it to the selected AMF in step 6.

After authentication, the AUSF provides the SEAF key to the AMF in step 8 and in parallel the UE creates the SEAF key and derives the AN key. The connection security between TWIF and AMF relies on the N2 security specified in TS 33.501 [2], section 9.2; therefore, no specific NAS security is used between the TWIF and AMF. The AMF selects therefore the NULL scheme for ciphering and integrity protection in the SMC in step 9.

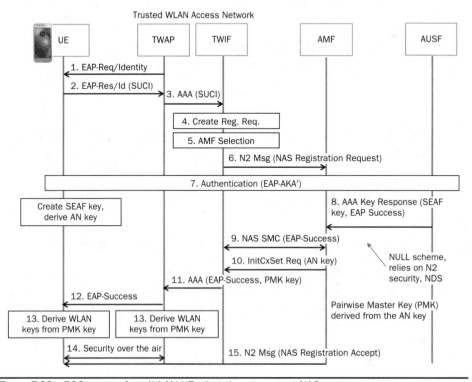

FIGURE 7.30   5GC access from WLAN UEs that do not support NAS

The AMF derives the AN key from the SEAF key and sends it to the TWIF in step 10. The TWIF now derives a pairwise master key (PMK) from the AN key and sends the PMK key and the EAP-success message to the trusted WLAN access point (step 11), which forwards the EAP-success to the UE (step 12). On the UE side, the PMK is derived from the AN key in the same way as in the TWIF. The PMK is used to secure the WLAN air-interface communication, i.e., the security on the air interface relies on Layer-2 security between UE and TNAP (security context to encrypt and integrity protect unicast and multicast traffic over the air). Since the TWIF is a trusted entity, no IPsec encryption is necessary between UE and TWIF, and NULL encryption is sufficient for the user plane and signaling, similar to the trusted non-3GPP access as described in Sec. 7.3.19.4.2.

7.4 IMS Security

IP multimedia services (IMS) is an access-independent multimedia feature that is used, e.g., for voice services from LTE onward and also is popular for voice-over WiFi over untrusted access, which is an (indoor) coverage extension of the mobile operators based on, e.g., local WiFi at home, without the need of additional RAN deployments. IMS requires IP connectivity and is a subscription feature. Therefore, a so-called IP multimedia services identity module (ISIM) can be installed on the UICC for authentication with the network, but it is also allowed to use the USIM if no ISIM is available.

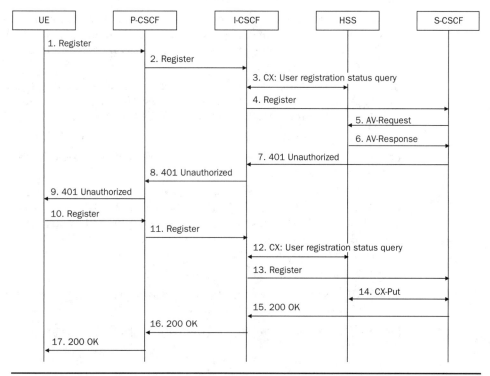

FIGURE 7.31 IMS authentication

The security features were described originally for 3G in 3GPP TS 33.203 [38], but they are applicable for any type of access. The detailed description of the different IMS functions and procedures can be found in Chap. 6 and in 3GPP TS 23.228 [39].

The security features comprise authentication of the subscriber and the network, re-authentication, as well as confidentiality and integrity protection of the SIP signaling between UE and P-CSCF. In the following, the authentication procedure is described.

As a prerequisite, the UE has to be registered in the network and need to have IP connectivity, i.e., a PDU session is established, as well as the discovered P-CSCF address as contact point for the registration. Figure 7.31 shows the IMS authentication procedures.

1. The UE sends an SIP REGISTER with its private and public user identity to the P-CSCF it discovered before.

2. The P-CSCF may be located in the serving network different to the home network. The P-CSCF checks the "home domain name" and discovers the I-CSCF in the home network and forwards the SIP REGISTER to the I-CSCF.

3. The I-CSCF queries the HSS whether the user is registered already in IMS and the HSS returns an indication whether the user is allowed to register via the P-CSCF in that serving network. The HSS also returns the S-CSCF address.

4. The I-CSCF forwards the SIP REGISTER to the S-CSCF.

5. The S-CSCF needs an authentication vector (AV) in order to start the authentication with the UE and requests an AV from the HSS.

6. The HSS generates several AVs for the S-CSCF and sends them back to the S-CSCF. Each AV consists of the following parameters: random number RAND, expected response XRES, cipher key CK, integrity key IK, and an authentication token AUTN.

7. The S-CSCF selects the next AV from the list and creates the authentication challenges inside an SIP 401 unauthorized response message including the RAND and AUTN for the UE as well as the CK, IK for the P-CSCF.

8. The I-CSCF forwards the SIP 401 unauthorized response to the P-CSCF.

9. The P-CSCF removes the keys CK, IK and stores them and forwards the rest of the message to the UE.

10. The UE checks the AUTN, which consists of a MAC and an SQN, and calculates the XMAC in order to check whether the SQN is in the correct range (see Sec. 7.3.9). The UE then calculates the authentication response RES and the keys CK, IK. The UE generates a new SIP REGISTER message including the authentication response RES toward the P-CSCF.

11. The P-CSCF forwards the message to the I-CSCF.

12. The I-CSCF performs a user registration status query and retrieves the same S-CSCF address as before.

13. The I-CSCF forwards the SIP REGISTER to the S-CSCF.

14. The S-CSCF compares the RES with the XRES from the HSS and authenticates the UE. The S-CSCF performs an S-CSCF registration notification procedure toward the HSS in order to inform the HSS about the successful authentication.

15. The S-CSCF indicates a successful authentication and IMS registration with an SIP 200 OK.

16. The I-CSCF forwards the SIP 200 OK to the P-CSCF.

17. The P-CSCF forwards the SIP 200 OK to the UE.

The SIP signaling message between UE and P-CSCF can be now integrity and confidentiality protected on IP level with IPsec ESP (RFC 4303 [40]).

References

1. 3GPP TS 33.401, "Technical Specification Group Services and System Aspects: 3GPP System Architecture Evolution (SAE) Security Architecture."
2. 3GPP TS 33.501, "Security Architecture and Procedures for 5G System."
3. Byeongdo Hong, Sangwook Bae, and Yongdae Kim, "GUTI Reallocation Demystified: Cellular Location Tracking with Changing Temporary Identifier," https://syssec.kaist.ac.kr/pub/2018/hong_ndss_2018.pdf
4. David Rupprecht, Katharina Kohls, Thorsten Holz, and Christina Pöpper, "Breaking LTE on Layer Two," https://alter-attack.net/
5. Ravishankar Borgaonkar, Lucca Hirschi, Shinjo Park, and Altaf Shaik, "New Privacy Threat on 3G, 4G, and Upcoming 5G AKA Protocols," https://eprint.iacr.org/2018/1175.pdf
6. Hongil Kim, Jiho Lee, Eunkyu Lee, and Yongdae Kim, "Touching the Untouchables: Dynamic Security Analysis of the LTE Control Plane," https://syssec.kaist.ac.kr/pub/2019/kim_sp_2019.pdf
7. 3GPP TS 24.501, "Non-Access-Stratum (NAS) protocol for 5G System (5GS); Stage 3."
8. 3GPP TS 36.331, "Radio Resource Control (RRC); Protocol specification."
9. R. T. Fielding, "Architectural Styles and the Design of Network-based Software Architectures," Doctoral dissertation, University of California, Irvine, 2000.
10. IETF RFC 7540, "Hypertext Transfer Protocol Version 2 (HTTP/2)."
11. IETF RFC 8259, "The JavaScript Object Notation (JSON) Data Interchange Format."
12. 3GPP TS 33.210, "3G Security; Network Domain Security (NDS); IP Network Layer Security."
13. 3GPP TS 33.310, "Network Domain Security (NDS); Authentication Framework (AF)."
14. IETF RFC 6749, "OAuth2.0 Authorization Framework."
15. IETF RFC 7519, "JSON Web Token (JWT)."
16. IETF RFC 7515, "JSON Web Signature (JWS)."
17. IETF RFC 7542, "The Network Access Identifier."
18. IETF RFC 7748, "Elliptic Curves for Security."
19. SECG SEC 1, "Recommended Elliptic Curve Cryptography," Version 2.0, 2009, http://www.secg.org/sec1-v2.pdf
20. SECG SEC 2, "Recommended Elliptic Curve Domain Parameters," Version 2.0, 2010, http://www.secg.org/sec2-v2.pdf

21. GSM 02.07 V7.1.0 (2000-03) Technical Specification, Digital Cellular Telecommunications System (Phase 2+); Mobile Stations (MS) Features (GSM 02.07 version 7.1.0 Release 1998).

22. Google support for Ciphering Indicator, https://issuetracker.google.com/issues/36911336

23. Syed Rafiul Hussain, Omar Chowdhury, Shagufta Mehnaz, Elisa Bertino, "LTEInspector: A Systematic Approach for Adversarial Testing of 4G LTE," Network and Distributed Systems Security (NDSS) Symposium 2018.

24. 3GPP TS 23.234, "3GPP System to Wireless Local Area Network (WLAN) Interworking; System Description," Mar. 2017.

25. 3GPP TS 23.401, "General Packet Radio Service (GPRS) Enhancements for Evolved Universal Terrestrial Radio Access Network (E-UTRAN) Access," Sep. 2019.

26. 3GPP TS 23.402, "Architecture Enhancements for Non-3GPP Accesses," Sep. 2019.

27. IETF RFC 5448, "Improved Extensible Authentication Protocol Method for 3rd Generation Authentication and Key Agreement (EAP-AKA')," 2009.

28. https://www.iana.org/assignments/enterprise-numbers/enterprise-numbers.

29. 3GPP TS 23.502, "Procedures for the 5G System; Stage 2, (Release-15)," Sep. 2019.

30. 3GPP TS 33.501, "Security Architecture and Procedures for 5G System," Sep. 2019.

31. IETF RFC 7296, "Internet Key Exchange Protocol Version 2 (IKEv2)," 2014.

32. 3GPP TR 33.807, "Study on the Security of the Wireless and Wireline Convergence for the 5G System Architecture," Sep. 2019.

33. 3GPP TS 33.402, "3GPP System Architecture Evolution (SAE); Security Aspects of non-3GPP Accesses."

34. IETF RFC 3748, "Extensible Authentication Protocol (EAP)."

35. IETF RFC 4187, "Extensible Authentication Protocol Method for 3rd Generation Authentication and Key Agreement (EAP-AKA)."

36. IETF RFC 7516, "JSON Web Encryption (JWE)."

37. A. Kunz and A. Salkintzis, "Non-3GPP Access Security in 5G," Journal of ICT Standardization, River Publishers, 2020.

38. 3GPP TS 33.203, "Access Security for IP-based Services," Mar. 2018.

39. 3GPP TS 23.228, "IP Multimedia Subsystem (IMS); Stage 2," Dec. 2019.

40. IETF RFC 4303, "IP Encapsulating Security Payload (ESP)."

41. 3GPP S3-194529, "Living doc for 5WWC."

Problems/Exercise Questions

1. What security features are new in 5G compared to previous generations?

2. What is the advantage of home control for authentication?

3. What are the reasons why privacy protection of the SUPI could be missing?

4. What is the minimum data rate for user-plane integrity protection?

5. What parameters are sent in the initial NAS message?

6. Which authentication methods can be used for primary authentication?

7. What are the representation formats of the SUPI?

8. How many NAS connections can one UE have?

9. For handover between EPS and 5GS, which functional entity is responsible for the security context mapping?

10. What is the reason for secondary authentication?

11. What is the relationship of IPX business and the interconnect security?

12. What is new for UP security?

13. How often is the K_{gNB} refreshed during mobility?

14. What is the difference between steering of roaming security and UE parameters update?

15. What is the difference between trusted and untrusted non-3GPP access?

8

Future Evolution of Mobile Systems

8.1   Current Evolution of 5G

In 3GPP, new studies for Release 17 were prioritized[1] based on system architecture working group 2 (SA WG2) proposal [1] supported by the participating companies. After the prioritization, five major topics were selected to be worked on which are explicitly described in the following sections. In addition, a short overview of the smaller study items that will be part of Release 17 is presented. The 3GPP system architecture working group 1 (SA WG1) which handles development of Stage 1 services and requirements has already started to discuss the Release 18 study proposals at a high level with potential new service requirements for the enhancement of the 5GS. The Stage 1 service requirements from SA WG1 are eventually fed into other 3GPP working groups, once completed, to become part of their work program. However, due to the limited progress on the Release 18 proposed study items at the time of writing this textbook, they are not described here.

Figure 8.1 illustrates the different features for 5G, starting with Release 15 up to current Release 17, following the ITU-T recommendations for the international mobile telecommunications (IMT) [2], which describe a service and feature set that should be supported for the year 2020 and beyond (see Chap. 4 for details). The 5G Phase 1 in Release 15 started with an initial set of features that are important for the first deployments. The radio interface for the new radio (NR) technology is specified in 3GPP TS 38.300 [3] (see Chap. 5 for details), the 5G core network architecture is specified in 3GPP TS 23.501 [4] (see Chap. 6 for details), and the security features are specified in 3GPP TS 33.501 [5] (see Chap. 7 for details).

8.1.1   Proximity-Based Services in 5GS

Proximity services (ProSe) is a relatively old feature standardized for EPS in 3GPP Release 12, which can be used for commercial and public safety services. The original idea was the introduction of device-to-device (D2D) communication in two different modes of operation (direct vs. indirect D2D) as well as two modes for the discovery of nearby devices (UE announcing, "I am here" versus UE asking, "who's there?") [6]. Due to legal interception issues, the D2D communication in direct mode was only allowed for public safety UEs. In Release 14, the direct mode of operation got enhanced for V2X services over LTE. This was acceptable from the legal point of view since those UEs are considered to be "machines" in the vehicle for sending/receiving warning or status messages. V2X was then introduced in Release 16 in NR/5GS in a similar way as for LTE/EPS.

The rationale behind studying ProSe in Release 17 [7] is that some features like direct mode D2D communication mechanisms including UE-to-network relay are not supported in 5GS, even if the mission critical services over 5GS may require them. Besides this, the objectives covering the commercial aspects will be studied, resulting from an SA WG1 study item [8] that comprises network-controlled interactive services

---

[1]The 3GPP working group meetings are announced at least 12-months in advance and available meeting time slots are assigned to individual work/study items. The 3GPP work program generally runs pretty full with limited free time slots available. This requires the 3GPP work program to be prioritized based on criticality of work/study items required to meet market demands. This prioritization exercise is done at the quarterly 3GPP plenary meetings. The work/study items that are de-prioritized due to lack of meeting time available are re-submitted by the interested parties for consideration in the next 3GPP Release.

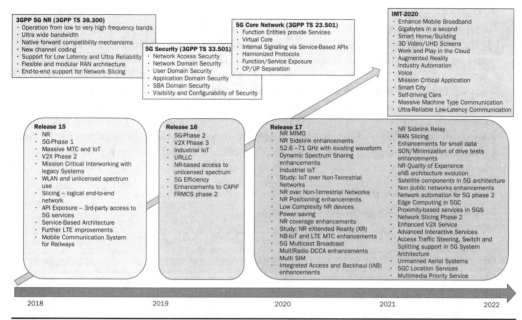

3GPP 5G NR (3GPP TS 38.300)
- Operation from low to very high frequency bands
- Ultra wide bandwidth
- Native forward compatibility mechanisms
- New channel coding
- Support for Low Latency and Ultra Reliability
- Flexible and modular RAN architecture
- End-to-end support for Network Slicing

5G Security (3GPP TS 33.501)
- Network Access Security
- Network Domain Security
- User Domain Security
- Application Domain Security
- SBA Domain Security
- Visibility and Configurability of Security

5G Core Network (3GPP TS 23.501)
- Function Entities provide Services
- Virtual Core
- Internal Signaling via Service-Based APIs
- Harmonized Protocols
- Function/Service Exposure
- CP/UP Separation

IMT-2020
- Enhance Mobile Broadband
- Gigabytes in a second
- Smart Home/Building
- 3D Video/UHD Screens
- Work and Play in the Cloud
- Augmented Reality
- Industry Automation
- Voice
- Mission Critical Application
- Smart City
- Self-driving Cars
- Massive Machine Type Communication
- Ultra-Reliable Low-Latency Communication

Release 15
- NR
- 5G-Phase 1
- Massive MTC and IoT
- V2X Phase 2
- Mission Critical Interworking with legacy Systems
- WLAN and unlicensed spectrum use
- Slicing – logical end-to-end network
- API Exposure – 3rd-party access to 5G services
- Service-Based Architecture
- Further LTE improvements
- Mobile Communication System for Railways

Release 16
- 5G-Phase 2
- V2X Phase 3
- Industrial IoT
- URLLC
- NR-based access to unlicensed spectrum
- 5G Efficiency
- Enhancements to CAPIF
- FRMCS phase 2

Release 17
- NR MIMO
- NR Sidelink enhancements
- 52.6 –71 GHz with existing waveform
- Dynamic Spectrum Sharing enhancements
- Industrial IoT
- Study: IoT over Non-Terrestrial Networks
- NR over Non-Terrestrial Networks
- NR Positioning enhancements
- Low Complexity NR devices
- Power saving
- NR coverage enhancements
- Study: NR eXtended Reality (XR)
- NB-IoT and LTE MTC enhancements
- 5G Multicast Broadcast
- MultiRadio DCCA enhancements
- Multi SIM
- Integrated Access and Backhaul (IAB) enhancements

- NR Sidelink Relay
- RAN Slicing
- Enhancements for small data
- SON/Minimization of drive tests enhancements
- NR Quality of Experience
- eNB architecture evolution
- Satellite components in 5G architecture
- Non public networks enhancements
- Network automation for 5G phase 2
- Edge Computing in 5GC
- Proximity-based services in 5GS
- Network Slicing Phase 2
- Enhanced V2X Service
- Advanced Interactive Services
- Access Traffic Steering, Switch and Splitting support in 5G System Architecture
- Unmanned Aerial Systems
- 5GC Location Services
- Multimedia Priority Service

2018   2019   2020   2021   2022

FIGURE 8.1 5G release overview

supporting, VR-based interactive service, Cloud/Edge/Split rendering for games, internet-of-everything (IoE)–based social networking services, and communication within network-controlled interactive service (NCIS) group.

8.1.2   5G Multicast-Broadcast Services

The MBMS (multicast/broadcast multimedia subsystem) feature was already developed in 2003 for UMTS networks in Release 6 [9] and was carried over to LTE/EPS later on. The goal was to enable group communication via video broadcasting and streaming services made possible with the support of higher available bandwidths of the radio interface. Unfortunately, there was limited commercial success for MBMS early on, with limited deployment by mobile operators. Later, mobile operators saw a deployment opportunity for MBMS when mission critical/public safety services were developed on top of the MBMS feature in Release 13/14. Also, vehicle-to-everything (V2X) and cellular IoT (CIoT) features started using the group communication feature provided by MBMS. To this end, a new study [10] is now in progress in 3GPP to look at the support of multicast requirements/use cases for CIoT, public safety, V2X, etc., and dedicated broadcasting requirements/use cases in the 5GS.

8.1.3   Network Automation for 5G

The area of data analytics is gaining importance in machine learning to find new techniques in drawing useful conclusions by examining huge amounts of data collected from a variety of sources. In support of this area, 3GPP in Release 15 introduced a new functionality in 5G core network (5GC), called "network data analytics function"

(NWDAF). The NWDAF is responsible for analyzing data of any part of the network and is a resource to be called upon by any network function (NF) in 5GC.

NWDAF was initially introduced to collect data on a particular network slice (NS) and provide the data analytics information to a consumer NF (e.g., for load-level information of a particular NS needed by PCF). The NWDAF was not aware of specific subscribers/UEs that were using the network slice and the data analytics was not subscriber specific. In Release 15, two NFs (PCF and NSSF) were the consumers for the information provided from the NWDAF, which may be used for policy decisions or slice selection. The framework for NWDAF in Release 16 was extended to enable data collection and provide data analytics to consumer NFs for various types of services, for example:

- QoS profile provisioning
- Traffic routing
- Performance improvement and supervision of mIoT terminals
- Future background data transfer
- Slice service-level agreements
- Support of northbound network status exposure
- Customized mobility management

The NWDAF is seen as a key function to enable artificial intelligence (AI) in the 5GS and the interaction between NWDAF and AI Model & Training Service is one new aspect in the current study in Release 17 [11]. In addition, further solutions for the NWDAF are under study to support network automation in various ways, including data analytics to operations support systems (OSS) and different deployment options.

8.1.4   Edge Computing in 5GC

The mobile edge computing (MEC), also known as multi-access edge computing in ETSI, is an open standardized environment to facilitate the integration of applications from third parties closer to mobile user's location. The MEC initiative is an industry specification group (ISG) within ETSI. The aim of the MEC-ISG is to unite the Telco and IT-cloud worlds, providing IT and cloud-computing capabilities within the RAN (radio access network). The MEC-ISG specifies the elements that are required to enable applications to be hosted in a multi-vendor multi-access edge computing environment.

MEC is a server providing computing resources and storage closer to where the user is located to provide quicker response times for applications running on devices. The MEC server can be part of the RAN or co-located with the RAN. A typical architecture is shown in Fig. 8.2.

The two MEC servers shown in the cloud computing environment are meant to do different tasks. For real-time applications and applications requiring fast response times (e.g., real-time streaming, V2X applications), the MEC server in the edge cloud computing environment will be used. For non-real-time applications, the MEC server in the central cloud computing environment will be used. The remote content server is shown to indicate that content being used can be cached in the MEC servers to be accessible on demand. The type of MEC server to be used is application dependent.

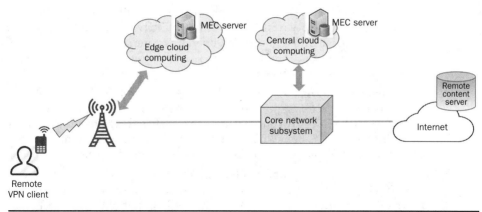

FIGURE 8.2   MEC architecture

The use of MEC technology, as defined by ETSI MEC-ISG, is in use in 3G and 4G mobile systems. Both ETSI and 3GPP work closely to leverage interactions between different network functions, aligning system operations with the NFV and SDN paradigms. In the 5GS, the MEC technology is under study (in Release 17) [12] to allow the provision of a 5G application function, to interact with the 3GPP 5GS to influence the routing of the edge applications' traffic and the ability to receive notifications of relevant events, such as mobility events, for improved efficiency and end user experience. Typical edge computing use cases in 5GS include, e.g., URLLC (ultra-reliable low-latency communication), V2X, AR/VR/XR, UAS, 5GSAT (satellite access in 5GS), and CDN (content delivery network).

8.1.5   Non-Public Networks

Non-Public Networks (NPN) are private networks or campus networks that are operated either by a network operator or by a private company. These networks may interwork with public mobile network, but could also operate in an isolated mode. It was first introduced in Release 16 within the scope of the "Vertical LAN" work in 3GPP SA WG2, which offered specific industry (verticals) solutions for companies/factories like redundant transmissions and URLLC. Private networks as such were already allowed in Release 15, but there were no specific features, except some special treatments, e.g., for authentication of subscribers. The current Release 17 study is building on the results of the Release 16 work and enhances the interworking between public networks and NPNs [13]. Further topics include the onboarding of UEs and their provisioning in NPNs or different deployment and managing issues.

8.1.6   Other Study Areas

There are also some smaller study items that are part of Release 17 and considered small in terms of 3GPP meeting time consumption; therefore, they are only introduced briefly here.

- Study on system enablers for multi-USIM devices (FS_MUSIM) [14]: This study item studies multi-USIM devices with respect to mostly paging/reachability optimizations. There are already commercially available phones that offer the possibility to insert two SIM cards/USIMs with two different subscriptions and potentially from different network operators. Since for cost reasons usually the radio and baseband components are shared among the USIMs, the UE would become unreachable on one USIM when the other one is in use.

- Study on enhancement of network slicing phase 2 (FS_eNS_ph2) [15]: This is another study looking into network slicing enhancements to identify the gap between GSMA 5GJA concept of a generic slice template and the support of slicing in the 5GS. Some more parameters like maximum number of users per slice or UL/DL data rate per slice may be introduced during this work.

- Study on enhanced support of industrial IoT–TSC/URLLC enhancements (FS_ IIoT) [17]: The basic time-sensitive communication (TSC) was studied in Release 16 of the vertical LAN study in TR 23.734 [16], but not all of the TSC aspects could be concluded. For these reasons, this study proposes enhancements in Release 17 on TSC aspects [17], e.g., support of industrial ethernet integration or new service requirements from audio-visual service production application.

- Study on supporting UAS connectivity, identification, and tracking (FS_ID_ UAS-SA2) [18]: This is a new topic in support of unmanned aerial systems (UAS), i.e., drones, which are using cellular connectivity to control unmanned aerial vehicle (UAV) as well as their supervision with an unmanned aerial system traffic management (UTM) for air traffic control.

- Study on architecture aspects for using satellite access in 5G (FS_5GSAT_ARCH) [19]: This was an earlier study proposal on architecture aspects for using satellite access in the 5GS as a major part in Release 16 feature development, but due to lack of time in the 3GPP RAN working groups and the dependencies there it could not be completed. Therefore, it is proposed again for Release 17, but will be granted only a relatively small amount of time to complete the feature, to consider use cases for provision of 5G services via satellite, in cases where it is not possible to provide 5G services over the terrestrial networks.

- Study on architecture enhancements for 3GPP support of advanced V2X services–phase 2 (FS_eV2XARC_ph2) [20]: V2X communication was studied already for several earlier releases in 3GPP and enhanced as part of 5GS Phase 2 during Release 16. The Release 17 study is now looking into studying specific functionality for the pedestrian type of UEs.

- 5G system enhancement for advanced interactive services (5G_AIS) [21]: This is another smaller study item looking at advanced interactive services in 5G with the focus on cloud gaming services. The UE sends sensor/pose data to the network in the uplink where the rendering takes place and the rendered data is sent back to the UE in the downlink, which requires high bandwidth with low latency of the audio and video data.

- Study extended access traffic steering, switch and splitting support in the 5G system architecture (FS_eATSSS) [22]: Access traffic steering, switch and splitting (ATSSS) study describes how traffic for a particular UE is routed over 3GPP and non-3GPP accesses, when the UE is connected simultaneously via both of them. The study item is an extension of the Release 16 work and looks especially on those aspects that could not be completed within Release 16.

- Study on enhancement to the 5GC location services–phase 2 (FS_eLCS_ph2) [23]: This study is a continuation of the Release 16 study on 5GC location services with several enhancements proposed for Release 17. Some key issues and requirements could not be concluded in Release 16 and thus proposed to be looked into in Release 17, such as service requirements for Industry IoT (IIoT) on very high accuracy and very low latency.

- Study on multimedia priority service (MPS) phase 2, stage 2 (FS_MPS) [24]: Finally, the last study for Release 17 is the next phase of MPS, which is looking for a priority for voice, video, and data communication for MPS users. Several use cases are part of this study with interworking different priorities and devices of the services and the study might mostly look at the policy control aspects.

8.2   Evolution toward 6G

As indicated in Fig. 8.3, every new generation of mobile network is expected to be deployed approximately every 10 years. Thus, even though 5G is still at a nascent stage with deployments being initiated in various geographies, researchers have already initiated work on investigating various technology trends and key performance indicators (KPIs) for the next generation of mobile networks (6G).

In this section, a review of the key trends is presented, in terms of technology and related performance requirements, use cases, and applications currently being discussed in literature. Since the technology trends cannot be objectively evaluated without considering the evolving business reality, an overview of the evolution of future financial projects and growth of the telecommunications industry in comparison to the so-called over-the-top service providers would also be considered.

8.2.1   Key Technology Trends and KPIs

An overview of the 6G system requirements, which could be used as possible research directions, using the framework of the three key use cases of 5G, is shown in Fig. 8.4 [26,27]. The detailed requirements indicated between the three edges of the triangle could be considered as ones which require a combination of the key use cases. Here it is assumed that enhanced mobile broadband would continue to be further improved in order to provide higher data rates. Massive machine-type communication (mMTC) could be logically extended to hyper-connectivity, where essentially all the objects that could be operated in an efficient manner with wireless connectivity could be connected to the mobile network in an efficient manner. The dimension related to ultra-reliability and low latency would also need to be extended to support hyper-reliability in 6G, with improved reliability and latency requirements as compared to 5G.

1980s	1990s	2000s	2010s	2020s	2030s	...

1G	2G	3G	4G	5G	6G	...

Figure 8.3   Approximated timeline for different generations (1G–6G) [25]

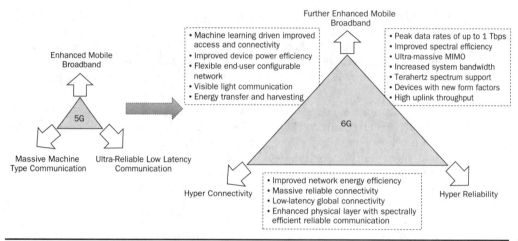

FIGURE 8.4   Overview of 6G system research directions [26,27]

In order to enable hyper-connectivity of devices with different form factors and features, the traditional static mechanisms for cell access, e.g., random access, and connectivity needs to be adapted to include machine learning–based predictive access and connectivity mechanisms. Based on analyzing the connectivity patterns of the devices, the network needs to adapt its behavior based on reasonably accurate predictions of future connectivity requirements. The device power efficiency would need to be further improved, especially taking into account the fact that the form factors for future devices could be significantly different from currently available smartphones.

Due to the wider anticipated proliferation of mobile networks—in factories, enterprise, and home environments—the need to improve the flexibility and end user configurability would gain higher prominence. The users in such environments would be significantly different in terms of skills and interests from mobile network operators, and would prefer simpler configurability of the network based on their requirements. Considering the limited amount of radio spectrum available where viable wireless network deployments are possible, visible light communication is expected to gain increasing prominence in the 6G era. Due to the low-power requirements and smaller form factor of future devices, wireless energy transfer and harvesting are considered to be integral requirements for the 6G system standards.

Traditionally, higher data rates have been one of the key requirements for a new generation of wireless network, and there are similar expectations from 6G with peak downlink data rate support increased to 1 Tbps as compared to 20 Gbps in 5G. The comparison of some of the key KPIs of 6G with respect to 5G are listed in Table 8.1 [27,28]. The higher data rates are expected to be enabled using further improvements in spectral efficiency as compared to 5G using novel physical layer techniques, and with the support of ultra-massive MIMO. The system bandwidth supported by 6G also would need to be enhanced as compared to 5G, in order to enable higher peak data rates. Due to the limited spectrum available even in the millimeter-wave frequency bands, terahertz frequency bands would need to be supported by the 6G system. Most of the applications in previous generations have been mainly designed considering increased

Generation/KPI		5G	6G
Peak data	UL	20 Gbps	1 Tbps
	DL	10 Gbps	0.5 Tbps
User experienced data rate	DL	100 Mbps	10 Gbps
	UL	50 Mbps	5 Gbps
Latency		1 ms	0.1 ms
Reliability (frame error rate)		10^{-5}	10^{-9}
Localization		10 cm	1 cm
Localization dimensionality		Two dimensional	Three dimensional

TABLE 8.1   5G and 6G KPI requirement comparison [27, 28]

downlink throughput. In 6G, uplink intensive services with significantly high uplink throughputs could also be supported using physical and other radio access network enhancements.

Energy efficiency has been one of the key requirements of 5G, with features such as lean carrier design and empowering RAN to make autonomous energy saving decisions, enabling significant improvements in terms of network power consumption. Since 6G would require higher data rates, increased availability, and reliability, energy efficiency improvements are expected to be an integral part of the system design. In 5G, it was inherently assumed that ultra-reliability would require significantly higher spectral resources, and hence would not be combined with massive connectivity. In 6G, system design enhancements would require the network to support higher number of devices with hyper-reliability with higher spectral efficiency. In terms of low latency, the requirements were mainly defined in 5G for over-the-air latency requirements, whereas the availability of low-latency global connectivity with the usage of a multitude of access and transport layer enhancements could be required in 6G. Hyper-localization in the three-dimensional space is also considered to be a key requirement for 6G, as compared to two-dimensional localization requirements for 5G.

8.2.2   Use Cases and Applications

In this section, some of the possible use case details and applications—on a wide range of industries—that could be enabled using 6G based on the technology features and requirements, and using the three key use cases as reference points, are discussed. An overview of the possible scenarios/use cases and related applications is shown in Fig. 8.5 [29,30]. The detailed overview of some of the use cases, along with the integration of AI with 6G networks for creating new services, is also presented. Detailed overview of some scenarios where the utilization of AI could create value in the context of media and entertainment and smart cities is also shown.

The tight integration of non-terrestrial networks into 6G networks would enable the global connectivity of devices, irrespective of geographic presence and the location of "home networks." It would also enable the provisioning of connectivity to remote areas with limited human presence, in a cost-efficient manner with possibilities of new value

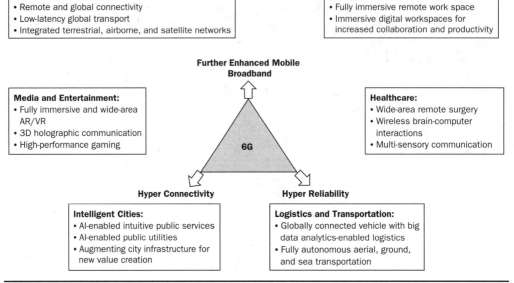

FIGURE 8.5   Possible scenarios and use cases [29, 30]

creation through increased economic activities in these areas. Integrated terrestrial, airborne, and satellite networks would enable provisioning of new services, particularly related to low-latency global transportation network connectivity.

Media and entertainment services have been dominating mobile networks in terms of resource utilization since 4G, and this trend is expected to continue in future as well with new use cases having requirements related to further enhanced mobile broadband and hyper-reliability. It is expected that 6G would be supporting fully immersive media content which requires the combination of low latency, ultra-reliability, and high throughput. The provisioning of immersive media content such as augmented/virtual reality (AR/VR) over a wide area is one of the key challenges that 6G networks would need to overcome. High-performance gaming and three-dimensional holographic communication—as compared to the video and audio services used for connectivity today—are some of the other key media- and entertainment-related applications.

Smart cities are expected to be one of the key use cases for 5G, which could be further enhanced to intelligent cities providing AI-enabled hyper-connectivity and new services to the public in 6G. The new services for intelligent cities could include intuitive public services and utilities to the residents of intelligent cities, which are augmented by AI in order to provide the applications in an efficient manner. The existing city infrastructure could be augmented using 6G to enable new value creation, with one use case related to the use of closed-circuit television (CCTV) cameras enabling new autonomous driving features discussed later in this section. As discussed earlier, the integration of terrestrial and non-terrestrial networks would enable new services in the area of logistics and transportation, with the enablement of globally connected vehicles which are augmented by AI-enabled real-time analytics. The provisioning of fully autonomous aerial, ground, and sea transportation is also considered to be one of the key applications of 6G networks.

Healthcare is one of the most important use cases for mobile networks, particularly due to the inherent support of reliability, low latency, and security for such networks. Some of the key use cases related to healthcare for 6G networks would be wide area remote surgery—essentially removing the current physical proximity requirements, wireless brain-computer interactions [30], and multi-sensory communication—which would require high throughput and low-latency transportation of information from the source to destination. Enterprise networks are also expected to be significantly transformed through 6G networks. The support for fully immersive remote workspaces could significantly transform the fixed investment costs for maintaining physical enterprise workspaces. Immersive digital workspaces could also act as a key enabler for increased collaboration and productivity on a global scale.

With the development of micro-operator networks [31], 6G networks could also act as a key driver for redefining the traditional *trust-relationship* between the end user and the network operator. Such changes could enable novel connectivity ecosystems, where the device layer and access/core network layer would be provisioned using global standards—similar to previous generations, with connectivity and access services provided by an application or shared data layer. Here, the trust-relationship between the end user and the network operator could be provided in a similar manner compared to the current internet services. The application/shared data layer could provide the same or enhanced amount of services that the current operators provide for the end user, along with the flexibility to configure the network. Building a global network of micro-operators could also enable shared and secure global access to such networks. An end-to-end overview of this new networking paradigm is shown in Fig. 8.6.

One of the key requirements for 6G is the support for 3D hyper-localization, which would enable network operators to provide new services to the end users. Such features being incorporated into 6G networks could also provide the provisioning of new services—such as three-dimensional spectrum allocation, as shown in Fig. 8.7. The support for higher frequency bands with inherent spectral isolation within indoor environments would enable network operators/spectrum owners to license their spectrum to enterprises and other private networks within 3D space, as compared to traditional 2D spectrum licensing considerations in previous generations. Such enhancements will enable new revenue streams to network operators/spectrum owners, while maximizing spectrum reuse.

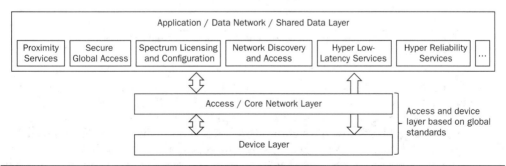

FIGURE 8.6 End-to-end network overview for enabling new connectivity ecosystems [31]

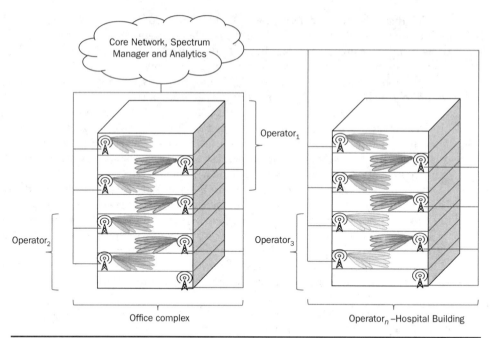

Figure 8.7   Dynamic three-dimensional spectrum allocation

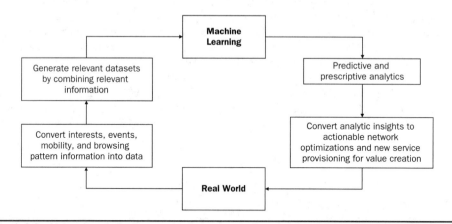

Figure 8.8   Network optimizations and enhancements for value creation using real world data [25]

One of the key themes of 6G is the unlocking of new value using AI within the network and through the services provided by the network [25, 30, 32]. While the usage of data for optimizing network operations is already a key topic in 5G, the availability of cheaper computing power and significant developments in the field of AI would enable improved optimizations and interworking with third-party services that would enable the network operators and infrastructure owners to provide new services to the end users. An overview of the steps involved in this process of converting real-world data into relevant datasets that machine learning models could utilize to generate predictive and prescriptive analytic insights is shown in Fig. 8.8. From the figure we can observe that the conversion of these insights into actionable network optimizations and provisioning of new services would be a key driver of new value creation.

In order to highlight the importance of generating insights and converting these insights into network optimizations in non-real-time, we consider the example of a realistic scenario where the network operator has access to the end user internet search history, content consumption history, and location history – as shown in Fig. 8.9. The available information is converted into datasets and using machine learning the probability of the user consuming certain types of content, such as popular sports events, live concerts, movies, is estimated. Based on this model generated using the available training dataset, the network operator can estimate the probability of the end user watching an upcoming popular event. By combining the data of all the users within the network, the operator could build powerful predictive analytics that enables future content consumption predictions with reasonable accuracy. The network operator could use this insight to optimize the positioning of the content for the upcoming event, e.g., a popular movie being released in a streaming site, sports events, etc., thereby minimizing the anticipated network load that could occur due to the simultaneous consumption of such content. Such optimizations could require the network operator to efficiently interwork with external content providers or content delivery network operators. This example indicates that relatively straightforward network optimizations that could be done in non-real-time could enable significant cost savings by avoiding over-dimensioning of the network while providing improved quality of service for the end users, thereby creating new value for the network operators and content providers.

Another example for the scenario where the combination of smart/intelligent city data generated using public utilities and prescriptive analytics in real time could enable

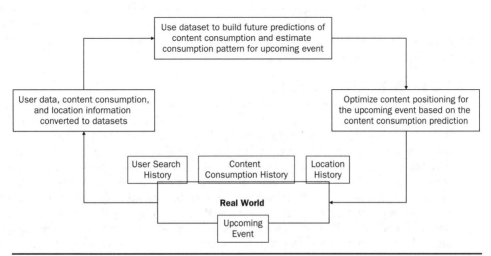

FIGURE 8.9 Example for combining physical world data and predictive analytics for non-real-time network operation optimization

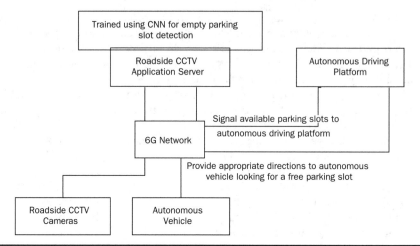

FIGURE 8.10   Example for combining intelligent city data and prescriptive analytics for real-time applications

enhanced autonomous driving features is shown in Fig. 8.10. In this example, it is shown that the roadside CCTV cameras that generate a significant amount of data which is currently utilized for optimizing public transporting and enabling public safety could also be utilized for detecting available free parking slots. It is assumed that the data is sent to the application server, using low-latency 6G network, which is trained using convolutional neural networks that are known to be optimal for image processing [33], so as to detect free parking spaces within the city. The application server could signal the free parking space information to the autonomous driving platform, which could then utilize this data to optimize the driving routes of autonomous vehicles. Such examples indicate the immense potential of combining the key features of the wireless network—such as support for high data rates and low latency could be combined with available public infrastructure to provision new services which could create significant value to the end users.

8.2.3   Evolving Business Trends

The main focus of this textbook has been on the end-to-end technology aspects of 5G mobile communication system and its evolution. But it is important to remember that technology evolution and adoption is tightly linked to the evolution of the overall industry from a business perspective. In order to highlight this aspect, the performance—quantified as the weekly stock prices[2]—of the telecommunication industry (Telcos), consisting of two European network infrastructure equipment vendors and network operators and three US-based network operators, is compared with the performance of some of the prominent internet companies (Webscales), and is shown in Fig. 8.11. The metric of weekly stock prices, normalized based on the aggregate stock price on Week-1 for each industry, provides an objective view from the perspective of the market in

---

[2]Source: http://finance.yahoo.com/

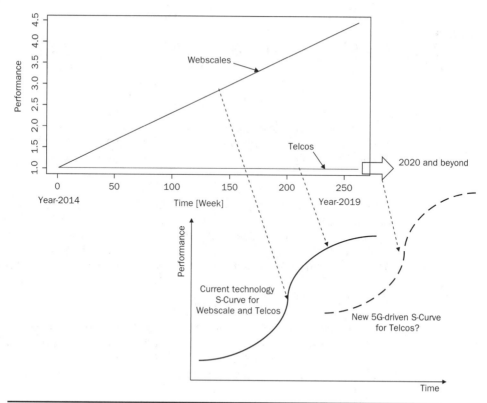

Figure 8.11   Overview of long-term industry trends and technology S-curves

terms of future growth prospects of these industries. In order to provide context to this data, it is compared with the innovation S-curves [34]. The innovation S-curve indicates various stages of growth of a company or industry—from early stage/adoption period with relatively low-growth, mid-stage/high-growth period, and final stage/maturity period. From the figure, an important aspect can be noted that while the Telcos growth/returns have remained relatively stable/flat, the Webscales that provide their services using the infrastructure provided by the network operators and infrastructure vendors to the end users have grown significantly. Based on the comparison with the S-curves, we can observe that while Telcos have reached the mature phase of the S-curve with performance staying relatively flat, the Webscales are on the growth phase while building innovative products that utilize the underlying communication infrastructure.

Based on these observations, the importance of innovation and providing services that create value for the end users and developing new capabilities for capturing at least part of this value needs to be an important objective for future mobile networks. Achieving this objective would be an essential requirement for enabling sustained innovation within the end-to-end mobile communication ecosystem. As shown in the figure, 5G—with its unique features and capabilities—could be a key enabler for a new S-curve for the telecommunications industry, and act as a key driver for future growth during 2020 and beyond.

References

1. 3GPP SP-191381, "Release 17 Prioritization for SA WG2, SA WG2 Chairman."
2. Recommendation ITU-R M.2083, "Framework and Overall Objectives of the Future Development of IMT for 2020 and Beyond," September 2015.
3. 3GPP TS 38.300, "NR; Overall description; Stage-2."
4. 3GPP TS 23.501, "System Architecture for the 5G System (5GS)."
5. 3GPP TS 33.501, "Security Architecture and Procedures for 5G System."
6. 3GPP TS 23.303, "Proximity-Based Services (ProSe); Stage 2."
7. 3GPP SP-190443, "Study on System Enhancement for Proximity-Based Services in 5GS (FS_5G_ProSe)."
8. 3GPP TR 22.842, "Study on Network Controlled Interactive Services."
9. 3GPP TS 23.246, "Multimedia Broadcast/Multicast Service (MBMS); Architecture and Functional Description."
10. 3GPP SP-190625, "Study on Architectural Enhancements for 5G Multicast-Broadcast Services (FS_5MBS)."
11. 3GPP SP-190557, "Study on Enablers for Network Automation for 5G—Phase 2 (FS_eNA_ph2)."
12. 3GPP SP-190185, "Study on Enhancement of Support for Edge Computing in 5GC (FS_enh_EC)."
13. 3GPP SP-190453, "Study on Enhanced Support of Non-Public Networks (FS_eNPN)."
14. 3GPP SP-190248, "Study on System Enablers for Multi-USIM Devices (FS_MUSIM)."
15. 3GPP SP-190931, "Study on Enhancement of Network Slicing Phase 2 (FS_eNS_ph2)."
16. 3GPP TR 23.734, "Study on Enhancement of 5G System (5GS) for Vertical and Local Area Network (LAN) Services."
17. 3GPP SP-190932, "Study on Enhanced Support of Industrial IoT—TSC/URLLC Enhancements (FS_IIoT)."
18. 3GPP SP-181114, "Study on Supporting UAS Connectivity, Identification, and Tracking (FS_ID_UAS-SA2)."
19. 3GPP SP-181253, "Study on Architecture Aspects for Using Satellite Access in 5G (FS_5GSAT_ARCH)."
20. 3GPP SP-190631, "Study on Architecture Enhancements for 3GPP Support of Advanced V2X Services—Phase 2 (FS_eV2XARC_ph2)."
21. 3GPP SP-190564, "5G System Enhancement for Advanced Interactive Services (5G_AIS)."
22. 3GPP SP-190558, "Study Extended Access Traffic Steering, Switch and Splitting Support in the 5G System Architecture (FS_eATSSS)."
23. 3GPP SP-190452, "Study on Enhancement to the 5GC LoCation Services—Phase 2 (FS_eLCS_ph2)."
24. 3GPP SP-190629, "Study on Multimedia Priority Service (MPS) Phase 2, Stage 2 (FS_MPS)."
25. NTT DOCOMO, INC, "White Paper: 5G Evolution and 6G," January 2020.
26. S. Dang, O. Amin, B. Shihad, et al., "What should 6G be?" Nat Electron 3, pp. 20–29 (2020). https://doi.org/10.1038/s41928-019-0355-6
27. E. Calvanese Strinati et al., "6G: The Next Frontier: From Holographic Messaging to Artificial Intelligence Using Subterahertz and Visible Light Communication," in IEEE Vehicular Technology Magazine, vol. 14, no. 3, pp. 42–50, Sept. 2019.

28. M Series, "IMT Vision–Framework and Overall Objectives of the Future Development of IMT for 2020 and Beyond," Recommendation ITU, 2015 Sep:2083-0. https://www.itu.int/dms_pubrec/itu-r/rec/m/R-REC-M.2083-0-201509-I!!PDF-E.pdf

29. V. Ziegler, T. Wild, M. Uusitalo, H. Flinck, V. Räisänen, and K. Hätönen, "Stratification of 5G Evolution and Beyond 5G," IEEE 2nd 5G World Forum (5GWF), Dresden, Germany, 2019, pp. 329–334.

30. W. Saad, M. Bennis, and M. Chen, "A Vision of 6G Wireless Systems: Applications, Trends, Technologies, and Open Research Problems," IEEE Network, doi: 10.1109/MNET.001.1900287.

31. A. Prasad, Z. Li, S. Holtmanns, M. A. Uusitalo, "5G Micro-Operator Networks—A Key Enabler for New Verticals and Markets," IEEE 25th Telecommunications Forum (TELFOR), Belgrade, Nov. 2017.

32. K. B. Letaief, W. Chen, Y. Shi, J. Zhang, and Y. A. Zhang, "The Roadmap to 6G: AI Empowered Wireless Networks," in IEEE Communications Magazine, vol. 57, no. 8, pp. 84–90, August 2019.

33. X. Wang, X. Wang, and S. Mao, "Deep Convolutional Neural Networks for Indoor Localization with CSI Images," in IEEE Transactions on Network Science and Engineering. doi: 10.1109/TNSE.2018.2871165

34. F. Hacklin, V. Raurich, and C. Marxt, "How Incremental Innovation becomes Disruptive: The Case of Technology Convergence," 2004 IEEE International Engineering Management Conference (IEEE Cat. No.04CH37574), Singapore, 2004, pp. 32–36, Vol. 1.

Problems/Exercise Questions

1. What are the three key use case dimensions currently considered for 6G? Explain why they are framed as extensions of 5G.

2. Explain in detail the system research directions currently being evaluated for 6G.

3. Compare the KPIs for 5G and 6G.

4. Explain in detail how the ultra-reliability and low-latency communication features are expected to be extended for 6G.

5. Explain in detail the use cases and applications being considered for 6G, and explain how they are different from 5G.

6. How is the relationship between end users and network operators expected to change in 6G? What kind of new connectivity ecosystem would this enable?

7. How can three-dimensional characteristics of the 6G network enable new spectrum licensing opportunities?

8. What role is AI expected to play in 6G?

9. Explain the example presented in the chapter related to non-real-time network optimizations in 6G.

10. Explain how the combination of intelligent city data and prescriptive analytics could enable new real-time applications in 6G. Could you describe any example for this, which is not covered in the textbook?

Glossary of Terms

Abbreviation	Term
3GPP	Third Generation Partnership Project
5G HE AV	5G Home Environment Authentication Vector
5G-GUTI	5G Globally Unique Temporary Identity
AAL2	ATM Adaptation Layer 2
ABBA	Anti-Bidding down Between Architectures
ADPCM	Adaptive Differential Pulse Code Modulation
AF	Application Function
AI	Artificial Intelligence
AIV	Air Interface Variants
AKA	Authentication Key Agreement
AMF	Access and Mobility Management Function
AMF	Authentication Management Field
AMPS	Advanced Mobile Phone Service
AN	Access Network
ANSI	American National Standards Institute
API	Application Programming Interface
AR	Augmented Reality
ARIB	Association of Radio Industries and Businesses (ARIB), Japan
ARPF	Authentication credential Repository and Processing Function
ARQ	Automatic Repeat Request
AS	Access Stratum
ATIS	Alliance for Telecommunications Industry Solutions (ATIS), North America
ATM	Asynchronous Transfer Mode
ATSSS	Access Traffic Steering, Switch and Splitting
AuC	Authentication Center
AUSF	Authentication Server Function
AUTN	Authentication Token
AV	Authentication Vector
BAIC	Barring of All Incoming Calls
BAOC	Barring of All Outgoing Calls
BBU	Baseband Unit

BIC	Barring Incoming Calls
BICC	Bearer Independent Circuit Switched Core
BOIC	Barring of Outgoing International Calls
BOIC-exHC	Barring of Outgoing International Calls except those directed to the Home PLMN Country
CAPEX	CAPital Expenditure
CBC	Cell Broadcast Centre
CBCF	CBC Function
CBE	Cell Broadcast Entity
CBS	Cell Broadcast
CCSA	China Communications Standards Association
CCTV	Closed-Circuit Television
CDMA	Code Division Multiple Access
CDR	Charging Data Record
CLI	Calling Line Identification
CMAS	Commercial Mobile Alert System
CMC	Connection Mobility Control
CN	Core Network
CoMP	Coordinated Multi Point transmission/reception
CP	Control Plane
CPRI	Common Public Radio Interface
CPU	Central Processing Unit
CS	Circuit Switched
CU	Central Unit
DCME	Digital Circuit Multiplication Equipment
DCN	Dedicated Core Network
DFT	Discrete Fourier Transform
DL	Downlink
DN	Data Networks
DN-AAA	Data Network Authentication Authorization and Accounting
DNS	Domain Name Server
DRA	Diameter Routing Agent
DRB	Data Radio Bearers
DU	Distributed Unit
ECIES	Elliptic Curve Integrated Encryption Scheme
EIR	Equipment Identify Register
eKSI	EPS KSI
EMSK	Extended Master Session Key
EN-DC	E-UTRA—NR Dual Connectivity
EPC	Evolved Packet Core
EPS	Evolved Packet System
ETSI	European Telecommunications Standards Institute
ETWS	Earthquake Tsunami Warning System
E-UTRA	Evolved Universal Terrestrial Radio Access
E-UTRAN	Evolved UMTS Terrestrial Radio Access Network
EWRS	Early Warning and Response System
FCC	Federal Communications Commission

GGSN	Gateway GPRS Support Node
GMSC	Gateway Mobile Switching Center
GPRS	General Packet Radio Service
GPSI	Generic Public Subscription Identifier
GSM	Global System for Mobile Communications
GTP	GPRS Tunneling Protocol
GUAMI	Globally Unique AMF Identifier
GUTI	Globally Unique Temporary Identifier
HARQ	Hybrid Automatic Repeat Request
HLR	Home Location Register
HPLMN	Home Public Land Mobile Network
HSS	Home Subscriber Server
HXRES*	Hash eXpected RESponse
ICIC	Inter-Cell Interference Coordination
IMS	IP Multimedia Subsystem
IMSI	International Mobile Subscriber Identity
IMT	International Mobile Telecommunications
IoE	Internet of Everything
IoT	Internet-of-Things
IP-SM-GW	IP Short Message Gateway
IPX	Internet Protocol (IP) Packet eXchange
ISDN	Integrated Services Digital Network
ISG	Industry Specification Group
ISIM	IP Multimedia Services Identity Module
ITU	International Telecommunication Union
ITU-T	ITU-Telecommunication Standardization Sector
JWS	JSON Web Signature
JTACS	Japan Total Access Communication System
KDF	Key Derivation Function
KPIs	Key Performance Indicators
LCP	Logical Channel Prioritization
LMR	Land Mobile Radio
LTE	Long-Term Evolution
LWA	LTE-WLAN Aggregation
LWIP	LTE-WLAN Radio Level Integration with IPsec Tunnel
MAC	Message Authentication Codes
MBB	Mobile BroadBand
MBMS	Multicast/Broadcast Multimedia Subsystem
MCC	Mission Critical Communications
MCC	Mobile Country Code
MCEF	Mobile Station Memory Capacity Exceeded Flag
ME	Mobile Equipment
MEC	Mobile Edge Computing or Multi-access Edge Computing
MeNB	Master eNB
MME	Mobility Management Entity
MMS	Multimedia Messaging Service
MNC	Mobile Network Code

MNO	Mobile Network Operator
MNRF	Mobile Station Not Reachable Flag
MNRG	Mobile Station Not Reachable for GPRS
MNRR-MSC	Mobile Station Not Reachable via the MSC Reason
MNRR-SGSN	Mobile Station Not Reachable via the SGSN Reason
MO	Mobile Originated
MPS	Multimedia Priority Service
MR-DC	Multi-RAT Dual Connectivity
MS	Mobile Station
MSA	Micro-Services Architecture
MSC	Mobile Switching Center
MSIN	Mobile Subscriber Identification Number
MSISDN	Mobile Subscriber Integrated Services Digital Network Number
MT	Mobile Terminated
MWD	Messages Waiting Data
N3IWF	Non-3GPP Inter-Working Function
NAS	Non-Access Stratum
NCC	Next Hop Chaining Count
NCIS	Network Controlled Interactive Service
NDS	Network Domain Security
NEF	Network Exposure Function
NF	Network Function
NFV	Network Function Virtualization
ngKSI	5G Key Set Identifier
NH	Next Hop
NPN	Non-Public Networks
NR	New Radio (5G Radio Technology)
NRF	Network Repository Function
NSI	Network Slice Instance
NSSAI	Network Slice Selection Assistance Information
NSSF	Network Slice Selection Function
NSSI	Network Slice Subnet Instance
NW	NetWork
NWDAF	Network Data Analytics Function
OFDM	Orthogonal Frequency Division Multiplexing
OPEX	Operating Expenditure
OSI	Open System Interconnection
OSS	Operational Support System
OTA	Over the Air
OTT	Over the Top
PCF	Policy Control Function
PCM	Pulse Code Modulation
PCRF	Policy and Charging Rules Function
PCS	Personal Communications Service (in USA)
P-CSCF	Proxy-Call Session Control Function
PDCP	Packet Data Convergence Protocol
PDG	Packet Data Gateway

PDU	Packet Data Unit
PGW	Packet Gateway
PHB	Per-Hop Behavior
PHY	PHYsical Layer (e.g., LTE PHY)
PLMN	Public Land Mobile Network
PMK	Pairwise Master Key
PNF	Physical Network Function
PP	Point to Point
ProSe	Proximity Services
PSA	PDU Session Anchor
PSAP	Public Safety Answering Point
PSTN	Public Switched Telephone Network
PTT	Push to Talk
PWS	Public Warning System
QCI	QoS Class Indicator
QoS	Quality of Service
RAN	Radio Access Network
RAND	RANDom challenge
RCLWI	RAN-Controlled LTE-WLAN Interworking
RLC	Radio Link Control
RNA	RAN-level Notification Areas
RNC	Radio Network Controller
RRC	Radio Resource Control
RRH	Remote Radio Head
RRM	Radio Resource Management
SA	Security Association
SA1	Service and Requirements group
SA2	System Architecture group
SBA	Service-Based Architecture
SBI	Service-Based Interfaces
SC	Service Centre
S-CSCF	Serving Session Control Function
SCTP	Stream Control Transmission Protocol
SDAP	Service Data Adaptation Protocol
SDN	Software Defined Network
SDU	Service Data Unit
SEAF	SEcurity Anchor Function
SEPP	Security Edge Protection Proxy
SgNB	Secondary gNB
SGSN	Serving GPRS Support Node
SGW	Serving Gateway
SIDF	Subscription Identifier De-concealing Function
SIM	Subscriber Identification Module
SIP	Session Initiation Protocol
SLA	Service-Level Agreement
SM	Session Management
SMC	Security Mode Command

SME	Short Message Entity
SMF	Session Management Function
SMS	Short Message Service
SMSoIP	SMS over IP
S-NSSAI	Single-Network Slice Selection Assistance Information
SOA	Service-Oriented Architecture
SoR	Steering of Roaming
SQN	Sequence Number
SRES	Signed RESponse
SS	Supplementary Service
SS7	Signaling System No. 7
SST	Slice/Service Type
STM	Synchronous Transfer Mode
S-TMSI	Serving Temporary Mobile Subscriber Identity
SUCI	SUbscription Concealed Identifier
SUPI	SUbscriber Permanent ID
TACs	Type Approval Codes
TACS	Total Access Communication System (1G Radio Technology)
TAI	Tracking Area Identifier
TAU	Tracking Area Update
TCP/IP	Transmission Control Protocol using IP
TDF	Traffic Detection Function
TDMA	Time-Division Multiple Access
TFP	Traffic Forwarding Policy
TIA	Telecommunications Industry Association (TIA), North America
TLS	Transport Layer Security
TMSI	Temporary Mobile Subscriber Identity
TNAN	Trusted Non-3GPP Access Network
TNAP	Trusted Non-3GPP Access Point
TNGF	Trusted Non-3GPP Gateway Function
TS	Teleservice
TSC	Time Sensitive Communication
TSDSI	Telecommunications Standards Development Society of India (TSDSI)
TTA	Telecommunication Technology Association (TTA), South Korea
TTC	Telecommunication Technology Committee (TTC), Japan
TTI	Transmission Time Interval
TWAP	Trusted WLAN Access Point
TWIF	Trusted WLAN Interworking Function
UAS	Unmanned Aerial Systems
UAV	Unmanned Aerial Vehicle
UDI	Unrestricted Digital Information
UDM	Unified Data Management
UDP	User Datagram Protocol
UDR	Unified Data Repository
UE	User Equipment
UICC	Universal Integrated Circuit Card
UL	UpLink

UMTS	Universal Mobile Telecommunications System
UP	User Plane
UPF	User-Plane Function
UPU	UE Parameters Update
URLLC	Ultra-Reliable Low-Latency Communication
URSP	UE Route Selection Policy
USIM	Universal Subscriber Identity Module
USSD	Unstructured Supplementary Service Data
UTM	Unmanned Aerial System Traffic Management
UTRAN	UMTS Terrestrial Radio Access Network
V2B	Vehicle to Business
V2H	Vehicle to Home
V2I	Vehicle to Infrastructure
V2P	Vehicle to Person
V2V	Vehicle to Vehicle
V2X	Vehicle to Everything
VLR	Visited Location Register
VM	Virtual Machine
VNF	Virtual Network Function
VoLTE	Voice over LTE
VPLMN	Visited PLMN
VPN	Virtual Private Network
VR	Virtual Reality
WAP	Wireless Application Protocol
WCDMA	Wideband Code Division Multiple Access
XMAC	Expected Message Authentication Code
XRES	Expected RESponse

Index